Applications of Solid-Phase Microextraction and Related Techniques

Applications of Solid-Phase Microextraction and Related Techniques

Guest Editor
Hiroyuki Kataoka

Basel • Beijing • Wuhan • Barcelona • Belgrade • Novi Sad • Cluj • Manchester

Guest Editor
Hiroyuki Kataoka
School of Pharmacy
Shujitsu University
Okayama
Japan

Editorial Office
MDPI AG
Grosspeteranlage 5
4052 Basel, Switzerland

This is a reprint of the Special Issue, published open access by the journal *Molecules* (ISSN 1420-3049), freely accessible at: www.mdpi.com/journal/molecules/special_issues/SJ3NJB64R2.

For citation purposes, cite each article independently as indicated on the article page online and using the guide below:

Lastname, A.A.; Lastname, B.B. Article Title. *Journal Name* **Year**, *Volume Number*, Page Range.

ISBN 978-3-7258-3378-8 (Hbk)
ISBN 978-3-7258-3377-1 (PDF)
https://doi.org/10.3390/books978-3-7258-3377-1

© 2025 by the authors. Articles in this book are Open Access and distributed under the Creative Commons Attribution (CC BY) license. The book as a whole is distributed by MDPI under the terms and conditions of the Creative Commons Attribution-NonCommercial-NoDerivs (CC BY-NC-ND) license (https://creativecommons.org/licenses/by-nc-nd/4.0/).

Contents

About the Editor vii

Hiroyuki Kataoka
Applications of Solid-Phase Microextraction and Related Techniques
Reprinted from: *Molecules* 2025, *30*, 827, https://doi.org/10.3390/molecules30040827 1

Nicolò Riboni, Erika Ribezzi, Federica Bianchi and Maria Careri
Supramolecular Materials as Solid-Phase Microextraction Coatings in Environmental Analysis
Reprinted from: *Molecules* 2024, *29*, 2802, https://doi.org/10.3390/molecules29122802 6

Alessandra Timóteo Cardoso, Rafael Oliveira Martins and Fernando Mauro Lanças
Advances and Applications of Hybrid Graphene-Based Materials as Sorbents for Solid Phase Microextraction Techniques
Reprinted from: *Molecules* 2024, *29*, 3661, https://doi.org/10.3390/molecules29153661 54

Hiroyuki Kataoka, Atsushi Ishizaki, Keita Saito and Kentaro Ehara
Developments and Applications of Molecularly Imprinted Polymer-Based In-Tube Solid Phase Microextraction Technique for Efficient Sample Preparation
Reprinted from: *Molecules* 2024, *29*, 4472, https://doi.org/10.3390/molecules29184472 82

Soyoung Ahn and Sunyoung Bae
Synthesis and Characterization of a Multi-Walled Carbon Nanotube–Ionic Liquid/Polyaniline Adsorbent for a Solvent-Free In-Needle Microextraction Method
Reprinted from: *Molecules* 2023, *28*, 3517, https://doi.org/10.3390/molecules28083517 103

Stefano Dugheri, Giovanni Cappelli, Niccolò Fanfani, Donato Squillaci, Ilaria Rapi and Lorenzo Venturini et al.
Vacuum-Assisted MonoTrap™ Extraction for Volatile Organic Compounds (VOCs) Profiling from Hot Mix Asphalt
Reprinted from: *Molecules* 2024, *29*, 4943, https://doi.org/10.3390/molecules29204943 116

Pingping Song, Bo Xu, Zhenying Liu, Yunxia Cheng and Zhimao Chao
The Difference of Volatile Compounds in Female and Male Buds of *Trichosanthes anguina* L. Based on HS-SPME-GC-MS and Multivariate Statistical Analysis
Reprinted from: *Molecules* 2022, *27*, 7021, https://doi.org/10.3390/molecules27207021 133

Sarathadevi Rajendran, Patrick Silcock and Phil Bremer
Volatile Organic Compounds (VOCs) Produced by *Levilactobacillus brevis* WLP672 Fermentation in Defined Media Supplemented with Different Amino Acids
Reprinted from: *Molecules* 2024, *29*, 753, https://doi.org/10.3390/molecules29040753 148

Stefano Dugheri, Giovanni Cappelli, Niccolò Fanfani, Jacopo Ceccarelli, Giorgio Marrubini and Donato Squillaci et al.
A New Perspective on SPME and SPME Arrow: Formaldehyde Determination by On-Sample Derivatization Coupled with Multiple and Cooling-Assisted Extractions
Reprinted from: *Molecules* 2023, *28*, 5441, https://doi.org/10.3390/molecules28145441 167

Yuyue Zang, Na Hang, Jiale Sui, Senlin Duan, Wanning Zhao and Jing Tao et al.
Magnetic Persimmon Leaf Composite: Preparation and Application in Magnetic Solid-Phase Extraction of Pesticides in Water Samples
Reprinted from: *Molecules* 2023, *29*, 45, https://doi.org/10.3390/molecules29010045 183

Ying Liu, Yong Zhang, Jing Wang, Kexin Wang, Shuming Gao and Ruiqi Cui et al.
Preparation of COPs Mixed Matrix Membrane for Sensitive Determination of Six Sulfonamides in Human Urine
Reprinted from: *Molecules* **2023**, *28*, 7336, https://doi.org/10.3390/molecules28217336 . . . . . . **199**

Stefano Dugheri, Donato Squillaci, Valentina Saccomando, Giorgio Marrubini, Elisabetta Bucaletti and Ilaria Rapi et al.
An Automated Micro Solid-Phase Extraction (SPE) Liquid Chromatography-Mass Spectrometry Method for Cyclophosphamide and Iphosphamide: Biological Monitoring in Antineoplastic Drug (AD) Occupational Exposure
Reprinted from: *Molecules* **2024**, *29*, 638, https://doi.org/10.3390/molecules29030638 . . . . . . **212**

Hiroyuki Kataoka, Saori Miyata and Kentaro Ehara
Simultaneous Determination of Tobacco Smoke Exposure and Stress Biomarkers in Saliva Using In-Tube SPME and LC-MS/MS for the Analysis of the Association between Passive Smoking and Stress
Reprinted from: *Molecules* **2024**, *29*, 4157, https://doi.org/10.3390/molecules29174157 . . . . . . **227**

About the Editor

Hiroyuki Kataoka

Hiroyuki Kataoka received a BS in Pharmaceutical Science from Nagasaki University, Japan (1977); an MS from Osaka University, Japan (1979); and a PhD from Tohoku University, Japan (1986). He was a Research Associate and Associate Professor at Okayama University from 1979 to 2003 and has since been a Professor at Shujitsu University. From 1998 to 1999, he collaborated with Professor Janusz Pawliszyn as a postdoctoral research fellow in the Solid Phase Microextraction Group at the University of Waterloo (Waterloo, Ontario, Canada), developing in-tube solid-phase microextraction. His research focuses on developing selective and sensitive methods for analyzing bioactive and potentially harmful compounds in living systems, foods, and the environment. His current projects include automated sample preparation techniques combined with LC-MS/MS or GC-MS for applications in environmental and biomedical research. Using these approaches, he develops methods to assess biological exposure to toxic chemicals and non-invasive techniques for analyzing stress- and fatigue-related biomarkers to enable early disease detection and health protection. His research aims to advance analytical technology and contribute to a healthier, more comfortable life. He has authored over 180 peer-reviewed scientific publications, with a Hirsch Index of 45 (as of January 2025). He has been recognized among the top 2% of scientists worldwide for six consecutive years, in a 2024 Stanford University survey. He serves on the editorial boards of *Analytica Chimica Acta*, *Journal of Chromatography A*, *Journal of Environmental & Analytical Toxicology*, *Journal of Analytical Methods in Chemistry*, *Chromatography*, and *Molecules*.

Editorial

Applications of Solid-Phase Microextraction and Related Techniques

Hiroyuki Kataoka

Laboratory of Applied Analytical Chemistry, School of Pharmacy, Shujitsu University, Nishigawara, Okayama 703-8516, Japan; hkataoka@shujitsu.ac.jp; Tel.: +81-86-271-8342

In recent years, various high-performance analytical instruments have been developed with improved sensitivity and performance. However, most still struggle to directly process complex matrices, such as biological materials, environmental samples, and food, making sample pretreatment essential. In the analytical process from sampling to data analysis, sample pretreatment is not only vital for extracting, separating, and concentrating target analytes from complex matrices, but also enhances detection, improves sensitivity and accuracy, and reduces instrument maintenance and operating costs, all of which significantly impact the reliability and practicality of analytical methods. Thus, developing environmentally friendly and efficient green sample pretreatment techniques, and integrating them into various analytical instruments, have become major challenges.

Liquid–liquid extraction, a classical sample pretreatment technique, requires large amounts of sample and harmful organic solvents, making it time-consuming, labor-intensive, and expensive. To address these drawbacks, solid-phase extraction (SPE) was developed, and has been widely adopted due to its relatively simple and efficient operation, lower cost, reduced organic solvent consumption, and high concentration capacity.

As part of green analysis strategies, various microextraction techniques have been developed to minimize sample sizes and solvent usage while improving efficiency and automation. These include micro solid-phase extraction (μSPE), dispersive-μ-SPE, magnetic solid-phase extraction (MSPE), solid-phase microextraction (SPME), stir bar sorptive extraction (SBSE), microextraction with packed sorbent (MEPS), in-needle microextraction (INME), pipette tip solid-phase extraction (PT-SPE), and disposable pipette extraction (DPX). Among these, SPME is a simple and convenient sample pretreatment technique that integrates sampling, extraction, preconcentration, and desorption into a single step, enabling automation, miniaturization, high-throughput processing, and direct online coupling with analytical instruments. Its advantages include improved extraction efficiency, reduced sample handling, and lower costs for solvents and disposal. Since its introduction by Arthur and Pawliszyn [1] in the early 1990s, more robust fiber assemblies and coatings with higher extraction efficiency, selectivity, and stability have been commercialized, making SPME widely applicable in fields such as environmental monitoring, food analysis, forensic science, pharmaceuticals, and bioanalysis [2,3].

Additionally, new extraction geometries utilizing capillary tubes, magnetic stirrer bars, or thin films instead of fibers, have been developed, along with advanced polymer coatings, offering superior absorption/adsorption capacity and selectivity [3,4]. In these microextraction techniques, the enrichment capacity, sample processing ability, and selective extraction of target compounds are highly dependent on the coating material. To enhance extraction efficiency and selectivity, new adsorbents including monoliths, ionic liquids (ILs)/polymer ILs, restricted access materials, molecularly imprinted polymers

Received: 8 February 2025
Accepted: 9 February 2025
Published: 11 February 2025

Citation: Kataoka, H. Applications of Solid-Phase Microextraction and Related Techniques. *Molecules* **2025**, *30*, 827. https://doi.org/10.3390/molecules30040827

Copyright: © 2025 by the author. Licensee MDPI, Basel, Switzerland. This article is an open access article distributed under the terms and conditions of the Creative Commons Attribution (CC BY) license (https://creativecommons.org/licenses/by/4.0/).

(MIPs), graphene/graphene oxide (GO), carbon nanotubes (CNTs), inorganic nanoparticles, metal–organic frameworks (MOFs), crystalline covalent organic frameworks (COFs), and amorphous covalent organic polymers (COPs) have been continuously developed.

This Special Issue, titled "Applications of Solid-Phase Microextraction and Related Techniques", features three peer-reviewed review articles and nine original research papers from groups worldwide. The topics include: studies on supramolecular materials such as MOFs and COFs [Contribution 1], hybrid graphene-based materials (GBMs) [Contribution 2], MIPs [Contribution 3] and other coating materials for microextraction; research utilizing gas chromatography–mass spectrometry (GC-MS) [Contributions 4–8], GC-electron capture detection (GC-ECD) [Contribution 9], high-performance liquid chromatography (HPLC) [Contribution 10], and LC-MS/MS [Contribution 11,12] as coupled analytical techniques. These also include applications such as GC-MS analysis of volatile organic compounds (VOCs) using multi-walled carbon nanotubes-ionic liquid/polyaniline adsorbents (MWCNT-IL/PANI) [Contribution 8] and MonoTrap^{TM} [Contribution 5] as adsorbents, headspace (HS)-SPME/GC-MS for VOC analysis using commercial fibers [Contribution 6,7], formaldehyde derivatization in water with SPME Arrow for GC-MS analysis [Contribution 8], MSPE GC-ECD analysis of pesticides in environmental water using magnetic persimmon leaf composites [Contribution 9], membrane solid-phase extraction (ME) HPLC analysis of sulfonamides in urine using a COP mixed matrix membrane [Contribution 10], biological monitoring of occupational exposures to anticancer drugs via µSPE coupled with ultra-performance liquid chromatography–tandem mass spectrometry (UHPLC-MS/MS) [Contribution 11], and simultaneous analysis of tobacco smoke exposure and stress biomarkers in saliva using in-tube (IT) SPME with a Supel-Q PLOT capillary for LC-MS/MS analysis [Contribution 12]. An overview of these studies is provided below.

SPME is a widely used sample preparation technique in environmental monitoring due to its simplicity, environmental friendliness, and extraction capacity. It is employed for the extraction, clean-up, and preliminary concentration of environmental pollutants. Supramolecular materials, with their high surface-to-volume ratio, controlled porosity, and tunable surface properties, have emerged as promising SPME coatings. These materials offer unique selectivity, three-dimensional structures, and flexible design, facilitating interactions between analytes and coatings through multiple oriented functional groups [Contribution 1]. By modifying linkers, nodes, and monomer units, 3D frameworks with tailored structures and surface chemistry can be designed, making them superior to other nanostructured materials such as alloys, silica, and carbon-based materials. Functionalization with phenyl, amino, and ionic groups has proven particularly effective in establishing multiple directional interactions, including hydrogen bonding, π–π interactions, electrostatic interactions, and van der Waals forces between analytes and coatings. As a result, extraction is driven by steric fit and complementarity between the sorbent and the target molecule, leading to enhanced selectivity and enrichment capabilities. Riboni et al. [Contribution 1] review the state of the art in SPME coatings based on MOFs, COFs, and supramolecular macrocycles such as cyclodextrins, calixarenes, and cavitands for environmental monitoring, while also discussing future challenges.

Graphene, a single-layer carbon allotrope, is widely used in energy generation, electronics, sensors, and other areas of materials science due to its flexibility, lightness, thermal and electrical conductivity, and mechanical resistance to high pressure [Contribution 2]. Its large surface area, honeycomb-patterned binding sites, delocalized π electrons, and monolayer structure make it particularly effective as an adsorbent in sample preparation techniques [Contribution 2]. GO can be chemically functionalized with various components, including ILs, silica derivatives, magnetic materials, MIPs, resins, deep eutectic solvents, and carbon-based biosorbents. These hybrid GBMs offer excellent versatility and modification potential, with their high surface area and functional groups making them ideal adsorbents for miniaturized extraction strategies. Their compatibility with environmentally friendly synthesis approaches, such as bio-based graphene materials, enables reduced adsorbent usage and the development of biodegradable materials, supporting green analytical strategies. Cardoso et al. [Contribution 2] review recent advances in hybrid GBMs for miniaturized solid-phase sample preparation techniques, discussing their typical characteristics, application trends, particularly in offline techniques such as SBSE, MEPS, PT-SPE, DPX, d-µ-SPE, and MSPE.

MIPs are stable, custom-made polymers with molecular recognition functions formed during synthesis, making them excellent "smart adsorbents" for selective sample preparation. The fabrication of MIPs involves three main steps: prepositioning recognition functions, polymerization, and template removal. First, a template-monomer complex is formed, followed by polymerization, which creates a polymer network with the template molecule embedded. In the final step, the template is removed, leaving behind cavities complementary to the target molecule [Contribution 3]. Meanwhile, IT-SPME [4] has gained attention as a "green extraction technique" due to its minimal use of organic solvents, reduced liquid waste, high throughput, compact size, online instruments compatibility, automation capabilities, and ability to operation continuously overnight. Kataoka et al. [Contribution 3] review recent advancements in MIP-based IT-SPME methods, which integrate the selectivity of MIPs with the efficiency of IT-SPME, and discuss their applications.

VOCs are present in various environments, and some have been identified as harmful. Ahn and Bae [Contribution 4] developed an HS-INME GC-MS method using a novel extraction device consisting of a needle inserted with an adsorbent-coated wire, enabling solvent-free sample extraction. A newly synthesized adsorbent, MWCNT–IL/PANI, produced by mixing aniline and MWCNTs in the presence of IL, followed by electrochemical polymerization, exhibits high thermal stability and can be reused up to 150 times without performance loss. This material offers an environmentally friendly, solvent-free extraction approach. Additionally, Dugheri et al. [Contribution 5] extracted VOCs using a vacuum-assisted HS-MonoTrapTM sampling system, which integrates a commercially available small monolith hybrid adsorption device, MonoTrapTM, into a vacuum-assisted extraction setup. The extracted VOCs were analyzed by GC-MS/olfactometry and applied to characterize VOC emission profiles from hot mix asphalt. Among the 35 hot mix asphalt samples analyzed, the major odor-active VOCs were primarily aldehydes, alcohols, and ketones.

Some VOCs produced by living organisms play important physiological roles. Song et al. [Contribution 6] analyzed the chemical composition and VOC content of female and male buds of *Trichosanthes anguina* L. (Cucurbitaceae), also known as *Trichosanthes cucunerina* L., using HS-SPME-GC-MS with a commercially available carboxen/polydimethylsiloxane/divinylbenzene (CAR/PDMS/DVB) fiber. Multivariate statistical analysis identified differences in VOC composition between female and male buds, revealing chemical distinctions in monoecious plants. These findings provide useful insights into plant sexual differentiation. Similarly, Rajendran et al. [Contribution 7] examined VOC production during the fermentation of the lactic acid bacteria *Levilactobacillus brevis* WLP672 (LB672) using HS-SPME-GC-MS with a CAR/PDMS/DVB fiber. They found that adding different amino acids, alone or in combination, to a specific medium influenced VOC production. These results contribute to the development of plant-based fermentation strategies for enhancing flavor in meat and dairy products.

Formaldehyde (FA) is widely used in manufacturing industries, including food processing, textiles, wood production, and cosmetics. However, human exposure to FA can cause both acute and chronic toxicity, necessitating the need for sensitive, economical, and specific monitoring tools. Dugheri et al. [Contribution 8] developed an on-sample derivatization and GC-MS method for FA analysis in water using an SPME Arrow PDMS fiber, which is a large-bore SPME probe with greater phase volume and mechanical durability than standard SPME fibers. This approach offers a greener and more efficient "one-pot" analytical method by integrating commercially available cooling-assisted SPME with multiplexed SPME techniques for fully automated monitoring.

Chemical pest control using pesticides is widely practiced in agriculture and forestry, but concerns remain about the impact of environmental water pollution on human health and ecosystems. Zang et al. [Contribution 9] developed an Fe_3O_4/persimmon leaf magnetic biomass composite for pollutant removal and detection in water. They also establishes a method for analyzing four pesticides (trifluralin, triadimefon, permethrin, and fenvalerate) in environmental water samples using MSPE combined with GC-ECD. This novel biomass composite serves as an effective, sustainable adsorbent, promoting green analytical methods by repurposing waste biomass.

Sulfonamides are widely used in meat-producing livestock to prevent bacterial infections due to their effectiveness, stability, and affordability. However, their excessive or prolonged use may lead to residue accumulations in animal products, posing health risks to humans, including allergic reactions, drug resistance, teratogenicity, carcinogenicity, and mutagenicity. Liu et al. [Contribution 10] developed COPs mixed-matrix membranes by incorporating porous COPs into polyvinylidene fluoride polymer using Schiff base chemistry. They also established an ME/HPLC method for analyzing of six sulfonamides in human urine. This novel COP-based membrane effectively preconcentrates trace organic compounds from complex matrices.

Antineoplastic drugs cause genetic damage not only to cancer cells, but also to healthy cells, posing significant risks to both patients undergoing treatment and healthcare workers exposed occupationally. Dugheri et al. [Contribution 11] developed a rapid, automated, highly sensitive method to determine urinary concentrations of the DNA alkylating agents cyclophosphamide and ifosfamide using μSPE followed by UHPLC-MS/MS for biological monitoring. This μSPE-bases approach has become a powerful tool for occupational health and aligns with green analytical strategies.

Passive exposure to environmental tobacco smoke not only increases the risk of lung cancer and cardiovascular disease, but may also act as a stressor linked to neuropsychiatric and other conditions. To investigate this relationship, Kataoka et al. [Contribution 12] developed an online automated analysis system for measuring tobacco smoke exposure biomarkers, including nicotine and cotinine, alongside stress-related biomarkers, such as cortisol, serotonin, melatonin, dopamine, and oxytocin in saliva samples. The method employs IT-SPME with a Supel-Q PLOT capillary as the extraction device, followed by LC-MS/MS analysis. This non-invasive, simple, and highly sensitive technique has confirmed that tobacco smoke exposure acts as a stressor for non-smokers.

Finally, as editor of this Special Issue, I hope that the articles published will inspire new perspectives and ideas for further research.

Acknowledgments: As Guest Editor of this Special Issue, I thank all of the authors for their contributions, and hope the contents of this publication will help readers to further develop their research.

Conflicts of Interest: The author declares no conflicts of interest.

List of Contributions

1. Riboni, N.; Ribezzi, E.; Bianchi, F.; Careri, M. Supramolecular Materials as Solid-Phase Microextraction Coatings in Environmental Analysis. *Molecules* **2024**, *29*, 2802. https://doi.org/10.3390/molecules29122802.

2. Cardoso, A.T.; Martins, R.O.; Lanças, F.M. Advances and Applications of Hybrid Graphene-Based Materials as Sorbents for Solid Phase Microextraction Techniques. *Molecules* **2024**, *29*, 3661. https://doi.org/10.3390/molecules29153661.
3. Kataoka, H.; Ishizaki, A.; Saito, K.; Ehara, K. Developments and Applications of Molecularly Imprinted Polymer-Based In-Tube Solid Phase Microextraction Technique for Efficient Sample Preparation. *Molecules* **2024**, *29*, 4472. https://doi.org/10.3390/molecules29184472.
4. Ahn, S.; Bae, S. Synthesis and Characterization of a Multi-Walled Carbon Nanotube–Ionic Liquid/Polyaniline Adsorbent for a Solvent-Free In-Needle Microextraction Method. *Molecules* **2023**, *28*, 3517. https://doi.org/10.3390/molecules28083517.
5. Dugheri, S.; Cappelli, G.; Fanfani, N.; Squillaci, F.; Rapi, I.; Venturini, L.; Vita, C.; Gori, R.; Sirini, P.; Cipriano, D.; et al. Vacuum-Assisted MonoTrapTM Extraction for Volatile Organic Compounds (VOCs) Profiling from Hot Mix Asphalt. *Molecules* **2024**, *29*, 4943. https://doi.org/10.3390/molecules29204943.
6. Song, P.; Xu, B.; Liu, Z.; Cheng, Y.; Chao, Z. The Difference of Volatile Compounds in Female and Male Buds of *Trichosanthes anguina* L. Based on HS-SPME-GC-MS and Multivariate Statistical Analysis. *Molecules* **2022**, *27*, 7021. https://doi.org/10.3390/molecules27207021.
7. Rajendran, S.; Silcock, R.; Bremer, P. Volatile Organic Compounds (VOCs) produced by *Levilactobacillus brevis* WLP672 Fermentation in Defined Media Supplemented with Different Amino Acids. *Molecules* **2024**, *29*, 753. https://doi.org/10.3390/molecules29040753.
8. Dugheri, S.; Cappelli, G.; Fanfani, N.; Ceccarelli, J.; Marrubini, G.; Squillaci, D.; Traversini, V.; Gori, R.; Mucci, N.; Arcangeli, G. A New Perspective on SPME and SPME Arrow: Formaldehyde Determination by On-Sample Derivatization Coupled with Multiple and Cooling-Assisted Extractions. *Molecules* **2023**, *28*, 5441. https://doi.org/10.3390/molecules28145441.
9. Zang, Y.; Hang, N.; Sui, J.; Duan, S.; Zhao, W.; Tao, J.; Songqing, L. Magnetic Persimmon Leaf Composite: Preparation and Application in Magnetic Solid-Phase Extraction of Pesticides in Water Samples. *Molecules* **2024**, *29*, 45. https://doi.org/10.3390/molecules29010045.
10. Liu, Y.; Zhang, Y.; Wang, J.; Wang, K.; Gao, S.; Cui, R.; Liu, F.; Gao, G. Preparation of COPs Mixed Matrix Membrane for Sensitive Determination of Six Sulfonamides in Human Urine. *Molecules* **2023**, *28*, 7336. https://doi.org/10.3390/molecules28217336.
11. Dugheri, S.; Squillaci, D.; Saccomando, V.; Marrubini, G.; Bucaletti, E.; Rapi, I.; Fanfani, N.; Cappelli, G.; Mucci, N. An Automated Micro Solid-Phase Extraction (µSPE) Liquid Chromatography-Mass Spectrometry Method for Cyclophosphamide and Iphosphamide: Biological Monitoring in Antineoplastic Drug (AD) Occupational Exposure. *Molecules* **2024**, *29*, 638. https://doi.org/10.3390/molecules29030638.
12. Kataoka, H.; Miyata, S.; Ehara, K. Simultaneous Determination of Tobacco Smoke Exposure and Stress Biomarkers in Saliva Using In-Tube SPME and LC-MS/MS for the Analysis of the Association between Passive Smoking and Stress. *Molecules* **2024**, *29*, 4157. https://doi.org/10.3390/molecules29174157.

References

1. Arthur, C.L.; Pawliszyn, J. Solid phase microextraction with thermal desorption using fused silica optical fibers. *Anal. Chem.* **1990**, *62*, 2145–2148. [CrossRef]
2. Kataoka, H.; Lord, H.L.; Pawliszyn, J. Applications of solid-phase microextrac-tion in food analysis. *J. Chromatogr. A* **2000**, *880*, 35–62. [CrossRef] [PubMed]
3. Kataoka, H.; Ishizaki, A.; Saito, K. Recent progress in solid-phase microextraction and its pharmaceutical and biomedical applications. *Anal. Methods* **2016**, *8*, 5773–5788. [CrossRef]
4. Kataoka, H. In-tube solid-phase microextraction: Current trends and future perspectives. *J. Chromatogr. A* **2021**, *1636*, 461787. [CrossRef]

Disclaimer/Publisher's Note: The statements, opinions and data contained in all publications are solely those of the individual author(s) and contributor(s) and not of MDPI and/or the editor(s). MDPI and/or the editor(s) disclaim responsibility for any injury to people or property resulting from any ideas, methods, instructions or products referred to in the content.

Supramolecular Materials as Solid-Phase Microextraction Coatings in Environmental Analysis

Nicolò Riboni *, Erika Ribezzi, Federica Bianchi * and Maria Careri

Department of Chemistry, Life Sciences and Environmental Sustainability, University of Parma, Parco Area Delle Scienze 17/A, 43124 Parma, Italy; erika.ribezzi@unipr.it (E.R.); careri@unipr.it (M.C.)
* Correspondence: nicolo.riboni@unipr.it (N.R.); federica.bianchi@unipr.it (F.B.)

Abstract: Solid-phase microextraction (SPME) has been widely proposed for the extraction, clean-up, and preconcentration of analytes of environmental concern. Enrichment capabilities, preconcentration efficiency, sample throughput, and selectivity in extracting target compounds greatly depend on the materials used as SPME coatings. Supramolecular materials have emerged as promising porous coatings to be used for the extraction of target compounds due to their unique selectivity, three-dimensional framework, flexible design, and possibility to promote the interaction between the analytes and the coating by means of multiple oriented functional groups. The present review will cover the state of the art of the last 5 years related to SPME coatings based on metal organic frameworks (MOFs), covalent organic frameworks (COFs), and supramolecular macrocycles used for environmental applications.

Keywords: solid-phase microextraction; metal organic frameworks; covalent organic frameworks; supramolecular receptors; cyclodestrins; environmental monitoring

1. Introduction

Since its development in 1989 by Arthur and Pawliszyn [1], solid-phase microextraction (SPME) has attracted increasing attention, being nowadays considered one of the most widely applied sample pretreatment techniques [2–4]. Its unique features, integrating sampling, extraction, preconcentration, and desorption into one single step, have provided noticeable advantages in terms of extraction efficiency, reduced sample handling, time consumption, and ease of use [5], thus enabling the development of fast, efficient, and high-throughput analytical methods [6,7]. Both the principles and the different geometries of SPME technologies and devices have been discussed in already-published reviews [2,3,8]. Environmental friendliness is another key feature of SPME, being that the reduced consumption of organic solvents, the prevention of waste generation, the automation, miniaturization, and possibility of in situ measurements are some of the criteria that perfectly comply with the principles of green analytical chemistry [9,10]. Therefore, SPME has been widely applied in different research fields, including environmental monitoring [6,11,12], food [13,14], forensics [15,16], pharmaceutical [15,17], and bio-analysis [4,15,18–20].

The performance of SPME in terms of selectivity and extraction efficiency is mainly related to the absorption/adsorption capacity of the coating [5,21]. Commercial coatings such as polydimethylsiloxane (PDMS), polydivinylbenzene (DVB), polyacrylate (PA), carboxen (CAR), and their combinations are usually characterized by low selectivity and extraction efficiency. To overcome these limitations, molecular hosts as well as nanostructured and porous materials able to promote an enhanced selectivity via size-exclusion mechanisms and oriented intramolecular interactions [6,21,22] have been proposed as novel SPME coatings, providing more selective and efficient solutions [5,7,23–25]. In this context, the functionalization of the material can both strengthen the interactions between the materials and the target compounds or improve analyte segregation.

Owing to the designed 3D architecture and adsorption capacity driven by chemical and size synergistic complementarity, supramolecular receptors have emerged as alternative coatings cable of strongly affecting both the selectivity and extraction capabilities of SPME. The aim of the present review is to cover the state of the art of the last 5 years, summarizing the main advances in the development of SPME coatings based on metal organic frameworks (MOFs), covalent organic frameworks (COFs), and supramolecular macrocycles, including cyclodextrins (CDs), calixarenes, and cavitands, for environmental monitoring. In vivo SPME applications are not discussed since reviews focused on the matter have been very recently published [26–28].

2. Supramolecular-Based Coatings: Deposition Methods

Different approaches have emerged to develop SPME coatings using supramolecular materials, depending on both the nature of the adsorbent material and on the sampling conditions. It is known that the proper planning of the synthetic strategy plays a key role in guaranteeing the availability of the active binding sites together with the thermal, mechanical, and chemical stability of the coating [7,29–31]. The first step to be performed is the activation of the supporting material: to this aim, different treatments have been proposed, including ultrasonic treatment using organic solvents, etching using mineral acids (HF, HCl, HNO_3), or deposition of a sol–gel coating bearing amino moieties, e.g., (3-aminopropyl) triethoxysilane (APTES). Both the ultrasonic treatment and acidic etching are useful to clean the support of the fiber, remove passive layers, and expose the metal ions/active groups on the support surface to strengthen the adhesion between the coating and the substrate. According to Omarova et al. [29], the coating approaches can be divided in two main groups: (i) in situ deposition during framework development, allowing the synthesis of the material directly on the surface of the fiber; (ii) two-stage deposition, including the synthesis of the adsorbent material and the post-synthetic deposition (Figure 1).

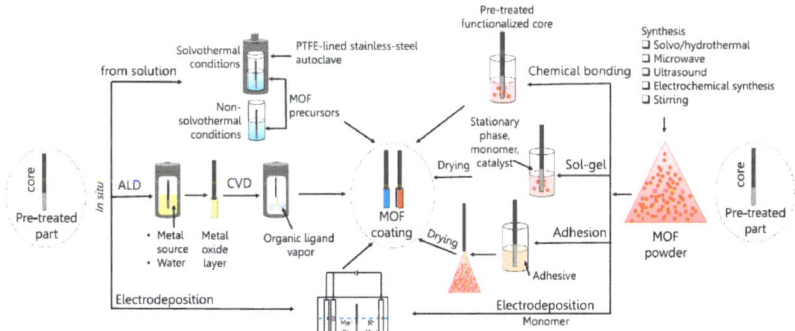

Figure 1. Schematic representation of the different approaches used for the development of supramolecular-based SPME coatings. Reprinted with permission from Omarova et al. [29].

In situ growth (or deposition from solution), chemical vapor deposition (CVD), and atomic layer deposition (ALD) are the most commonly applied methodologies of in situ deposition. The main advantages of these approaches rely on the development of coatings containing only the supramolecular frameworks, thus increasing the extraction selectivity.

In situ growth is based on the immersion of the pretreated fiber in the reacting solution, allowing for the synthesis of the material directly onto its surface [7,29–32]. In this context, stainless-steel (SS) supports are preferred to fused silica and quartz substrates, being that these fibers very fragile. Compared to the other in situ approaches, this technique is simpler and does not require the use of expensive equipment. However, it is time-consuming and several parameters, including reaction time, concentration of reagents, and agitation speed, need to be optimized to obtain stable and reproducible coatings [29,30].

CVD involves the deposition of the coating onto a heated substrate surface via the chemical reactions of evaporated precursors [33]. The structure, morphology, dimensions, and orientation of the coating can be finely tuned by changing the deposition conditions. This process is highly reproducible and can be performed at higher rates compared to the solution deposition [29]. Atomic layer deposition (ALD, or ALD–molecular layer deposition) is a nanotechnological modification of CVD, involving a vapor-phase layer-by-layer deposition of the coating materials onto a solid support exploiting gas–surface reactions [34]. The deposition process is divided into sequential reaction stages between the precursors and substrate, leading to the consecutive deposition of reactants (Figure 2): (i) in the first step a precursor (metallic center, polymeric material, or organic linker) is pulsed into a vacuum-heated chamber containing the fiber; (ii) after the deposition of the precursor layer, the unfunctionalized reactants are removed from the chamber; and (iii) the other precursors are purged onto the already-formed layer, leading to a formation of a second layer of the coating. The process is repeated until the desired coating is obtained. ALD provides a unique control on thin film formation in terms of conformality and uniformity, with atomic-level accuracy [29,34].

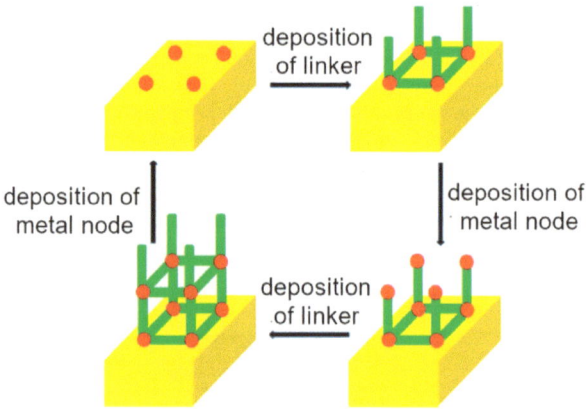

Figure 2. Schematic representation of the ALD process, edited from Zhou et al. [34].

As for dual-stage deposition, different approaches have been proposed, including physical adhesion, the sol–gel technique, and chemical cross-linking [7,29,30,32].

The physical adhesion of a pre-synthesized material is the most commonly applied procedure [7,29–32]. Besides fiber pretreatment, the coating process consist of several steps: (i) the inert surface is immersed in an adhesive material (e.g., silicone-based sealants or epoxy glues) or pre-coated with a polymeric layer (usually polyacrylonitrile—PAN); (ii) the fiber is dipped in the fine powdered supramolecular material to obtain a coating with the intended thickness (multiple dipping cycles can be required); (iii) an additional layer of adhesive material can be applied to increase the mechanical stability of the fiber; and (iv) the fiber is dried and thermally conditioned to promote coating hardening, curing, and allowing for the evaporation of both the residual solvents and organic sub-products. Steps i–iii can be merged when the pretreated fiber is directly immersed into a solution containing the suspended material without the need of an adhesive medium [29–31]. In this case, the immersion–heating stages are usually performed multiple times to obtain the intended coating thickness. In physical adhesion, the adhesive (or the suspending solvent) used for developing the coating has a major impact on the SPME performance, affecting the thermal and mechanical stability of the material, as well as the resistance to organic solvents. The main advantage of this approach relies on its simplicity, cost-effectiveness, and versatility. Since the synthesis of the material is not affected by the deposition, microwave- or ultrasound-assisted reactions can be applied to obtain the supramolecular material [30]. The main drawback of this approach is related to the presence of adhesives or polymeric substances that could clog the pores or change the surface chemistry of the adsorbent, thus reducing the surface area or affecting the extraction efficiency and selectivity.

Concerning the sol–gel technique, it involves the embedding of the supramolecular materials into a gel network for coating deposition [29,30,35,36]. The pre-synthesized supramolecular material can be either dispersed in a sol–gel suspension or deposited after the gel formation onto the SPME support. The produced coatings are characterized by high homogeneity and uniformity as well as enhanced thermal, mechanical, and chemical resistance compared to the use of adhesives. In addition, the sol–gel process usually occurs under very mild conditions, and the thickness, composition, and morphology of the coating can be finely tuned by changing the concentration of the precursors, the dipping rate, the number of dipping cycles, and the drying temperature. The main drawback of this approach relies on the encapsulation of the supramolecular framework within the gel network, which can lead to the pores clogging and possible interactions between the silica gel and the compounds present in the sample.

Chemical cross-linking is based on the chemical bonding of the pre-synthesized supramolecular material with the fiber surface [25,29,30]. The fiber is pre-treated to obtain carboxylic or amino groups at the surface to be used as binding sites for the immobilization of the material. In this context, the frameworks are linked to the fiber with strong covalent bonds, thus providing high thermal and mechanical stability. Since no additive is required, selectivity is improved, and the extraction/desorption kinetics are higher than those achieved by the coatings obtained using other dual-stage deposition processes. However, chemical cross-linking could be a complex procedure and the amount of loaded material and the coating thickness could present a serious challenge to be tuned during coating deposition.

Finally, supramolecular coatings onto SPME fibers have been also obtained by using the electrochemistry approach: [29,31,37,38] in this case, both in situ and two-stage electrodeposition have been applied, depending on the synthetic approach, i.e., electrochemical reduction/oxidation or deposition of pre-synthesized supramolecular sorbents or by using conductive polymers such as polyaniline (PANI), polypyrrole (PPy), or polythiophene. The main advantages of this process rely on its high efficiency, repeatability, and ease of control, thus allowing the acquisition of coatings characterized by uniform thicknesses and tuned porous structures. By contrast, the main drawbacks are related to the use of a polymeric network embedding the adsorbent material and the possible redox reaction that could modify the properties of the supramolecular material.

3. Metal–Organic Frameworks

MOFs are crystalline, porous hybrid materials, based on the self-assembly of metal ions (or metal clusters) and organic ligands via coordinative bonds, leading to the development of 3D coordination polymers [7,17,25,29,31]. Their structure could be summarized considering the metals as the nodes of a highly porous crystal lattice, coordinated by rigid organic linkers, creating the scaffolds of the 3D framework.

3.1. Synthetic Procedures

Different procedures have been proposed to synthesize MOFs, having a great impact on their structure and physical properties, namely morphology, lattice dimension, porosity, and surface area [31,38]. The main strategies used to produce MOF-based coatings rely on hydro/solvothermal methodology, microwave-assisted synthesis, sonochemistry, mechanochemical, and electrochemistry, as depicted in Figure 3.

Both the hydrothermal and the solvothermal approaches are based on the dissolution of metal salts and organic ligands in either water (hydrothermal) or an organic (solvothermal) medium, followed by reaction at high temperatures in pressurized reactors. The obtained material is then cooled, washed, and vacuum-dried to remove the solvent and sub-products from the pores of the framework. Although these approaches are simple, long reaction times and the use of high volumes of organic solvents or harsh conditions are usually required. Hydro/solvothermal synthesis has been frequently used for the in situ growth of MOFs [17,29–31,38]. To fasten the nucleation process, microwave-assisted synthesis has been proposed, obtaining MOF particles with a homogeneous size distribution in a very short time. Similarly, the ultrasound-assisted method has been proposed as a valid alternative to the previously described approach, being able to synthesize MOFs in shorter times, under milder conditions, and with high reaction yields. In fact, the acoustic cavitation provides a high local temperature and pressure, followed by high cooling rates, which facilitate both the reaction and the nucleation processes [38]. Microwave- and ultrasound-assisted synthesis are usually not compatible with in situ deposition; therefore, they are mostly proposed using the two-stage processes [31,38]. Recently, mechanochemical synthesis (MC), based on the use of a mechanical force to break the intramolecular bonds and promote chemical transformation, has emerged as a valuable green approach [39]. Being solvent-free, improvements in terms of reaction time, energy conversion efficiency, and reduction in organic waste, can be achieved. Finally, the electrochemical approach can be used to synthesize the MOFs directly onto the SPME fiber: the metal ions are obtained by anodic dissolution of the metal salts and react with the dissolved organic linkers.

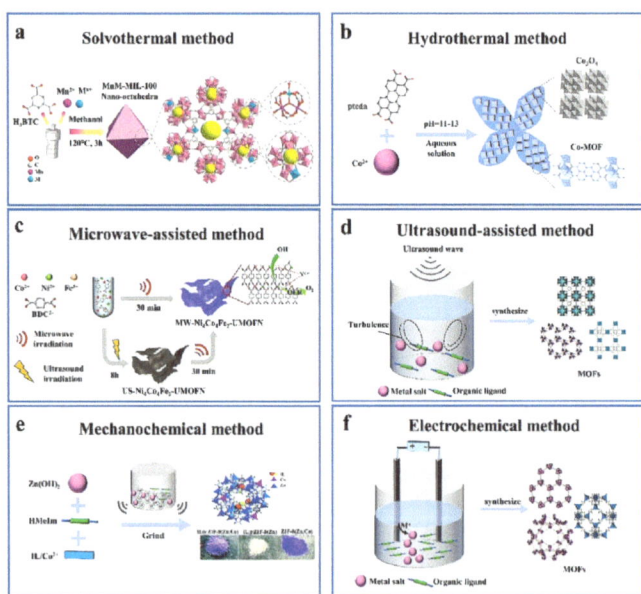

Figure 3. Schematic representation of the different approaches for MOF synthesis. (**a**) Solvothermal method; (**b**) hydrothermal method; (**c**) microwave-assisted synthesis; (**d**) ultrasound assisted method; (**e**) mechanochemical method; (**f**) electrochemical assisted approach. Reprinted with permission from Zhang et al. [38].

3.2. MOF Features for SPME Extraction

Since their first application as SPME coatings in 2009 [40], MOFs have demonstrated excellent capability for the extraction of compounds belonging to different chemical classes due to their high surface area, good thermal and chemical stability, controlled pore size, and tunable surface chemistry by selecting metal ions, organic ligands, or synthetic conditions [25,29,31]. In particular, adsorption studies have demonstrated that after activation MOFs present surface areas characterized by Brunauer–Emmett–Teller (BET) values up to 7000 m^2/g [29,31], higher than commonly applied SPME coatings such as activated carbon (~1500 m^2/g), zeolites (300–600 m^2/g), silica-based (300–800 m^2/g) [41,42], or commercially available coatings, e.g., DVB (750 m^2/g) [43]. The proper design of the MOF framework can also play a key role in enhancing selectivity by strengthening the steric complementarity between the host and the guests or by enhancing MOF–analyte interactions via hydrogen bonding, π-π, CH-π, electrostatic, van der Waals, hydrophobic/hydrophilic interactions, or via the coordination of the free metal sites present in the lattice [25,44].

3.3. MOFs as SPME Coatings for Environmental Applications

SPME based on MOF coatings has been widely applied for environmental monitoring, combining the simplicity, miniaturization, automation, and greenness of the extraction technique with the enhanced extraction capabilities and selectivity of 3D designed frameworks. These aspects are of paramount importance to meet the demand for analytical methods able to detect pollutants at trace levels in complex matrices [6,11]. The main application of MOF-based SPME coatings for environmental applications are discussed in the next subsections and summarized in Table 1.

3.3.1. Extraction of Benzene, Toluene, Ethylbenzene, and Xylenes

Benzene, toluene, ethylbenzene, and xylenes (BTEX) are volatile non-polar organic compounds mostly derived from petrochemical fuel combustion and used as solvents in the pharmaceutical and dye industries. The International Agency for Research on Cancer (IARC) set benzene in group 1, ethylbenzene in group 2B, and toluene and xylenes in group 3. The main challenge in the detection of these compounds is related to their presence at trace levels in environmental samples, generally interfered with by the overwhelming amounts of aliphatic hydrocarbons [45,46]. Several MOFs have been proposed to improve the extraction efficiency and selectivity, including Zeolitic Imidazolate Frameworks (ZIFs) and Matériaux de l'Institut Lavoisier (MIL). These materials demonstrated both π-π stacking and size complementarity between the analytes and the framework.

ZIFs are a class of MOFs that feature divalent metal nodes coordinated by imidazole organic ligands to obtain a material with a tetrahedron structure characterized by enhanced thermal and water stability. These materials have been applied as SPME coatings since 2010 [47]; however, in 2019 Maya and coworkers proposed the use of the single-crystal ordered macroporous (SOM) ZIF-8 as an SPME coating for BTEX extraction from wastewater samples [48]. The material was obtained by synthesizing a polystyrene monolith and a ZIF-8 MOF was grown in its interstices. Then, the polymer was dissolved using dimethylformamide, obtaining the SOM-ZIF-8 material, which was deposited onto SS using a silicone sealant. The obtained coating showed an ordered array of macropores and was characterized by 2.5–3.1-fold times higher intensities than both the crystalline ZIF-8 and the commercial PDMS fibers. The HS-SPME-GC-FID-based method was validated, obtaining limits of quantitation (LOQs) in the low ng/L level, allowing for BTEX detection in environmental samples. ZIF-8 was also the starting material for developing a superhydrophobic MOF composite material ($_{NS}ZIF-8^{Si}$) [49]: after synthesis of Mn_xO_y nanosheets, in situ growth of a ZIF-8 MOF was performed on the nanomaterial, followed by deposition of PDMS on its surface. The SPME fiber coating was obtained by physical adhesion using a silicone sealant. The developed material showed excellent thermal stability (up to 450 °C), superhydrophobic behavior, and significantly higher extraction efficiencies than commercial 30 μm PDMS due to the combined hydrophobic and π-π stacking effects.

Recently, Hu and coworkers [50] used the MIL-101-NH$_2$ framework, a moisture-sensitive MOF, as a template to develop a carbonaceous coating. After solvothermal synthesis, cobalt and thiourea were encapsulated inside the MOF structure and the material was annealed, obtaining urchin-like nanoporous carbon. The final coating was obtained by physical deposition using a silicone sealant. The extraction performance of the urchin-like nanoporous carbon-based SPME coating was better than that obtained by using commercial PDMS and MIL-derived nanoporous carbon not functionalized with cobalt and thiourea, achieving LOQ values in the 0.28–1.2 ng/L range. Finally, the material provided a very high sampling rate for all the analytes, reaching extraction equilibrium within 2 min. Kardani and coworkers developed a polyacrylonitrile/MIL-53(Al)MOF@SBA-15/4,4′-bipyridine hybrid nanocomposite [51]. The material was based on mesoporous silica obtained by the sol–gel procedure, functionalized with a MIL-53(Al) MOF by solvothermal reaction. The coating was obtained by electrospinning PAN onto an SS fiber in the presence of the hybrid nanocomposite material. A very porous structure with a high degree of cross-linking and increased adsorption/desorption rates was achieved. Finally, LOQs in the 7.6–20 ng/L range allowed the ultra-trace determination of BTEX in wastewater samples.

BTEX extraction from air samples using a MOF-199-based coating deposited in situ via a solvothermal approach was proposed by Omarova et al. [52]. Different thicknesses of the MOF-199 coating were tested for the extraction of 20 low-molecular-weight VOCs including BTEX and 3 high-molecular-weight VOCs. For most of the substituted monoaromatic compounds, the extraction capabilities of the developed coating were significantly higher than those achieved by using a PDMS/DVB fiber. Finally, LOQs in the range of 0.09–0.31 µg/m^3 proved the reliability of the MOF-199 coating for air monitoring purposes.

3.3.2. Extraction of Polycyclic Aromatic Hydrocarbons

Polycyclic aromatic hydrocarbons (PAHs) are a broad class of compounds consisting of two or more fused aromatic rings, exhibiting a linear, angular, or cluster arrangement. These compounds are characterized by extremely low water solubility, very different vapor pressures, high toxicity, and high bioaccumulation potential; therefore, they are considered as priority pollutants by both the United States Environmental Protection Agency (US EPA) and the European Environment Agency [53]. The main analytical challenge is related to their detection at sub-trace levels in the presence of overwhelming amounts of interfering compounds.

ZIF-based coatings were developed and tested for the extraction of PAHs using different deposition approaches or post-synthetic functionalization of the MOFs. In 2019, Kong and coworkers synthesized ZIF-8 MOFs by different approaches, namely solvothermal, stirring, and ball-milling methods, and deposited the MOF-based material onto an SS wire using a dual-stage deposition via the sol–gel technique [54]. The obtained coatings were tested for the direct immersion (DI)—SPME—gas chromatography—mass spectrometry (GC—MS) analysis of 16 US EPA PAHs and 11 nitro-PAHs. The best enrichment capabilities were obtained by using the solvothermally synthesized material, which was also characterized by the highest BET surface area (1390 m^2/g). Under optimal conditions, LOQs in the 1.1–90.0 ng/L range were obtained, thus assessing the reliability of the method for the detection of PAHs at sub-trace levels. Finally, the analysis of environmental samples highlighted high levels of contamination of river and lake water and industrial wastewater, with PAHs detected in the 30.0($\pm$1.1)–1509($\pm$83) ng/L range. A ZIF-8-based SPME coating was synthetized in situ onto metal alloy wires by CVD and then solvothermally grown by Rocio-Bautista and coworkers [55]. The coating was applied for the extraction of acenaphthene, fluorene, and pyrene as model PAHs together with homosalate, ethylhexylsalicilate, methyl-antralinate, padimate-O, and ethylhexyl 4-methoxycinnamate as personal care product representatives. The DI-SPME-GC-FID method was characterized by limit of detections (LODs) in the 0.6–2 µg/L range and relative standard deviations (RSDs) $\leq$ 23%; however, a strong matrix effect was observed when wastewater samples were analyzed. Lian and coworkers studied the in situ electrochemical deposition of ZIF-8 onto SS wires pretreated using dopamine under alkaline conditions [56]. A homogeneous porous coating characterized by very high thermal stability (up to 460 °C) was obtained. The coating was tested for the HS-SPME-GC-FID analysis of eight low-molecular-weight PAHs obtaining LOQs in the 35–179 ng/L range, allowing for the detection of acenaphthene, acenaphthylene, and fluoranthene in lake water at concentration levels in the 200 $\pm$ 13–360 $\pm$ 33 ng/L range. Zeng et al. developed ZIF-8-based coatings with tunable properties in terms of polarity, porosity, surface area, and conductivity, thus allowing the extraction of either polar or apolar compounds [57]. SS wires coated with PANI deposited by means of cyclic voltammetry were used as an SPME support. Aligned ZnO nanorods were deposited using a dipping -seed-mediated method and then a hydrothermal method was applied for ZIF-8 in situ growth. The developed coating was tested both for the electroenhanced SPME extraction (EE-SPME) of aromatic amines and ionic drugs in spiked standards and for the DI-SPME-GC-FID analysis of six low-molecular-weight PAHs in sewage water samples. Electrochemical deposition was applied for a two-step synthesis to obtain a nickel/titanium alloy coated with cobalt and for the in situ growth of a ZIF-67 coating [58]. The material was then annealed to obtain a Co@ZIF- 67-derived coating to be used for the selective DI—SPME—liquid chromatography—UV detection (LC—UV) of six PAHs. The proposed material was characterized by a higher extraction capability and selectivity than commercially available materials and other developed coatings, with LODs in the 5–42 ng/L range, good linearity, and precision (RSD $\leq$ 7%). Finally, the method was applied for the analysis of snow, lake and river water, and wastewater samples, detecting a total concentration (Σ) of six PAHs in the 1.47–6.34 µg/L range.

Metal azolate frameworks (MAFs) were also tested as SPME coatings by Liu et al. [59] and Wang et al. [60] for the extraction of 7 and 16 PAHs in water, respectively. In both studies, Zn was used as the metal ion and the coatings were obtained by a two-stage methodology consisting of a preliminary step based on the solvothermal synthesis of the MOFs, followed by the dipping of etched SS fibers in a methanolic solution containing the dispersed MOFs. In Liu et al. [59], the synthesis of MAF-66 was performed at room temperature and the layer-by-layer deposition was completed after 10 dipping cycles. The SPME fiber was then used for the HS-SPME-GC-FID analysis of low-molecular-weight PAHs, obtaining LODs in the 0.1–7.5 ng/L range. Two different materials, namely MAF-5 and MAF-6, were tested by Wang et al. [60] for the extraction of the 16 US EPA priority pollutant PAHs from aqueous samples, showing that MAF-6 had the highest extraction capability. Finally, the DI-SPME-high-performance liquid chromatography-fluorescence detection (HPLC-FLD) method was characterized by LOQs at the sub-µg/L level for all the investigated PAHs, except for fluoranthene (LOQ 1.8 µg/L) and benzo(k)fluoranthene (LOQ 1.5 µg/L).

Sun et al. developed a HKUST-1-based SPME coating using both the in situ growth and the two-stage process using neutral silicone gel as the adhesive [61]. Both coatings were tested for the HS-SPME-GC-MS analysis of eight PAHs from aqueous samples: the in situ grown coating provided a better extraction capability compared to the two-stage coating due to the pore size and radial alignment of channels. In addition, in situ growth is a less time-consuming procedure. The method was characterized by LOQs in the 0.4–33 ng/L range and proved to be suitable for the extraction of eight PAHs from lake water samples, allowing the detection of medium-weight PAHs in both sites, with concentrations in the 15.3–64.4 ng/L and 19.6–56.9 ng/L range, respectively.

Bianchi et al. tested the extraction capability of a triple-catenated Zn^{2+}-pillared MOF, named PUM-210, for the extraction of 16 US EPA PAHs [62]. Compared to the commonly applied homoleptic materials, this framework incorporates two different ligands, i.e., 2,6-naphthalenedicarboxylic acid and a pyridine-functionalized biphenylene ligand, for the complexation of Zn(II) ions. The MOF was obtained by solvothermal synthesis and the coating was deposited onto a silica fiber by means of epoxy glue. The developed coating provided low quantitation limits, in the 1–7 ng/L range, and very high enhancement factors in the 300–14,950 range. Finally, the addition of BTEX as interfering compounds did not affect the quantitation of PAHs due to the presence of additional π–π interactions, higher hydrophobicity, and size complementarity between the MOF and the analytes.

Qiu et al. developed a sheathed MOF-based SPME coating through the in situ heteroepitaxial growth of copper-2,5-diaminoterephthalate crystals (Cu-DAT) on copper wires, followed by dip-coating to obtain a PI sheath [63]. The coating was deposited by a multi-step synthesis directly onto the copper wire (Figure 4): (A) copper hydroxide nanotubes were grown onto the fiber to provide a substrate for the in situ growth of the MOF; (B) the fiber was immersed in the ligand solution; (C) the material was annealed; and (D) the PI sheath coating was obtained by dip-coating. The obtained fibers were tested for the extraction of 10 model PAHs from both water samples and fish muscle by using both

HS and DI extraction, obtaining LOQs in the range of 0.3–2.1 ng/L and 4.0–18.9 ng/g, respectively. Finally, the reliability of the developed procedure was assessed both for the detection of PAHs in river water and for their in vivo monitoring in tilapia dorsal muscle, obtaining results comparable to conventional liquid extraction.

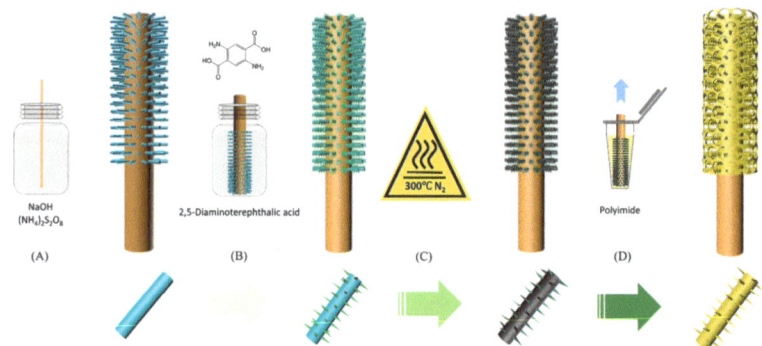

Figure 4. Schematic representation of the in situ heteroepitaxial growth of copper-2,5-diaminoterephthalate crystals on copper wires, followed by dip-coating: (**A**) immersion of copper wire in the conversion solution; (**B**) immersion of the treated copper wire in the saturated ligand solution; (**C**) coating annealing; (**D**) dip-coating using polymide solution. Reprinted with permission from [63].

3.3.3. Extraction of Organophosphorus and Organochlorine Pesticides

Organophosphorus (OPP) and organochlorine (OCP) pesticides are widely employed in agriculture for pest control. Being toxic and persistent compounds, they can exert a harmful impact on both ecosystems and human health, thus requiring strict monitoring to guarantee compliance with the current regulation [64]. To face this challenge, selective, sensitive, and reliable SPME methods based on MOF coatings have been proposed [65–67].

Electrospinning was applied by Amini and coworkers to deposit a polyacrylonitrile/nickel-based MOF (PAN/Ni-MOF) nanocomposite coating on an SS wire for the extraction of diazinon and chlorpyrifos from environmental water samples [65]. The obtained hydrophobic coating exploited the capability of the nitrile backbone of PAN to promote hydrogen bonding and π-π interactions with the OPPs. Corona discharge-ion mobility spectrometry (CD-IMS) was applied for the detection of diazinon and chlorpyrifos, obtaining LOQs of 0.5 and 1 µg/L, respectively. Finally, the reliability of the developed method was assessed by analyzing river water (no pesticide detected), agricultural wastewater (Σpesticides 24.6 µg/L), and groundwater samples near rice fields (Σpesticides 20.0 µg/L).

Mohammadi et al. developed a flexible/self-supported ZIF-67 film for thin-film microextraction of OPPs from agricultural wastewater and underground water [66]. Firstly, electrospinning followed by calcination was used to obtain cobalt oxide fibers; in a second step, functionalization with the MOF was carried out by hydrothermal reaction. The developed film was used for the extraction of ethion using secondary electrospray ionization–ion mobility spectrometry (SESI-IMS) as a detection technique, requiring only 2 min of detection time per sample. In addition, the thermal desorption of the film was in accordance with the principles of analytical green chemistry. Under optimized conditions, a LOQ of 500 ng/L was obtained, allowing for the detection of ethion in one agricultural wastewater sample (1.1 µg/L).

A novel MOF coating was developed by Gong et al. using a two-stage deposition process: a NU-1000 MOF was synthesized by solvothermal reaction and deposited onto SS wires using a silicone sealant [67]. The major features of the framework were the very high surface area (BET 1966 m^2/g) and the presence of mesoporous channels able to increase mass transfer inside the coating. The material was tested for the extraction of six model OCPs at ultra-trace levels from water samples. Although the enrichment factors (EFs) were higher than those obtained by using either commercial 65 μm PDMS/DVB or 85 μm PA fibers were achieved, selectivity issues were observed. In fact, when PAH extraction was carried out, a high adsorption capability was observed, with EFs in the 9800–11,000 range, due to the π-π and CH-π interactions between the polyaromatic compounds and the coating. Under optimized conditions, LOQs in the range of 37–190 ng/L were obtained; thus, the DI-SPME-GC-MS method was suitable for ultra-trace analysis of OCPs in river and pond water, detecting only hexachlorobenzene at a concentration of 1.86 ± 0.08 ng/L and 2.08 ± 0.13 ng/L, respectively.

Xu et al. developed a lotus-like Ni@NiO–embedded porous carbon (Ni@NiO/PC) SPME coating for the extraction of chlorobenzenes from aqueous samples [68]. The coating was prepared following a two-stage deposition process: after solvo- and hydrothermal synthesis of cellulose nanocrystals functionalized with MOF-74, the material was pyrolyzed to obtain the Ni@NiO/PCs, which was deposited onto SS wires by means of a silicone sealant. The obtained coating was able to adsorb the chlorobenzenes more effectively compared to both the pristine materials and commercial coatings. The target analytes could diffuse through the surface of the material due to the presence of meso- and micropores with a size of 1.26 nm, interacting with the active sites of the coating via dipolar (thanks to the Ni-O groups) and π-π and hydrophobic interactions. The coating was tested for the DI-SPME-GC-MS analysis of eight model chlorobenzenes in tap and surface water: despite the LOQs ranging from 0.017 to 0.118 ng/L, no chlorobenzene was detected in tap water, whereas 1,3,5-trichloro-benzene (15.7, 28.9 ng/L), 1,2,3-trichloro-benzene (14.4, 10.9 ng/L), and 1,2,4,5-tetrachloro-benzene (1.4 ng/L) were detected in river water samples.

3.3.4. Extraction of Poly- and Perfluoroalkyl Substances

Poly- and perfluoroalkyl substances (PFASs) are a wide class of emerging pollutants, accounting for thousands of synthetic compounds with linear, branched, or cyclic structures, characterized by a hydrophilic terminal, such as a carboxylic or sulfonate group, and a hydrophobic fluorinated alkyl chain. PFASs have been widely used in manufacturing, packaging, the textile and semiconductor industry, or as foams and detergents [69]. Owing to the chemical and biological stability of the C-F bonds, these compounds are extremely persistent in the environment and can be easily accumulated in the tissues of living organisms, thus being considered as a major threat to human life and ecosystems [69]. Both EU and US EPA have set stringent regulations for PFASs, limiting their concentration at trace levels in drinking water and environmental samples. Analytical monitoring of these compounds is a challenging issue due to their unique hydrophobic and oleophobic nature [69]. In this context, MOFs have been proposed as valuable coating materials for PFAS extraction [70–73].

Four different water-resistant MOFs, namely, ZIF-8, UiO-66, MIL88-A, and $Tb_2(BDC)_3$, were investigated as coatings for probe extraction of perfluorooctanoic acid (PFOA) from environmental water, being that this compound is one of the most widely spread PFASs [70]. The coating was obtained by in situ growth of the MOFs onto poly(dopamine) (PDA)-coated SS needles. ZIF-8 and UiO-66 were able to adsorb PFOA inside the pores and exhibited similar binding energies, whereas MIL88-A and $Tb_2(BDC)_3$ ZIF-8 were characterized by surface adsorption, being that their pores were too small for hosting the analyte. Due to the lower production costs and higher reproducibility, only ZIF-8 was selected for PFOA quantitation. Finally, a DI-SPME extraction followed by direct analysis by nanoelectrospray ionization–mass spectrometry (nanoESI-MS) method was validated, allowing for very rapid analysis (10 min per sample, including pretreatment and MS analysis) and achieving a LOD value of 11 ng/L. Therefore, the reliability of the method was demonstrated for the purpose of high-throughput screening of PFOA at trace levels. The validated method was applied for the quantitation of PFOA in contaminated samples of tap water, rainwater, and seawater. A strong matrix effect was observed, especially when seawater samples were analyzed, obtaining a sensitivity (7.75×10^8 M^{-1}) higher than that observed in ultrapure water (2.43×10^8 M^{-1}), probably as a consequence of the alkalinity and ion abundance of seawater.

Another strategy to improve the enrichment capability of SPME through electrostatic interactions and hydrogen bond formation between the amine group and the acidic group of PFAAs was based on the functionalization of MOFs with amino groups [71–73]. In this context, the extraction of eight model PFASs using an amino-functionalized ZIF-8 (NH$_2$-ZIF-8) coating followed by HPLC-MS/MS analysis was proposed by Gong and coworkers [71]. The MOF was synthesized by solvothermal reaction and deposited onto quartz fibers by means of PAN. The developed coating showed a higher performance than pristine ZIF-8 and commercial fibers, namely 65 µm PDMS/DVB and 85 µm PA fibers. After optimization and validation, the DI-SPME-HPLC-MS/MS method was applied for the analysis of river water, seawater, and wastewater treatment plant effluent samples, allowing for the detection of the target analytes at ng/L levels. In 2022, Ouyang et al. proposed a modification of UiO-66(Zr) introducing both amino groups on the surface and the end-capping of the material to avoid any interaction with water molecules, thus enhancing the hydrophobic interactions [72]. The NH$_2$-UiO-66(Zr) MOF was obtained in a one-step solvothermal synthesis, capped using phenylsilane, and deposited onto the fiber by means of PAN. The presence of capped nanocrystals (NH$_2$-UiO-66(Zr)-hp) resulted in higher hydrophobicity, faster extraction kinetics and higher enrichment capacity compared to the uncapped coating. The DI-SPME-HPLC-MS/MS method was validated, obtaining LOQs in the range of 0.21–1.8 ng/L, and applied for the analysis of tap, river, and pond water samples detecting ΣPFASs of 29.3, 58.7, and 45.0 ng/L, respectively. A similar approach was proposed by Jia et al., who developed an SPME coating based on MIL-101(Cr) functionalized with amino and fluorinated groups to enhance the interaction towards PFASs [73]. As depicted in Figure 5, the bifunctional MOF was synthesized by the hydrothermal method, aminated by diethylenetriamine (DETA), and fluorinated using perfluorooctanoyl chloride. Finally, the coating was obtained by dipping the SS needle coated with epoxy resin in the developed material. The dual-functionalized MIL-101(Cr) (MIL-101-DETA-F) SPME coating was applied for the extraction of nine model PFASs, obtaining EFs in the 70–112 range: these values were higher than those obtained by using UiO-66(Zr), MIL-101(Cr), the mono-amino-functionalized MOF, and the commercial PDMS/DVB coatings. In addition, the pore size of MIL-101-DETA-F was close to the molecular size of perfluorododecanoic acid, allowing for the size exclusion of interfering compounds. Compared to the monofunctionalized MOF, fluorination enabled F−F interactions, improving both the selectivity and extraction capability of SPME. Finally, the DI-SPME-ultra-high-performance liquid chromatography-tandem mass spectrometry (UHPLC-MS/MS) method was validated for PFAS extraction in aqueous samples, achieving extremely low LOQ values, in the range of 0.01−0.40 ng/L, a three order of magnitude linear range, and RSDs $\leq$ 12%. The ΣPFASs was 5.5 and 27.2 ng/L for tap and river water, respectively, and in the 263.8–2790.7 ng/L range for municipal, dyeing, and mining wastewater samples.

3.3.5. Extraction of Pharmaceutical and Personal Care Products

Pharmaceutical and personal care products (PPCPs) are emerging pollutants with potential detrimental effects on human health, the environment, and ecosystems. Their widespread use, inappropriate disposal, and incomplete removal by conventional wastewater treatment procedures have led to their ubiquitous presence in all the environmental compartments. Due to their presence at trace or sub-trace levels in various environmental matrices and their different physical and chemical properties, the development of a reliable method for their monitoring is still a challenging issue [74].

In this context, new MOF-based coatings for the detection of nonsteroidal anti-inflammatory drugs and anti-cancer drug residues have been reported by the research group of Liu and Khodayari [75,76]. In particular, a zirconium-based MOF and graphene oxide nanocomposite coating (Zr-MOF@GO) was synthesized by hydrothermal reaction and deposited onto SS wires by means of an aluminum chromium phosphate binder and applied for the DI-SPME-GC-FID determination of ibuprofen and diclofenac [75]. The extraction mechanism of the hybrid material was based on free-metal site coordination, hydrogen bond formation, and π–π interactions, achieving LOQs of 10 and 100 ng/L, respectively. Finally, the method was applied for the quantitation of the investigated drug residues in a Yellow River water sample, obtaining similar results to those achieved using a commercial PA fiber, i.e., an ibuprofen concentration of 27 ng/L and a diclofenac concentration below the LOQ. Khodayari and coworkers [76] developed electrospun polyfam/Co-MOF-74 composite nanofibers for the thin-film SPME extraction of sorafenib, dasatinib, and erlotinib, used as anti-cancer drug representatives. The Co-MOF-74 particles were obtained by the hydrothermal procedure and deposited onto the SS fibers by electrodeposition using polyfam as a conductive polymer. The incorporation of the MOF within the polyfam fiber network produced a significant increase in both the surface area and total pore volume compared to the pure polyfam electrospun fibers (28.3 m^2/g and 0.244 cm^3/g vs. 8.4 m^2/g and 0.044 cm^3/g), thus improving the extraction efficiency toward the anti-cancer drugs. The optimized thin-film-SPME-HPLC-UV method was validated, obtaining EFs in the 24.2–37.1 range and allowing for the quantitation of the analytes at the sub-µg/L level. Finally, the analysis of environmental water samples revealed a high level of contamination, being that the concentration levels of the detected analytes were always higher than 10 µg/L.

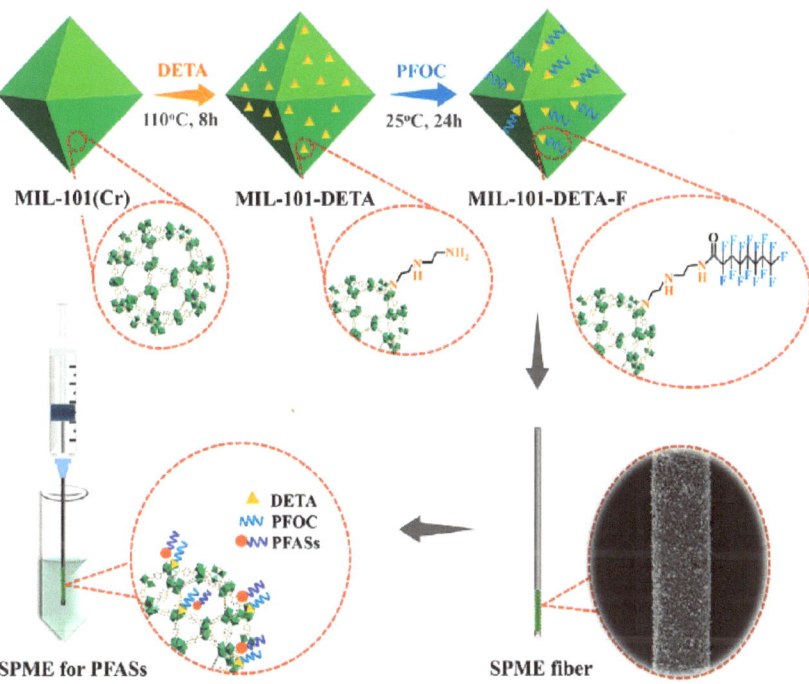

Figure 5. Schematic representation of the synthesis and deposition of dual-functionalized MIL-101-DETA-F onto SPME fibers, reprinted with permission from [73].

Linear and cyclic methylsiloxanes are another class of contaminants of emerging concern, being used in the formulation of PPCPs to improve their physical properties and as precursors of silicone polymers. To improve the detection of these pollutants in wastewater samples, both a MIL-101-based and a CIM-80(Al)-based coating were recently proposed by Zhang et al. and by González-Hernández and coworkers [77,78]. The MIL-101-based fiber was obtained by SS fiber dip-coating in a solution containing MIL-101 and polysulfone [77] and an HS-SPME-GC-MS method was validated for the extraction of octamethylcyclotetrasiloxane, decamethylcyclopentasiloxane, and dodecylcyclohexasiloxane from environmental water samples. Under optimal conditions, the MIL-101-based method provided LOQs in the 140–270 ng/L range and RSDs ≤ 4%. The extraction performance of the developed coating was better or equal to those obtained by using a commercial 65 µm PDMS/DVB fiber, which was affected by high background contamination. Finally, the validated method was applied to quantify the target analytes in wastewater samples collected at different treatment stages from a wastewater treatment plant in Guangzhou, China, obtaining the highest contamination levels in the aeration unit (1.8–85.1 µg/L) and barscreen (0.7–3.3 ng/L). Regarding the CIM-80(Al)-based coating, it was obtained on APTES-pretreated nitinol wires by in situ growth [78]. The fibers were used for the HS-SPME-GC-MS analysis of six linear and cyclic methylsiloxanes and seven musk fragrances as model PCPs. The performance of the MOF-based coating was compared with that achieved by using the commercially available 65 µm PDMS/DVB-coated fibers: the developed coating provided lower or comparable LOQs for cyclic methylsiloxanes compared to the commercial one (in the 0.2–0.7 µg/L vs. 0.5–0.6 µg/L range), whereas an opposite behavior was observed for linear methylsiloxanes (0.1–0.4 µg/L vs. 0.05–0.06 µg/L). As for musk fragrances, LOQs in the 1.2–3.5 µg/L and 0.1–0.4 µg/L ranges were obtained for the MOF-based and commercial coating, respectively. However, the CIM-80(Al)-based coating demonstrated slightly better precision, a larger linear range, and reduced cross-contamination compared to the commercial fibers. Finally, the HS-SPME-GC-MS method was applied for the analysis of three seawater and three wastewater samples, detecting cashmeran (15 ± 2 and 1.4 ± 0.2 µg/L) and galaxolide (46.9 ± 0.2 and 6 ± 1 µg/L) in two out of three wastewater samples, whereas no analyte could be quantified in the analyzed seawater samples.

A superhydrophobic amino-functionalized UiO-66(Zr) coating was proposed by Liu et al. for the extraction of semi-volatile UV filters from environmental water samples [79]. The superhydrophobic MOF was synthesized in solvothermal conditions and deposited onto SS wires by means of a neutral silicone sealant. By capping the NH_2-UiO-66(Zr) MOF with phenylsilane groups, the adsorption of water molecules within the MOF lattice was strongly reduced, thus increasing the enrichment capability of the coating up to six and eight times higher compared to PDMS and unfunctionalized MOFs, respectively. After optimization and validation, the HS-SPME-GC-MS method, characterized by LOQs in the range of 0.7–7.1 ng/L and RSDs ≤ 6%, was applied for the quantitation of UV filters in river and pond water samples, detecting concentration levels in the 40.1–193.7 ng/L and 26.1–250.1 ng/L ranges, respectively.

A ZIF-8-based coating was developed for the in-tube (IT)-SPME-HPLC-FLD detection of five model fluoroquinolones in tap water, river water, and wastewater samples [80]. Firstly, a porous monolith was obtained by copolymerization of 4-vinylbenzoic acid with ethylenedimethacrylate in a fused silica capillary. Then, the ZIF-8 MOF was obtained by in situ layer-by-layer deposition using $ZnNO_3$ and imidazole solutions, allowing both a controlled self-assembly of the MOF and a good column-to-column reproducibility (RSDs < 10%). The IT extraction allowed for the on-line SPME-HPLC-FLD analysis of the aqueous solutions containing the analytes, obtaining LOQs in the 0.48–1.8 ng/L range and EFs in the 255–296 range. Finally, the reliability of the method was assessed by detecting low concentration levels of fleroxacin, enrofloxacin, and sarafloxacin in the investigated samples.

3.3.6. Extraction of Polychlorinated Biphenyls

Polychlorinated biphenyls (PCBs) are a class of persistent organic pollutants accounting for more than 200 synthetic organic compounds. PCBs are listed among the top five priority hazardous substances by the US EPA due to their toxicity, persistency, ubiquitous presence, and bioaccumulation [81]. Despite their production being banned in the 1970s, they have been detected in all the environmental compartments and ecosystems, including air, water, soil, and in wildlife, plants, animal tissues, and food products.

Considering that PCBs are listed by the IARC in group 1 (carcinogenic to humans) and are recognized endocrine disruptors, their monitoring in the environment is of pivotal importance: to this aim, different MOF-based SPME coatings have been devised [82–84].

A coating with nitrogen-doped carbon nanotube cages (N-CNTCs), derived from ZIF-67 obtained by calcination of the MOF and deposited onto SS wires by means of an epoxy resin, was successfully proposed for PCB determination [83]. The MOF morphology was preserved, obtaining a thin N-doped carbon nanotube assembled structure with uniform hollow cages. The material showed better extraction performance than that achieved using either commercial coatings (75 μm CAR/PDMS and 65 μm PDMS/DVB fibers) or the non-hollow material, which also required prolonged time to reach equilibrium. The extraction capability of the N-CNTCs was attributed to π-π interactions between the analytes and the interconnected graphitic framework, as well as to the N doping. The DI-SPME-GC-MS method was characterized by LOQs in the 0.33–0.72 ng/L range, a four order of magnitude linear range, and good precision. The method was applied for the determination of PCBs in six surface water samples, quantifying PCBs in the 1.9–26.2 ng/L range. Similarly, hollow carbon nanobubbles (HCNBs) were obtained from a ZIF-8 MOF [84]. The developed material maintained the morphology of the framework, also featuring an internal hollow structure able to provide short diffusion distances and increased extraction capability. The SPME coating was tested for the extraction of different classes of persistent pollutants, namely BTEX, PAHs, and PCBs, always showing higher extraction capabilities than commercial PDMS and PDMS/DVB fibers. As in Guo et al. [83], the presence of a high content of nitrogen in the structure increased the adsorption sites and surface polarity, resulting in the efficient extraction of PCBs. The HS-SPME-GC-MS method was optimized and validated, obtaining LOQs in the 0.0017–0.0042 ng/L range, a five order of magnitude linear range, and RSDs $\leq$ 14%. Finally, the method was tested for the extraction of five model PCBs from rainwater, pond, and river water samples, obtaining ΣPCBs of 37.1, 64.1, and 100.2 ng/L, respectively.

3.3.7. Extraction of Other Compound Classes

A hollow zirconium–porphyrin-based MOF (HZ-PMOF) was proposed as an SPME coating for the extraction of 1-naphthol and 2-naphthol, which are extremely toxic environmental contaminants, widely used in industry as precursors for the production of dyes, pesticides, and pharmaceuticals [85]. The MOF was synthetized by a solvothermal approach and physically deposited by means of a polyimide sealing resin. Comparing the non-hollow with the hollow MOF material, different morphologies were obtained: the HZ-PMOF exhibited mesopores and a BET surface of 1585 m^2/g, with benefits in terms of mass transfer efficiency and specific surface area, whereas the non-hollow material was characterized only by micropores and a surface of 997 m^2/g. The developed HS-SPME-GC-MS/MS method based on the HZ-PMOF-coated fiber was validated, obtaining low LODs (1.0 ng/L) for both napthols and, applied to the analysis of environmental water samples from five cities in China, detecting the analytes in two out of five samples (37.7 ± 5.6 and 15.0 ± 3.4 ng/L for 1- naphthol and 48.9 ± 4.3 and 8.9 ± 2.1 ng/L for 2- naphthol, respectively).

Darabi and Ghiasvand developed a PPy/chromium-based MOF nanocomposite, PPy@MIL-101(Cr), which was obtained by the hydrothermal method and deposited by electrochemical polymerization of pyrrole onto SS wires [86]. The coating was tested for the extraction from soil samples of methyl tert-butyl ether, a persistent environmental contaminant used as an octane enhancer in gasoline [87]. The novel material showed a higher extraction capability than commercially available materials and pure electrodeposited PPy due to the larger sorptive surface area and stronger interactions. The validation of the HS-SPME-GC-FID method demonstrated a LOQ value of 0.5 ng/g and a linear response over four orders of magnitude. As a main drawback, a low intra-fiber reproducibility was observed, with RSDs up to 26% due to the scarce robustness of the electropolymerization process. Finally, the validated method was applied for the quantitation of methyl tert-butyl ether in six soil samples collected from an oil refining company and different gas stations, detecting the analyte in the 0.92(±0.48)–2.72(±0.50) µg/g range.

In 2019, Wei et al. developed a porous carbon SPME coating performing both hydrothermal growth of MOF-74 and in situ carbonization on SS wires [88]. The MOF-74-C coating was tested for the extraction of five odorous organic contaminants, namely, 2-chlorophenol, 2,4,6-trichloroanisole, 2-isobutyl-3-methoxypyrazine, thiophenol, and 4-methylthiophenol from water samples. The developed material showed an extraction efficiency 2.0–16.0 times higher than PDMS and PA fibers, whereas similar results were obtained using the PDMS-DVB fiber. Compared to the commercial polymeric materials, the MOF-74C coating was characterized by enhanced selectivity due to the presence of micropores with sizes ranging from 0.59 to 1.71 nm, very close to the dynamic diameters of the odorants, thus resulting in a sieve and micropore-filling effect. The DI-SPME-GC-MS method was validated, obtaining LOQs in the 0.03–300 ng/L range, a three-to-five order of magnitude linear range, recoveries in the 90.1–107.3% range, and RDSs always lower than 8.8%. The method was finally applied to the analysis of wastewater, tap, and river water samples, detecting 2-isobutyl-3-methoxypyrazine in the surface water sample at trace levels (0.2 ng/L).

UiO-66-NH$_2$ can be considered as an excellent MOF platform to obtain hybrid materials [89–91]. In 2019, Ni et al. developed zirconium and nitrogen co-doped ordered mesoporous carbon (Zr/N-OMC) SPME coatings to be used for the extraction of substituted phenolic compounds [89]. UiO-66-NH$_2$ was obtained by solvothermal synthesis, dispersed in a solution containing phenolic resin precursors, and the composite resin/UiO-66-NH$_2$ material was synthesized by solvent-evaporation-induced self-assembly. Finally, Zr/N-OMC was obtained by carbonization of the composite and deposited onto SS wires using a neutral silicone sealant. Zr/N-OMC was characterized by faster equilibrium rates and a higher extraction efficiency compared to non-doped and N-doped OMC, despite the similar pore structure and lower BET surface (775, 720, and 583 m^2/g for OMC, N-OMC, and Zr/N-OMC, respectively). Zr doping increased the adsorption rates and the extraction amounts of phenols, benefiting from Lewis acid–base interactions between the coating and the analytes. The hybrid coating was applied for the extraction of six model phenols from water samples and the HS-SPME-GC-MS method was validated, achieving LODs in the 0.21–1.7 ng/L range, a three order of magnitude linear range, and a fiber-to-fiber precision with RSDs $\leq$ 10%. The method was applied for the extraction of phenols from surface water: all the analytes except 2,4,6-trichlorophenol were detected in the 20.0(±1.0)–315.0(±10.0) and 20.0(±1.2)–330(±12.0) ranges in river and pond samples, respectively. UiO-66-NH$_2$ was also used to develop a new SPME coating that combines the extraction capability of MOF with ionic liquids (IL) [90]. The coating was deposited by treating the SS fibers with hydrofluoric acid and hydrogen peroxide to obtain hydroxyl groups for in situ MOF growth. The functionalization with the 1-hydroxyethyl-3-vinylimidazole chloride was performed in situ, exploiting the interaction between the anion and the zirconium nodes (Figure 6); the final IL/MOF was used for the DI-SPME extraction of phthalate esters (PAEs) from water samples. Interestingly, the incorporation of the IL inside the MOF increased the hydrophobicity of the channels while promoting hydrogen bond formation and π-π interactions between the coating and the PAEs. The developed coating provided a higher extraction efficiency than commercial 100 µm PDMS, 75 µm CAR/PDMS, and 65 µm PDMS/DVB fibers, and the non-ionic coating. Under optimal conditions, the validated DI-SPME-GC-MS method was showed by LOQs in the 0.6–1.2 ng/L range, a three order of magnitude linearity, and precision, with RSDs $\leq$ 12%. Finally, the method was tested for the extraction of PAEs from spiked surface and bottled water samples, obtaining good recoveries in the 90.3–102.4% range.

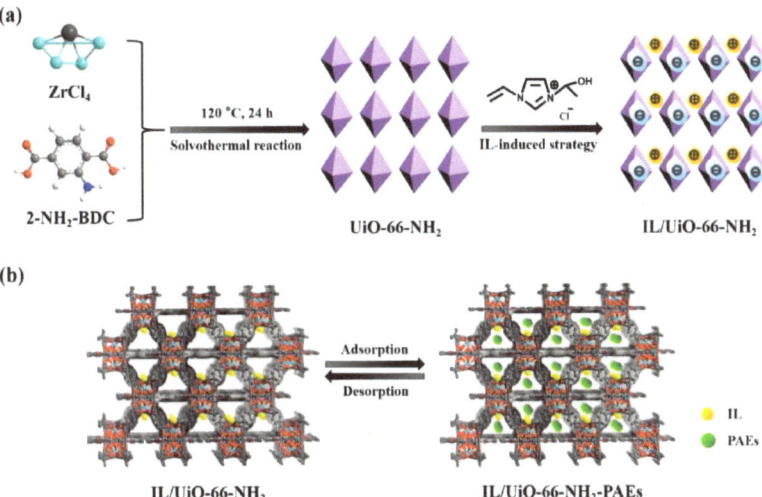

Figure 6. Schematic of the fabrication of IL/UiO-66-NH 2 (a) and the extraction and desorption process for PAEs: (a) fabrication of the coating; (b) extraction/desorption process for PAEs. Reprinted with permission from [90].

Finally, an interesting study was carried out by Xu and coworkers, who developed an aptamer-functionalized MOF (PAN/UiO@UiO2-N3-aptamer), based on the UiO-66-NH$_2$ structure [91], for the selective LC-MS recognition of Microcystin-LR at trace levels. The UiO-66-NH$_2$ seed crystals were prepared using solvothermal conditions and electrospun with PAN to obtain a PAN/UiO SPME coating. The MOF was grown in situ by a seeding procedure and functionalized to obtain the azide intermediate to be used as binding site for the aptamer. The specific affinity recognition of the material toward the target analyte was demonstrated by comparing the performance of the PAN/UiO@UiO2-N3-aptamer with that of the intermediate materials and the MOF functionalized with a control oligonucleotide for the extraction of Microcystin-LR in the presence of two different structural analogs. The results highlighted the superior performance of the PAN/UiO@UiO2-N3-aptamer coating, characterized by a high extraction capability, selectivity, and recovery up to 95.2%. By contrast, no significant difference in terms of adsorption capability was observed when the other materials were used, thus indicating that non-specific interactions were responsible for the adsorption of the compounds. The DI-SPME-LC-MS method was validated, obtaining a LOQ of 8 ng/L, a three order of magnitude linear range, and good precision (RSDs < 13%). Finally, the method was successfully applied for the detection of Microcystin-LR in tap, pond, and river water samples.

Table 1. MOF-based coatings used for the SPME extraction of environmental samples.

Analyte	Material	Deposition Method	Matrix	Extraction Mode	Platform	LOD (ng/L)	EFs	References
BTEX	SOM-ZIF-8	physical adhesion	wastewater	HS	GC-FID	1.0–12	-	[48]
BTEX	NSZIF-8Si	physical adhesion	river water	HS	GC-MS	0.02–0.21	-	[49]
BTEX	MIL-101-NH$_2$ derived urchin-like nanoporous carbon	physical adhesion	pond water and river water	HS	GC-MS	0.08–0.36	-	[50]
BTEX, styrene, and trimethylbenzene	PAN/MIL-53(Al)@MOF@SBA-15/4,4′-bipyridine hybrid nanocomposite	in situ electrodeposition	tap water, mineral water, well water, and wastewater	HS	GC-FID	2.3–3.6	318–385	[51]
BTEX + 14 VOCs	MOF-199	in situ growth	air	DI	GC-MS	0.03–0.09 ^{a}	-	[52]
16 PAHs and 11 nitro-PAHs	ZIF-8	sol–gel deposition	tap water, surface water, and wastewater	DI	GC-MS	0.3–27.0	-	[54]
3 PAHs and 5 PPCPs	ZIF-8	CVD deposition and in situ growth	wastewater	DI	GC-FID	600–2000	-	[55]
8 PAHs	ZIF-8	in situ electrodeposition	lake water	HS	GC-FID	10–54	-	[56]
6 PAHs	PANI/ZnO nanorods/ZIF-8	in situ growth	sewage water	EE-SPME/DI-SPME	GC-FID	8.2–134	-	[57]
6 PAHs	Co@ZIF-67	in situ growth and electrodeposition	snow, lake water, river water, and wastewater	DI	HPLC-UV	5–42	-	[58]
7 PAHs	MAF-66	physical adhesion	lake water and food	HS	GC-FID	0.1–7.5	127–3108	[59]
16 PAHs	MAF-5 and MAF-6	physical adhesion	wastewater and milk products	DI	HPLC-FLD	6–540	-	[60]
8 PAHs	HKUST-1 membrane	in situ growth and physical adhesion	lake water	HS	GC-MS	0.1–9.9	-	[61]
16 PAHs	PUM-210	physical adhesion	contaminated water	DI	GC-MS	0.50–3.7	300–14,950	[62]
10 PAHs	PI(Cu-DAT)	in situ growth and dip-coating	river water and fish muscle	HS/DI	GC-MS	0.3–2.1, 4.0–18.9	-	[63]
diazinon and chloropyrifos	PAN/Ni-MOF	post-synthetic electrodeposition	river water, farm water, groundwater, and beverages	HS	CD-IMS	200–300	-	[65]
ethion	ZIF-67 film	in situ electrodeposition	underground water and agricultural wastewaters	DI	SESI-IMS	100	-	[66]
6 OCPs	NU-1000	physical adhesion	river water and seawater	DI	GC-MS	0.011–0.058	972–2275	[67]
8 chlorobenzenes	Ni@NiO/PCs	physical adhesion	tap water and river water	DI	GC-MS	0.07–0.165	-	[68]

Table 1. Cont.

Analyte	Material	Deposition Method	Matrix	Extraction Mode	Platform	LOD (ng/L)	EFs	References
PFOA	ZIF-8, UiO-66, MIL88-A, and Tb$_2$(BDC)$_3$	in situ growth	contaminated tap water, rainwater, and seawater	DI	Direct-MS	11	-	[70]
8 PFASs	NH$_2$-ZIF-8	physical adhesion	river water, seawater, and wastewater	DI	HPLC-MS/MS	0.15–0.75	-	[71]
11 PFASs	NH$_2$-UiO-66(Zr)-hp	physical adhesion	tap water, river water, and pond water	DI	HPLC-MS/MS	0.035–0.616	6.5–48	[72]
9 PFASs	MIL-101-DETA-F	physical adhesion	tap water, river water, and wastewater	DI	UHPLC-MS/MS	0.004–0.12	70–112	[73]
Ibuprofen and diclofenac	Zr-MOF@GO	physical adhesion	river water	DI	GC-FID	1–30	-	[75]
3 PPCPs	Polyfam/Co-MOF-74 composite nanofibers	post-synthetic electrodeposition	wastewater and biological fluids	Thin film-SPME	HPLC-UV	30–200	24–37	[76]
3 PPCPs	MIL-101	physical adhesion	municipal wastewater	HS	GC-MS	4–60	-	[77]
6 methylsiloxanes and 7 musk fragrances	CIM-80(Al)	in situ growth	wastewater and seawater	HS	GC-MS	100–3500 ^{b}	-	[78]
5 UV filters	NH$_2$-UiO-66(Zr)	physical adhesion	river water and pond water	HS	GC-MS	0.2–2.1	865–3321	[79]
5 fluoroquinolones	ZIF-8	in situ deposition	tap water, river water, and wastewater	IT	HPLC-FLD	0.14–0.61	255–296	[80]
7 PCBs	ZIF-67 derived N-CNTCs	physical adhesion	river water	DI	GC-MS	0.10–0.22	-	[83]
5 PCBs	ZIF-8 derived HCNBs	physical adhesion	river water, pond water, and rainwater	HS	GC-MS	0.0017–0.0042	-	[84]
1-naphthol and 2-naphthol	HZ-PMOF	physical adhesion	urban water samples	HS	GC-MS/MS	1.0	-	[85]
methyl tert-butyl ether	PPy@MIL-101(Cr)	post-synthetic electrodeposition	soil	HS	GC-FID	0.01 ^{c}	-	[86]
5 odorous organic compounds	MOF-74-C	in situ growth	tap water, lake water, and wastewater	DI	GC-MS	0.01–100	520–3000	[88]
6 substituted phenolic compounds	NH$_2$-UiO-66(Zr) derived Zr/N-OMC	physical adhesion	river water and pond water	HS	GC-MS	0.21–1.7	-	[89]
8 PAEs	IL/UiO-66-NH$_2$	in situ growth	river water, lake water, and bottled water	DI	GC-MS	0.2–0.4	-	[90]
Microcystin-LR	PAN/UiO@UiO$_2$-N$_3$-aptamer	post-synthetic electrodeposition	tap water, pond water, and river water	DI	LC-MS	3	-	[91]

^{a} µg/m^3; ^{b} LOQ; LOD not provided; ^{c} ng/g.

4. Covalent Organic Frameworks (COFs)

In recent years, COFs [92] have emerged as a novel type of ordered crystalline and tunable porous materials: unlike MOFs, they consist of light elements (H, B, C, N, and O) connected by covalent bonds in which the pores are obtained by linking different groups in a cyclic manner. Although the number of developed MOF structures is higher compared to the number of developed COFs, one major advantage in the use of COFs relies on the absence of metallic elements in their structure, thus making them more stable when exposed to water or solvents. In addition, the lack of a metallic center increases the surface/weight ratio and the greenness of the materials. Moreover, COFs can form either 2D or 3D constructs, whereas MOFs usually present a 3D architecture. Owing to their unique features in terms of thermal and chemical stability, high specific surface area, permanent porosity, and tunable frameworks, COFs are considered as promising coatings for the SPME extraction of different classes of compounds from highly complex matrices [30,32,93].

4.1. Synthetic Procedures

Similarly to MOFs, the morphology and properties of COFs strongly depend on both the organic building blocks and synthetic conditions, including reaction medium, time, temperature, and type of catalyst [30,32,93,94]. According to the dimensions of the building blocks, they can be classified into 2D structures and 3D consolidated networks. Two-dimensional COFs are based on rigid conjugated planar macrocyclic molecules, such as porphyrin, thiophene, phthalocyanine, and tetrazolium, presenting strong interlayer interactions, mainly π-π stacking. Both 2D and 3D COFs include different active functional groups, namely, B-O, C=N, C–N, and C=C [30,32,93]. They can be synthesized via boron or nitrogen condensation, triazine-based trimerization, metal-catalyzed coupling reactions, and Schiff base reactions between organic monomers.

Boron-based COFs consist mainly of boronated ester COFs, obtained through covalent interaction between catechol and polyfunctional boric acid. The main features of these materials are low density, high specific surface, and multiple binding sites; however, they suffer from low chemical stability, being that these COFs are usually unstable in humid environments and in air [12,93,95]. COFs based on Schiff base linkages encompass different organic ligands with substituents on the amine moiety, including imines, hydrazones, azines, and β-ketoenamines. Among them, imine-based COFs are based on the formation of C=N linkage though the condensation reaction between amino groups and aldehydes in the presence of a Lewis acid catalyst. These COFs are stable both in common organic solvents and water, providing superior stability compared to boron-based COFs. Imide-based COFs are based on the condensation between an amine and an anhydride at high temperatures. Amine-type COFs are less stable and require a laborious synthesis, which largely limits their application. Recently, novel β-ketoenamine-linked COFs have emerged, featuring high chemical and thermal stability owing to the presence of irreversible bonds, which cannot be hydrolyzed in water, acidic or basic environments [96,97]. Another important class is the triazine-based COFs, which are characterized by surface areas up to 2390 m^2/g and high thermal and chemical stability. Finally, C=C linkage-based COFs have been developed, exploiting less reversible or irreversible poly-condensations processes: 2D sp^2 C=C COFs have been obtained by condensation of aldehydes and benzyl cyanides in the presence of a base catalyst [32,93].

Up to date, different synthetic strategies have been proposed to produce COF-based SPME coatings. The most applied technique is solvothermal synthesis, due to the use of high-pressure conditions to promote the formation of the crystalline framework. Additional approaches have been suggested, among which microwave-assisted and sonochemical syntheses are gaining increasing interest, reducing the reaction time and preserving the porosity of the framework. Ionothermal synthesis has been also proposed for the preparation of triazine-based COFs: despite the green and solventless approach, this strategy requires very harsh reaction conditions and has limited applications due to the poor control of the crystallinity [32]. Finally, mechanochemical (MC) synthesis is a new mild, simple, and rapid approach to synthetize COFs by grinding reaction: high yields, great thermal stability, and promising crystallinity are the key features of this synthetic strategy, although limitations in terms of porosity and superficial area have been demonstrated [20,32,93].

4.2. COF Features for SPME Extraction

Among the interesting features of COFs, their low density, large surface area, and adjustable pore size and structure, as well as customizable properties, are the most attractive features that make them exceptional SPME materials. As previously reported, the thermal and chemical stability of COFs depend on their crystalline structure, particularly on both the bond strength and the electrostatic forces between the interlayers. Different methodologies have been proposed to enhance the stability of COFs, including the application of highly planar building blocks able to increase the interlayer force, the linkage conversion, and the conversion from reversible (e.g., imide-linked) to irreversible (e.g., amide-linked) COFs [14,30,95].

The most common typologies of COFs are 2D and 3D COFs. The 2D COFs exhibit intralayer covalent bonds and interlayer non-covalent interactions, such as π-π stacking. Adsorption usually occurs via hydrophobic interaction with planar aromatic analytes between the layers. In contrast, 3D COFs present a framework linked with strong covalent bonds and accessible open binding sites with a high surface area characterized by a hierarchical intrinsic nanoporosity. This feature increases the extraction selectivity and enrichment performance of the SPME coating compared to 2D frameworks due to the requirements in terms of steric complementarity [98,99]. Recent studies have introduced heteroporous COFs as an alternative solution for the simultaneous extraction of molecules with different sizes, thus enhancing both the selectivity and the extraction performance of the coating material.

Pore dimensions can be tuned by changing the reaction conditions or the utilized building blocks; however, it has to be considered that micropores and small mesopores generally limit mass transfer and accessibility to the inner surface of the framework. Post-functionalization of the binding sites can also be performed to induce the formation of specific analyte–COFs interactions, improving both the extraction performance and selectivity [95]. This strategy can involve either the modification of the organic groups already present in the original framework or the incorporation inside the COF of predesigned structures able to host target analytes [20,95]. Finally, the use of a protective, thin layer mostly made by a polyimide polymer proved to be useful in improving both the mechanical strength and life span of COF-based coatings [100].

4.3. COFs as SPME Coatings for Environmental Applications

COFs as SPME coatings, prepared by physical adhesion, the sol–gel method, in situ growth, and chemical cross-linking, have recently attracted considerable attention for the analysis of organic pollutants, including polycyclic aromatic hydrocarbons [94,101–112], phenols [97,113–115], polybrominated diphenyl ethers [98,100,116–120], polyhalogenated biphenyls [99,121,122], pesticides [123–128], and PFASs [129–131] in environmental samples. The COF-based coatings used for the extraction of targeted compounds in environmental samples are summarized in Table 2.

4.3.1. Extraction of Polycyclic Aromatic Hydrocarbons and Nitroaromatic Compounds

COFs have been proposed as novel SPME coatings for the extraction of PAHs from water and soil samples, based on either HS or DI modes. Recent studies investigated the HS extraction performances of different COF coatings toward low-molecular-weight PAHs [94,101,105–107]. Huo and coworkers developed a hydrazone-based covalent organic framework (BTCH-PTA-COF) coating by the sol–gel technique [101]. The adsorption of PAHs within the framework was mainly attributed to the π-π and hydrophobic interactions between the target compounds and the benzene rings and C=N bonds present within the COF. The material was characterized by a high surface area (1460 m^2/g), mesoporous structure, and high hydrophobicity. Under the optimal extraction conditions, the BTCH-PTA-COF-coated fiber demonstrated a high extraction efficiency for PAHs in a water sample, with EFs in the 767–1411 range, LOQs in the 0.10–0.15 µg/L range, RSDs always lower than 9.6%, and recovery rates between 88.5 and104.8%, allowing for the sensitive analysis of target pollutants in river water samples. A facile and layer-by-layer self-assembly was proposed by Tian et al. to develop under very mild conditions a thin SPME film based on imine COFs (TFPA-TAPP-COF) [105]. The TFPA-TAPP-COF coating fiber consisted of tetra(4-aminophenyl) porphyrin and tris (4-formyl phenyl) amine: the use of very mild synthetic conditions together with good mechanical strength, remarkable thermal stability (up to 260 °C), and high extraction efficiency towards PAHs were the main features of the proposed material. The synthetic procedure allowed a close control over both the thickness of the coating and the number of assembled layers. The HS-SPME extraction performance of the novel COF material was assessed by extracting five model PAHs from aqueous samples, achieving LODs in the 6–24 ng/L range and RSDs < 10%. Finally, the HS-GC-FID method was applied for the analysis of spiked river water samples, obtaining recoveries in the 80.3–102.6% range. Another porphyrin-based COF was synthesized through a Schiff base reaction by Yu and coworkers [106] and deposited onto SS wires using a silicone sealant. An average narrow pore size of 8.7 nm was achieved, leading to an improvement in the selectivity of the material. PAH extraction was mostly related to π-π stacking, van der Waals interactions, and weak hydrogen bond formation between the amino group of the COF and the hydrogen atoms in the PAHs. The developed HS-SPME method showed LOQs in the range of 0.5–10 µg/L and good inter-day and fiber-to-fiber repeatability, with RSDs < 9%. Finally, the HS-SPME-GC-FID method was applied for the extraction of five model low-molecular-weight PAHs from both lake water and soil samples, obtaining recoveries in the 68–99% and 41–105% ranges, respectively. Recently, Yang et al. developed a COF based on 1,3,5-tris(4-aminophenyl)benzene and trimesoyl chloride (TAPB-TMC-COF) by amide coupling at room temperature. The coating was obtained by the physical deposition of the material onto etched SS wires and used for the HS-SPME-GC-MS determination of low-molecular-weight PAHs in water samples [107]. Analyte adsorption was mostly due to π-π stacking, with EFs in the 819–2420 range. The developed method exhibited good sensitivity (LODs in the 0.29–0.94 ng/L range) and linearity over three orders of magnitude. The HS-SPME-GC-MS method was applied for the analysis of PAHs from spiked river, pond, and wastewater samples, obtaining RSDs < 8% and recoveries in the 79–105% range, although a significant matrix effect was demonstrated for the wastewater samples. An imine-based COF-SCU1 material was prepared through a one-step synthesis and packed into a stainless-steel needle to be used as a needle trap device (NTD) for the extraction of PAHs from soil samples [94]. Compared to classical fiber extraction, the NTD proved to be a more robust microextraction tool, characterized by a higher sorption capacity and extraction efficiency. The validation of the COF-packed NTD-GC-FID method allowed for the achievement of LODs in the 0.01–0.05 ng/g range, excellent linearity over five orders of magnitude, and RSDs always lower than 13%. Finally, a ΣPAH concentration in the 0.86–2.3 µg/g range was obtained when soil samples collected from the neighboring ground of gas stations were analyzed.

Another interesting solution for the extraction of PAHs was based on the development of hybrid MOF/COF materials, which combine the outstanding features of MOF and COF structures, integrating the rigidity and high surface area of MOFs with the high flexibility and tunability of the COF surface. Different MOF/COF SPME coatings were obtained for the HS-SPME-GC-FID determination of PAHs from contaminated soil samples [108–110]. Koonani and coworkers developed a Zn-MOF/COF coating by solvothermal synthesis of the MOF precursor, followed by a further functionalization using ethylenediamine and a hydrothermal treatment with melamine and 1,4-benzenedicarboxylate to prepare the new MOF/COF hybrid material [108]. A similar approach was followed by Nouriasl at al. to obtain a porous Cu-MOF/COF coating using an amino-functionalized Cu-MOF as the starting material [109]. The coatings were obtained by the physical deposition of the material onto SS fibers using an epoxy glue. These synthetic strategies resulted in the COF assembly within the pores of the amino-MOF by in situ condensation using the amino groups present on the MOF surface as linkers to obtain stable covalent bonds between the MOF and the COF. The hybrid MOF/COF coatings showed higher extraction efficiency than both commercial fibers and individual materials, being able to interact with PAHs via π-π, van der Waals, and hydrophobic interactions. In addition, the presence of nitrogen atoms in the framework was exploited to promote electron donor–acceptor interactions with π-donor aromatic compounds, whereas the Zn or Cu metal sites could interact with the Lewis bases of the condensed aromatic rings. The HS-SPME-GC-FID method proposed by Koonani et al. was validated for the analysis of six PAHs, obtaining LODs below 1 ng/g and RSDs < 12%. The method was then applied for the analysis of contaminated soil samples, detecting the analytes up to 70 ng/g [108]. Compared to the previous study, the HS-SPME-GC-FID method proposed by Nouriasl et al. was characterized by lower LODs (in the 0.1–0.5 ng/g range), and higher precision, with RSDs below 8%, allowing for the determination of the investigated analytes in environmental samples at concentration levels up to 839 ng/g [109]. A hybrid COF/MOF (2DTP/MIL-101-Cr) coating was developed by Maleki et al. to be used for the extraction of BTEX and six model PAHs [110]. The MIL-101 MOF was obtained by hydrothermal synthesis, whereas the 2D COF was synthesized through metal-assisted and solvent-mediated synthesis. The hybrid material was obtained by ultrasonically mixing the two components, resulting in the embedding of the MOF within the COF framework. The material was physically deposited onto an SS needle to be used as an SPME coating for the in-syringe, vacuum-assisted HS-SPME (IS-VA-HS-SPME). The VA-HS-SPME relies on the application of a reduced pressure to improve the transfer rate of high-boiling-point components from the sample matrix to the headspace, effectively increasing the extraction efficiency of semi-volatile compounds (Figure 7). The hybrid material exhibited higher extraction performances than both the individual materials and commercially available PDMS and CAR/PDMS fibers. The optimized VA-HS-SPME-GC-FID method was validated, obtaining LOQs in the 7.1–17 ng/g and 0.23–3.7 ng/g ranges for BTEX and PAHs, respectively, with RSDs below 16%. When applied to the analysis of contaminated soil samples, both BTEX and PAHs were quantified at sub-trace levels, in the 0.26–3.6 μg/g and 0.02–1.8 ng/g ranges, respectively.

As for the DI-SPME extraction of PAHs from aqueous samples, novel COF-based coatings have been proposed [102–104,111,112]. Li and coworkers synthesized by ultrasonic-assisted reaction a COF based on 1,3,5-triformylphloroglucinol (Tp) and benzidine (BD), which was grafted on etched SS fibers using APTES [111]. The electroetching procedure resulted in an array of nanopores on the SS surface, promoting the grafting of the COF in the microwells. The SPME fibers were tested for the extraction of seven model PAHs from tap and lake water samples followed by GC-FID analysis. Validation resulted in LODs in the 1–5 μg/L range, RSDs < 10%, and recoveries in the 88–163% range. Feng et al. developed a triazine-based COF under solvothermal conditions, subsequently deposited onto SS wires using an epoxy resin. Finally, the SS wires were inserted in a peek tube to be used for the IT-SPME-LC-UV determination of different contaminants in environmental water samples [112]. The good extraction capability towards eight model PAHs, four estrogens, and eight plasticizers was assessed by achieving EFs in the 542–942, 190–929, and 68–1235 ranges, respectively. After optimization of the extraction conditions, the IT-SPME-LC-UV method was validated for the analysis of PAHs, obtaining LOQs in the 13–33 ng/L range, a fiber-to fiber reproducibility with RSDs $\leq$ 14%, and a wide linear range. The method was then applied for the analysis of tap, river, and rainwater samples; however, no analyte could be detected in unspiked samples. In another study, COF TpPa-1 was used as a template framework to develop a novel carbonaceous material [102]: the derived N-doped porous carbons (TpPa-1–1000) were deposited onto SS wires by physical deposition and used for the extraction of PAHs from soil samples. The TpPa-1–1000 material featured a high nitrogen content (5.62% atomic), a specific surface area of 435.6 m^2/g, a uniform pore size distribution (3.9 nm), and a porous layered structure, thus increasing the number of accessible binding sites for promoting the interaction of the adsorbent with the target analytes. Compared to commercially available materials, the coating provided enhanced EFs (in the 1686–41,294 range) due to both the high degree of graphitization and the formation of electron-donor–acceptor interactions. The validation of the DI-SPME-GC-MS method resulted in LOQs in the range of 10.4–28.5 µg/kg of soil sample, and RSDs lower than 12%. A DI-SPME-GC-MS method was also proposed by Zang and coworkers, who tested three different COF-modified graphitic carbon nitride (g-C_3N_4) materials for the extraction of PAHs from aqueous samples [103]. The g-C_3N_4@TpBD coating provided the best extraction performances toward the investigated PAHs. The developed coating was characterized by higher extraction performances than commercial 85 µm PA, 60 µm PDMS/DVB, and 100 µm PDMS fibers, probably due to the π-conjugated structure of g-C_3N_4, allowing for π-π stacking and hydrophobic and electrostatic interaction with the target analytes. The DI-SPME-GC-MS method showed LOQs in the 0.07–0.17 µg/L range, good linearity (over a three orders of magnitude range), and RSDs always lower than 11%. Finally, the method was applied for the analysis of river, lake, well, and rainwater samples and melted snow, allowing the detection of naphthalene and acenaphthene at the sub-µg/L level.

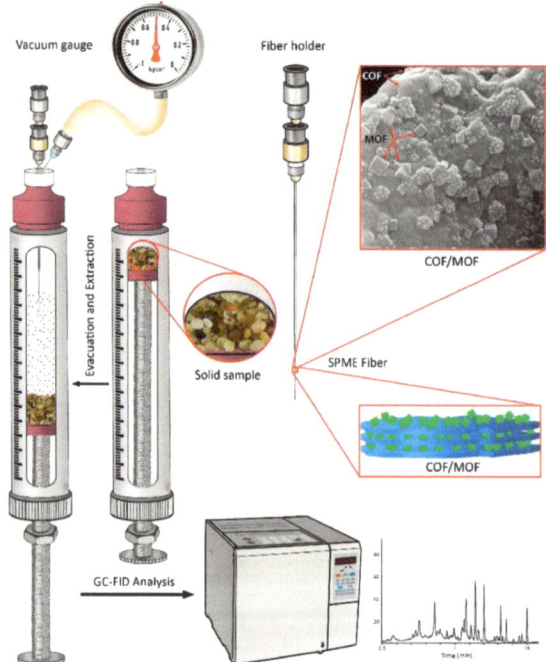

Figure 7. Schematic representation of ISV-HS-SPME extraction using hybrid MOF/COF coating, reprinted from [110].

A multi-component coating was developed by Sun and coworkers: carbon fibers were coated with titania nanorod arrays (NARs) and functionalized with either biochar nanospheres or COF nanospheres [104]. The obtained hybrid fibers were introduced in a PEEK tube to be used for the IT-SPME extraction of PAHs before LC-UV analysis. The functionalization with either the biochar or COF nanospheres proved to be useful in increasing the BET surface from 10.07 m^2/g to 15.97 m^2/g and 32.75 m^2/g, respectively. The extraction capability of the different materials was tested for the extraction of PAHs, estrogens, and bisphenols, resulting in the selection of TiO_2NARs–CFs, biochar nanosphere– TiO_2NARs–CFs, and COF nanosphere–TiO_2NARs–CFs as the best coatings for the extraction of the different classes of compounds. The IT-SPME-LC-UV method was validated obtaining LODs in the 5–10, 1–5, and 5–100 ng/L ranges for PAHs, estrogens, and bisphenols, respectively, and RSDs always lower than 15%.

Hydrogen-bonded organic frameworks (HOFs) are other emerging SPME coatings consisting of organic or metal–organic architectural units connected by hydrogen bonds. They are gradually developing into an attractive new type of porous crystalline material for the determination of nitroaromatic compounds. To increase the extraction selectivity toward this class of analytes, Hu and coworkers developed an SPME coating based on the self-assembly of melamine and 1,3,6,8-tetra(4-carboxylphenyl) pyrene to obtain a MA/PFC-1 HOF [132]. The use of melamine promoted the formation of hierarchical micropores with dimensions in the 5–20.0 Å range, a surface area of 1550($\pm$26) m^2/g, and strong hydrogen bonds between the constituting units. The HOF coating significantly improved the extraction performance of five model nitroaromatic compounds via polar, electrostatic, hydrogen bonding, π–π stacking, and C–H$\cdots$π interactions between the analytes and the adsorbent material. The HS-SPME-GC-MS method was validated, obtaining LOQs in the 4.3–20.8 ng/L range and RSDs $\leq$ 9%, allowing for the detection of *p*-nitrochlorobenzene and *p*-nitrotoluene and p-nitrochlorobenzene in water samples at the sub-µg/L level.

4.3.2. Extraction of Phenols and Derivatives

Phenols, phenolic compounds, and synthetic derivatives are characterized by very different chemical and physical properties. The main challenge in their detection is related to their hydrophilicity and low concentrations in environmental matrices. In this context, COFs as SPME coatings could be promising materials to address both enrichment and selectivity issues [97,113–115].

Recent studies investigated the extraction capabilities of two different COFs, namely a spherical TPB-DMTP-COF [113] and a piperazine-linked, copper-doped phthalocyanine metal covalent organic framework (CuPc-MCOF) [114], to be used as SPME coatings for the extraction of phenols and chlorophenols (CPhs). The spherical TPB-DMTP-COF was synthesized in an organic solvent at room temperature and deposited onto etched SS wires by means of a silicone sealant. A uniform spherical superstructure (1–2 µm) with a strong π-π stacking architecture between the layers, a pore size of 2.54 nm, and a BET surface of 1560 m^2/g, higher than that of solvothermal synthesized COFs, were the main features of the developed coating [113]. The SPME fibers were applied for the HS extraction of phenols and CPhs, obtaining EFs in the 1741–4265 range. The adsorption was mostly driven by hydrophobicity and steric hindrance effects, with the formation of π–π, van der Waals interactions, and hydrogen bonds between the analytes and the material. Validation demonstrated LOQs and recoveries in the 0.016–0.050 ng/L and in the 81–116% ranges, respectively, with RSDs $\leq$ 10%. When the HS-SPME-GC-MS/MS method was applied for the analysis of environmental samples, no analyte was detected in drinking and reservoir water, whereas underground water samples were contaminated by CPhs (in the 0.6–22.7 ng/L range). Wang and coworkers synthesized a CuPc-MCOF coating via Buchwald–Hartwig cross-coupling and deposited it onto etched SS fibers using a polyimide resin. The coating was used for the EE-SPME of CPhS from water and biological samples, namely oysters and prawns [114]. The high hydrophobicity of the COF-based material was ascribed to its conjugated structure and abundant fluorine atoms, whereas the presence N and of Cu atoms within the framework provided free metal sites for coordination binding. During EE-SPME, the analytes underwent mass transfer, thus increasing the enrichment of chlorophenols onto the fiber coating, acting as an electrode. The extraction performance of the developed coating was better than that of commercially available SPME fibers, with EFs in the 339–988 range. The EE-SPME-GC-MS/MS method was validated, obtaining LOQs at low ng/L levels, linearity over three orders of magnitude, and good precision (RSD < 9%). Finally, the method was applied for the analysis of seawater samples, detecting trichlorophenols in a 6.8–30.1 ng/L range with a recovery rate ranging from 90% to 113%. Similarly, when oysters and prawns were analyzed, 2,4-dichlorophenol,

2,4,5-trichlorophenol, 2,4,6-trichlorophenol, and 2,4,5,6-tetrachlorophenol were detected at low ng/L levels.

Guo and coworkers developed a TpBD COF based on an MC approach and the coating was obtained either by in situ growth or physical deposition onto SS wires [115]. Depending on the deposition procedure, different morphologies were obtained: in situ deposition resulted in a highly porous structure with a uniform section, whereas a less uniform coating with a smooth cross-section characterized by a less porous structure was observed for physical deposition, probably due to pore obstruction by the silicone adhesive. Both coatings were tested for the HS-SPME-GC-MS analysis of 2-nitrophenol, 2,6-dimethylphenol, 2,4-dimethylphenol, 2,4-dichlorophenol, and 2,4,6-trichlorophenol. The extraction performance of the in situ grown coating was superior to both the physically deposited COF and 85 µm PA commercially available fibers, mainly due to the formation of π–π interactions and hydrogen bonding. The reliability of the method for the determination of the investigated analytes was attested at trace level, obtaining LOQs in the 1.3–2.4 ng/L range with RSDs < 8%. The feasibility of the method was further demonstrated by analyzing water and soil certified reference materials, obtaining phenol concentrations in compliance with the reference value.

Finally, Gao and coworkers developed a COF–graphene oxide composite material (COF-GO) deposited onto a glass fiber using a silicone sealant to be used as an SPME coating for the analysis of bisphenol A [97], a well-recognized endocrine-disrupting compound used in the polymer and packaging industries. The hybrid material was synthesized by a three-step synthesis involving: (i) the preparation of the amino-functionalized GO using APTES, (ii) the addition of Tp between the layers of the material, and (iii) the in situ growth of the COF. The functionalization of GO with the COF increased the BET surface, pore volume, and the interlayer spacing from 353 m^2/g, 0.16 g/cm^3 and 8.4 Å to 824 m^2/g, 0.46 g/cm^3, and 14.7 Å, respectively, thus providing more active sites and faster mass transfer processes. The obtained SPME fibers were used for the CFDI-MS determination of bisphenol A in water samples, obtaining an extraction performance 2.2–4.7 times higher than TpBD and GO and 2.1–4.9-fold higher than 65 µm PDMS/DVB-, 85 µm PA-, and 100 µm PDMS-coated fibers. The validation of the DI-SPME-CFDI-MS method resulted in LOD and LOQ values at sub-trace levels (22.2 and 73.9 ng/L, respectively) and RSDs < 6%. When applied to the analysis of river and seawater samples, bisphenol A was detected in all samples in the 0.1222($\pm$0.0077)–4.794($\pm$0.072) range, with recovery rates in the 95–107% range.

The same research group developed different COF-based coatings for the analysis of tetrabromobisphenol A (TBBPA) and its analogs in environmental matrices using ambient mass spectrometry [98,119,120]. These pollutants have gained great attention due to their worldwide distribution, persistence, bioaccumulation, immunotoxicity, and neurotoxicity. TpBD-COF and multicomponent COFs were tested for the extraction of TBBPA and its analogs from environmental water samples. In an early study, a TpBD-COF was tested as an SPME coating for the DI-SPME–constant flow desorption ionization–mass spectrometry (CFDI-MS) analysis of TBBPA in water samples, obtaining LOD and LOQ values of 0.92 and 3.1 ng/L [119]. To improve the extraction performance of the SPME, different multi-component COFs containing Tp, Pa, and BD were tested for the analysis of TBBPA derivatives in water samples. The main feature of the multicomponent COF is the pore size tunability by changing the ratio of the organic linkers: the best performance was obtained using the TpPaBD$_{50}$ framework, exhibiting extraction efficiencies up to four times higher than TpPa and TpBD COFs [120]. A further study demonstrated that the best analytical performance could be obtained using porous TpBD, synthesized using a polystyrene sphere (PS) template-assisted method (Figure 8) [98]. Porous TpBD was characterized by a hierarchical porous structure with a BET surface and pore volume higher than those of the conventional TpBD (797 m^2/g and 0.75 cm^3/g vs. 638 m^2/g and 0.41 cm^3/g). The extraction capabilities of the COF were strongly related to its hierarchical porous structure, which featured micro-, meso-, and macropores, promoting the diffusion/mass transfer processes. The CFDI-MS-based method for the analysis of four TBBPA analogs showed good performance, with LOQs in the 0.4–3.2 ng/L range and good precision, with RSDs < 8%. When applied to the analysis of environmental samples, contamination was assessed for all the river water samples, with a concentration up to 51 ng/L and, for two out of three seawater samples, concentration levels up to 66 ng/L.

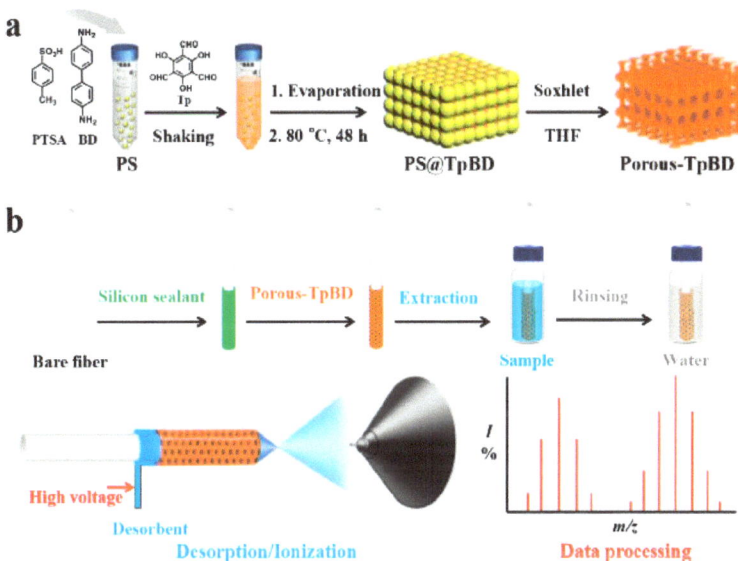

Figure 8. Schematic illustration of the (**a**) synthesis and (**b**) coating and SPME-CFDI-MS procedure for analysis of TBBPA analogs, reprinted with permission from [98].

4.3.3. Extraction of Polybrominated Diphenyl Ethers and Polyhalogenated Biphenyls

Polybrominated diphenyl ethers (PBDEs) and polybrominated biphenyls (PBBs) have been widely used as flame retardants in manufacturing since the mid-1990s; now, their use is strictly regulated due to their toxicity, high persistence in the environment, and impact on ecosystems [133]. Owing to their potential ecotoxicological effects on organisms, sensitive methods based on gas chromatography-negative chemical ion mass spectrometry (GC-NCI-MS) and COF-based SPME coatings have been developed to increase both selectivity and extraction efficiency, allowing for the detection of these pollutants at trace levels.

In this context, a TpPa-based coating was developed by Liu and coworkers [116] using a solvothermal reaction followed by physical adhesion on the fiber support by means of a silicone sealant. The material showed a flower-shaped morphology, high porosity, and a BET surface of 625 m^2/g. The coating was tested for the DI-SPME-GC-NCI-MS analysis of five model PBDEs, allowing for the extraction of the analytes via hydrophobic, van del Waals, and π–π interactions, with EFs in the 2035(±22)–6859(±67) range. The selectivity of the coating was explained on the basis of the size complementarity between the pores of the material (approximately 1.25 nm) and the PBDEs (average size 0.81 nm). Validation of the optimized method resulted in LOQs in the 0.019–0.074 ng/L range and RSDs < 10%. The analysis of environmental water samples did not show the presence of the analytes in groundwater and drinking water, whereas concentration levels in the 0.50–2.18 ng/L range were reported in pond water. Song et al. developed a coating based on polyimide and a TAPB-DMTP COF (PI@TAPB-DMTP) supported on SS wires for PBDE extraction from water samples [100]. The etched SS fibers were pretreated with APTES to obtain an amino-functionalized surface to be used for the grafting of DMTP, followed by in situ growth of the COF at room temperature. A PI protective layer was finally deposited onto the fiber (Figure 9). The characterization of the coating revealed a very high BET surface (1880 m^2/g) and the presence of mesopores with an average diameter of 2.2 nm, suitable for hosting the analytes. The PI@TPB-DMTP coating exhibited a high extraction performance with EFs in the 1470–3555 ng/L range, benefiting from strong hydrophobic effects, π−π stacking, enhanced surface area, and porosity. The optimized DI-SPME-GC-NCI-MS method was validated for the analysis of six PBDEs, obtaining LOQs in the low ng/l level, i.e., 0.027–0.063 ng/L, with RSDs lower than 9%. Despite the sub-trace detection limits, all the investigated PBDEs were below the LODs in the analyzed river, lake, and wastewater samples.

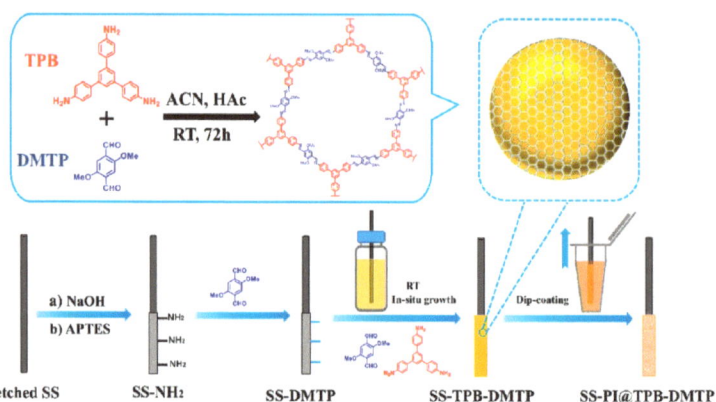

Figure 9. Schematic representation of the synthesis of the PI@TPB-DMTP coating, reprinted with permission from [100].

The use of poly(ionic liquid)s (PILs) to functionalize the surface of COFs was proposed by Su et al. [117]: two vinyl-decorated COF analogs (COF-β and COF-γ) were synthesized by solvothermal treatment, and then photo-initiated polymerization was performed to synthesize two series of poly(1-vinyl-3-methylimidazolium bis ((trifluoromethyl) sulfonyl) imide)-based hybrids (COF-PIL) with different mass ratios of IL/COF. Finally, the SPME coating was obtained by physical deposition using a silicone sealant. The use of co-polymerization to obtain covalently composited hybrids resulted in stable PIL@COF bonds. Interestingly, the COF-β and COF-γ hybrids showed different morphologies, namely, spherical and sheet stacking with a hierarchical porous structure, with PIL decorating the pores and interlayer volume. Despite the presence of PIL decreasing the BET surface area of the materials, the functionalization generally improved the extraction performance of the coating. This behavior was ascribed to the increased affinity between the COF surface and the adsorbed PBDEs, able to interact via weak hydrogen bonding, π-π stacking, and Van der Waals forces. The COF-γ-PIL exhibited a superior extraction capability compared to the COF-β-PIL: this behavior could be explained considering that the sheet-like stacking morphology featured shorter diffusion paths and broader pore size distribution, improving the accessibility of adsorption sites and mass transfer processes. The COF-γ-PIL-based coating also demonstrated a higher extraction capability when compared to commercially available coatings, namely, 100 μm PDMS, 65 μm PDMS/DVB, and 75 μm CAR/PDMS fibers, providing EFs in the 913–3625 range. Finally, the DI-SPME-GC-MS method was validated, obtaining LOQs in the 0.0070–0.046 ng/L range and excellent precision (RSDs < 8%), and successfully applied for the analysis of river, lake, and seawater samples.

Zhou and coworkers developed a coating based on TAPB—2,5-dimethoxyterephaldehyde (DMTP) COF using 2,5-dimethoxybenzaldehyde (DBA) as a modulator for the extraction of PBBs [118]. The material was fabricated at room temperature and deposited onto etched SS wires by means of a silicone sealant. The use of a modulator allowed the optimization of the COF microstructure in terms of specific surface area (BET values in the 1545–1611 m^2/g range, using different modulator percentages), surface hydrophobicity (contact angles higher than 90° with DBA higher than 40%), and morphology (in terms of particle size). The material containing 40% DBA showed EFs higher than those obtained using the unmodulated COF (in the 4400–11,360 range). Adsorption occurred through π-π interactions, hydrophobic effects, and electrostatic interactions between the material and the analytes. The optimized HS-SPME-GC-MS method was validated, achieving LOQs in the range of 0.12–0.94 ng/L, precision with RSDs < 8%, and recoveries in the 80–120% range, thus allowing the determination of 3-bromo-1,1′-biphenyl and 4,4′-dibromo-1,1′-biphenyl at concentration levels in the 0.33–1.1 and 2.6–3.0 ng/L ranges, respectively.

COF-based SPME coatings have also been developed for the selective extraction of PCBs [99,121,122]. Lv and coworkers developed a TpPa-functionalized PAN membrane (PAN-SiO$_2$@TpPa) [121]: the COF was grown in situ on the PAN-SiO$_2$ electrospun nanofiber membrane, which was deposited onto etched SS fibers using a resin glue. Functionalization provided the COFs with improved properties in terms of BET area (33 m^2/g vs. 108 m^2/g for PAN-SiO$_2$ and TpPa@PAN-SiO$_2$, respectively), average pore size (35.1 ± 5.1 vs. 9.8 ± 1.3 nm) and pore volume (0.009 vs. 0.104 cm^3/g). The TpPa@PAN-SiO$_2$ coating showed extraction capabilities up to three times higher than the unfunctionalized membrane for the six investigated PCBs. The adsorption was obtained by the synergistic effect of hydrophobic interaction, π-π stacking, and hydrogen bonding. Under optimized conditions, the HS-SPME-GC-ECD method exhibited LOQs in the 0.1–5 ng/L range, EFs in the 2714–3949 range, and precision with RSDs < 9%. No PCB could be detected in river, lake, and seawater samples, whereas the analysis of spiked matrices resulted in recoveries in the 73–126% range. Su et al. synthesized a chlorine-functionalized COF coating onto etched SS wires pretreated with APTES [122]: the COF was grown in situ using Tp, 2,5-dichloro-1,4-phenylenediamine, mesitylene, and 1,4-dioxane, obtaining a chlorinated analogue of TpPa-1. Using the chlorine-functionalized COF, EFs in the 699–4281 range were achieved, higher than those obtained using the TpPa-1 and commercially available polymeric coatings (100 µm PDMS and 50/30 µm DVB/CAR/PDMS fibers). This behavior could be ascribed to π-π stacking and hydrophobic interaction between the target molecules and the coating, presenting mesopores able to accommodate PCB molecules. The HS-SPME-GC-MS method exhibited excellent performance for the determination of 17 PCBs, with LOQs in the 0.005–0.029 ng/L range, a linearity over five orders of magnitude, and good precision, with RSDs ≤ 9%. Finally, the method was applied for the analysis of sea, river, and reservoir water, detecting PCB-209 in all samples at sub-ng/L concentration levels. Finally, Lu et al. developed an SPME coating using a 3D COF [99]: four different 3D COFs were synthesized using tetra (*p*-aminophenyl) methane (TAM) and aldehyde reagents (Tp, 1,3,5-triformylbenzene, terephthalaldehyde and 1,1′-biphenyl-4,4′-dicarbaldehyde) under solvothermal conditions and each coating was deposited onto etched SS fibers using an epoxy glue. The TpTAM coating showed a higher BET surface area (537.2 m^2/g), larger pore volume (0.32 cm^3/g), sharper pore size distribution (0.68 nm), and higher EFs (6940–10,305) compared to the other developed materials. The adsorption of both planar and non-planar PCBs was enhanced by π-π stacking, hydrophobic interactions, halogen bonds, and steric complementarity between host and guest. Finally, the selectivity of the material towards the investigated analytes was proved by the achievement of EFs lower than those calculated for PCBs and biphenyl when structural PCBs analogs, including *o*-dichlorobenzene, naphthalene, anthracene, and pyrene were considered. An excellent HS-SPME-GC-MS method performance was obtained, with LOQs in the 0.004–0.066 ng/L range, recoveries in the 85–117% and 85–115% ranges for water and soil, respectively, and precision, with RSDs < 11%.

4.3.4. Extraction of Pesticides and Insecticides from Water Samples

The use of COFs as key materials for SPME devices has been proposed as a valid alternative to the MOF-based coatings (described in Section 3.3.3), and also for the selective extraction of OCPs and OPPs at trace levels in environmental matrices [123–126]. Xin and coworkers developed a ketoenamine COF (Tp−Azo−COF) to be used as an SPME coating for the extraction of OCPs [123]. The COF was obtained by solvothermal treatment and deposited onto etched SS wires using an epoxy resin. The adsorbent material showed a high percentage (5.8%) of electronegative N atoms, facilitating the formation of halogen bonds between the analytes and the coating, a BET area of 1218 m^2/g, and high porosity, with a total pore volume of 0.689 m^3/g. The Tp−Azo−COF-coated fiber exhibited higher EFs compared to commercial fibers, ranging from 1061 to 3693 for the five investigated OCPs. DI-SPME sampling using this fiber followed by GC-MS/MS analysis of OCPs resulted in LOQs in the 0.005–0.26 ng/L range and good precision (RSDs < 11%). Green tea, milk, tap, and well water samples were analyzed by the validated DI-SPME-GC-MS/MS method: no analyte was detected in real samples, whereas recoveries in the 83.4−101.6% range were obtained on spiked matrices. A novel porous COF coating synthesized by condensation between cyanuric chloride, 4,4′-ethylendianiline, and 3,4,9,10-perylenetetracarboxylic dianhydride was devised by Tabibi and Jafari [124]. The COF was deposited onto SS wires using a silicon glue solution. The DI-SPME extraction was coupled with GC-CD-IMS analysis. The coating was used for the extraction of trifluralin (a common herbicide) and chlorpyrifos from agricultural wastewater and vegetables. Validation resulted in LOQs of 0.45 and 0.50 μg/L, and EFs of 1950 and 2123 for trifluralin and chlorpyrifos, respectively. A good precision was calculated, with RSDs below 11%. Finally, chlorpyrifos was detected in agricultural wastewater (1.2 μg/L) and both analytes were quantified in cucumber (2.4 and 5.6 μg/L for trifluralin and chlorpyrifos, respectively). A magnetic COF nanohybrid ($NiFe_2O_4$@COF) was developed for the SPME extraction of triclosan and methyltriclosan in environmental water and human urine samples [125]. Magnetic $NiFe_2O_4$ particles were synthesized by hydrothermal synthesis and the COF was grown in situ under ambient conditions by aldehyde-/amine-based condensation using terephthaldicarboxaldehyde and 1,3,5-tris (4-aminophenyl) benzene. The $NiFe_2O_4$@COF hybrid nanocomposite was physically deposited onto fused silica fibers. The core-shell nanohybrid-based coating was characterized by a pore diameter of 3.9 nm and BET surface (169.7 m^2/g) larger than those of a pristine COF (1.2 nm and 58.4 m^2/g, respectively). The extraction selectivity of the $NiFe_2O_4$@COF coating was demonstrated by comparing the EFs of the target analytes with those of five pyrethroid pesticides: EFs of 279 and 334 were calculated for triclosan and methyltriclosan, respectively, whereas values in the 76–147 range were obtained for the other pyrethroids. The adsorption of the analytes was ascribed to hydrophobic and strong π-π interactions between the COF shell and phenyl groups of the OCPs, leading to a higher extraction performance compared to commercially available fibers. The DI-SPME-GC-ECD method exhibited LOQs of 23 and 3.3 ng/L for triclosan and methyltriclosan, respectively, allowing for the detection of the investigated analytes in tap, river, and barreled water samples at the sub-μg/L level. Yang and coworkers developed an ultrastable crystalline, quinoline-linked 2D COF (COF-CN) for the selective extraction of 14 OCPs from environmental water samples [126]. The COF was synthesized based on a cycloaddition approach using TPB and 2,5-dimethoxyterephthalaldehyde as building blocks, followed by post-synthetic functionalization using 4-ethynylbenzonitrile. The SPME coating was fabricated by a two-step deposition using a PDMS layer for the physical deposition of the COF. Although post-synthetic functionalization led to a reduction in both the BET surface and pore volume of the material (700 m^2/g and 1.2 nm for the COF-CN vs. 1983 m^2/g and 2.5 for the unfunctionalized COF), the COF-CN material showed higher thermal stability and increased hydrophobicity. The developed COF exhibited EFs higher than commercially available coatings: values in the 540–5065 range were calculated, benefitting from suitable pore size, π-π stacking, and a larger specific surface area. The HS-SPME-GC-MS/MS method was validated for 14 OCPs, obtaining LOQs in the 0.0030–45.13 ng/L range and good precision (RSDs < 12%). Finally, the validated method was applied for the determination of OCPs in water samples from various cities in Henan Province, detecting six OCPs in all samples, hexachlorobenzene and aldrin in all samples except one, and heptachlor and 4,4′-DDT in some urban samples.

Trace detection of insecticides is another key issue for environmental monitoring due to their worldwide use, high stability, and high toxicity, resulting in a strong impact on the aquatic ecosystems [127,128]. Song et al. developed an SPME coating based on a trifluoromethyl-grafted COF named COF-$(CF_3)_2$ for the extraction of benzoylurea insecticides from water samples [127]. The COF scaffold was synthesized under solvothermal conditions using 2,5-Dibromoterephthaldehyde and 1,3,5-tris(4-aminophenyl) benzene, post-synthetically functionalized through the Suzuki–Miyaura cross-coupling reaction under mild condition and deposited onto SS wires using a silicone glue. Although the functionalization with the trifluoromethyl group decreased both the BET surface and pore size from 1991 m^2/g to 1636 m^2/g and from 2.5 nm to 2.0 nm, respectively, a strong increase in the hydrophobicity as well as in the chemical and thermal stability was observed. The COF-$(CF_3)_2$ coating showed higher EFs (in the 44–105 range) compared to unfunctionalized COF and up to three times higher than those calculated using commercially available fibers, due to strong hydrogen bonds and fluorine–fluorine interactions between the COF coating and the analytes. In addition, hydrophobic, π-π stacking interactions, C–H$\cdots$π interactions, and C–F$\cdots$π interactions could be established. The DI-SPME-UHPLC-MS/MS method was validated, obtaining LOQs in the 0.20–1.70 ng/L range, recoveries in the 76.2–107.6% range, and good precision (RSDs $\leq$ 11%). The method was applied for the analysis of wastewater, river, lake, pond, and farmland water: the insecticides were detected in all the samples except for river water, with concentrations up to 33.2 ng/L. Eleven pyrethroid insecticides were extracted by Han and coworkers using a sea-urchin-like COF (COF$_{TDBA-TTL}$) [128]. The COF was synthesized under solvothermal conditions and deposited onto SS wires using a silicone adhesive. The coating interacted with the investigated pyrethroid insecticides through electrostatic, hydrophobic, and π-π interactions, and hydrogen bonding, showing extraction performance up to 446 times higher than commercially available coatings. The DI-SPME-GC-MS method was validated, obtaining LOQs in the 0.720–5.61 ng/L range, EFs in the 2584–7199 range, and RSDs $\leq$ 13%. Finally, tefluthrin, cyhalothrin, acrinathrin, permethrin, and etofenprox were detected in Pearl River water samples at a concentration of 7.54, 9.96, 7.70, 10.3, 17.2, and 11.0 ng/L, respectively.

4.3.5. Extraction of per- and Polyfluorinated Alkyl Substances

The use of COFs as SPME coatings for the extraction of PFASs has been proposed as a valid alternative to the use of MOFs, featuring a highly hydrophobic and π electron-rich structure [129–131]. In this context, a novel dioxin-linked COF (TH-COF) was synthesized by microwave-promoted nucleophilic substitution of 2,3,5,6-tetrafluoro-4-pyridinecarbonitrile with 2,3,6,7,10,11- hexahydroxy triphenylene [129]. The COF was then deposited onto SS wires using an epoxy resin and the coating was applied for the DI-SPME-UHPLC-MS/MS analysis of perfluoroalkyl carboxylic acids (C_4–C_{10}), perfluorobutanesulfonic acid, and perfluorooctanesulfonic acid. The use of microwave conditions strongly increased the BET surface of the material (from 576 m^2/g to 1254 m^2/g), obtaining pores with a diameter of 2 nm and a reticulate morphology. PFASs were adsorbed via van der Waals forces and strong hydrogen bonding between the pyridine N atoms of TH-COF and the analytes. Method validation resulted in very low LOQs (0.0066–0.015 ng/L), recoveries in the 89.5–105% range, and good precision, with RSDs $\leq$ 8%, allowing for the ultra-trace analysis of pollutants in river, drinking, and underground water samples: perfluoropentanoic acid and perfluorobutanesulfonic acid were detected only in the river water samples at a concentration of 0.480 ng/L and 0.451 ng/L, respectively. Song and coworkers developed a multifunctionalized COF, featuring an amide group and perfluoroalkyl-chain-functionalized chains (COF-NH-CO-F9), to be used for the extraction of anionic, cationic, and zwitterionic PFASs [130]. The material was obtained under solvothermal conditions and deposited onto

SS wires using silicone elastomer. A BET surface of 681 m^2/g, an interlayer distance of 4.8 Å, and an average pore size of 24 Å were observed, allowing for a size complementarity with the target analytes. In addition, the functionalization with fluoroalkyl chains increased the hydrophobicity of the surface and was beneficial for the selective enrichment of PFASs, which interacted with the coating via electrostatic, hydrophobic, fluorine−fluorine, and π−CF interactions, as well as through hydrogen bonding between N−H···F and N−H···O groups. The EFs for the investigated compounds were in the 66−160 range, up to 57 times higher than commercially available coatings, the non-fluorinated COF analogue, and the COF-(CF$_3$)$_2$ coating described in Section 4.3.4. By coupling DI-SPME with UHPLC-MS/MS, LOQs in the 0.01−0.65 ng/L range, recovery rates in the 77.1−108% range, and RSDs ≤ 11% were obtained. The method was then applied for the analysis of tap, river, lake, pond, and farmland water, and wastewater samples, detecting total PFAS concentrations in the 14.5−96.3 ng/L range. A triazine-core-based, F-functionalized COF (COF-F-1) was synthesized by Hou and coworkers at room temperature and physically deposited onto an SS probe to be used for the nanoESI-MS analysis of PFASs [131]. After the COF deposition, a chitosan polymeric layer was applied, enhancing the hydrophilicity, biocompatibility, and stability of the SPME probe. The COF-F-1 presented hydrophobic channels, fluorine-rich subunits, and electron-deficient groups containing nitrogen, able to interact with PFASs via fluorine−fluorine and electrostatic interactions, as well as hydrogen bonding. A good complementarity between the pore dimensions of the coating (2.8 nm) and the molecular sizes of the investigated C$_7$−C$_{18}$ PFASs (9.4−23.8 Å) was observed. The nanoESI-MS method was applied for the analysis of C$_7$−C$_{10}$ perfliuoroalkylsulfonic acids and C$_7$−C$_{18}$ perfluoroalkyl carboxylic acids, obtaining EFs in the 105−4538 range, indicating a high enrichment capacity of the developed coating towards PFAS. The highest values obtained for the perfluoroalkylsulfonic acids were ascribed to the hydrogen bonding between the oxygen atoms of the guest molecules and the protonated nitrogens within the COF. LOQs in the 0.06−3 ng/L range were obtained, together with recoveries in the 84−113% range, and RSDs below 12%, allowing for the detection of C$_8$−C$_{18}$ PFASs in the environmental sample with concentrations in the 5.4−38.6 ng/L range.

4.3.6. Extraction of Other Compound Classes by COF-Based SPME

Phthalic acid esters (PAEs) are a class of plasticizers widely used in the polymer industry. Different compounds have been included in the list of priority substances by the US EPA due to the potential endocrine-disrupting activity; therefore, the development of new analytical methods able to detect these compounds at trace levels in environmental and biological matrices is demanded [134,135]. To improve their enrichment and selective extraction, Yu and coworkers developed different β-ketoenamine-linked COF coatings [96]. Four different COFs were tested, all synthesized by MC to obtain a clay-like material that was deposited onto APTES-pretreated SS fibers, obtaining covalently bonded COF coatings. The best performance was obtained using Tp and 4,4'-diamino-p-terphenyl (TpTph-COF). The material showed high hydrophobicity, a BET surface of 267.6 m^2/g, and an average pore diameter of 2.52 nm. The analytes could be adsorbed via hydrogen bonding, π-π and hydrophobic interactions, together with a size-matching effect to guarantee a good extraction selectivity. The DI-SPME-GC-MS/MS method was validated, obtaining LOQs in the 0.2−0.5 ng/L range, EFs in the 1140−3720 range, and RSDs always lower than 10%. The analysis of environmental samples resulted in the detection of dipentyl phthalate (0.27 ng/L) and butyl benzyl phthalate (11.62 ng/L), diisohexyl phthalate (0.33 ng/L), dipentyl phthalate (1.65 ng/L), and butyl benzyl phthalate (2.34 ng/L) in lake water.

Porphyrin-based COFs were proposed as novel materials to be used for the EE-SPME extraction of PAEs [136]: the coating was obtained by in situ electropolymerization onto etched SS fibers, obtaining a fine tuning for both the thickness and morphology of the coating. The final COF exhibited a good crystallinity, with stacked 2D-COF sheets, and good conductivity due to the conjugated macrocycle of the porphyrin ligand. The use of EE-SPME resulted in improved selectivity towards PAEs, even in the presence of interfering compounds, including vitamin C, bovine serum albumin, hemoglobin, aromatic amines, PAHs, and hydroxyl-PAHs. The application of a constant positive voltage increased the extraction of negatively charged PAEs while oxidizing hydroxyl-PAHs to aldehydes. The $\pi-\pi$ stacking of porphyrin between the 2D adjacent layers produced a high value of charge carrier mobility, thus improving the extraction capability, whereas the amino groups interacted with the analytes by hydrogen bonding. The EE-SPME-GC-MS method was validated for six model PAEs, obtaining LOQs in the 0.2–10 ng/L range and RSDs $\leq$ 7%. When applied to the analysis of environmental water samples, diisobutyl phthalate, di-n-butyl phthalate, and dihexyl phthalate were detected in lake water in the 45.9–208.6 ng/L range, whereas di-n-butyl phthalate and dihexyl phthalate were detected in the 40.6–178.7 ng/L range in industrial wastewater samples.

Wen and coworkers tested the extraction capability of TpPa-1 for the extraction of synthetic musks from water samples [137]. These compounds are contaminants of emerging concern due to their extensive use, persistence, and bioaccumulation, leading to their widespread presence in environmental and biological matrices. The novel COF coating was synthesized using the solvothermal approach and deposited onto etched SS wires using silicone glue. The material was tested for the DI-SPME-GC-MS/MS analysis of six model synthetic musks, obtaining EFs in the 1214–12,487 range, higher than those achieved using commercial fibers due to hydrophobic interactions and π-π stacking. After validation, LOQs in the 0.04–0.31 ng/L range and RSDs < 10% were obtained. Finally, the DI-SPME-GC-MS/MS was applied for the analysis of tap, underground, and river water, determining contamination levels from 0.72 to 571.1 ng/L. A novel templating strategy based on SiO$_2$ nanofibers (SiO$_2$ NFs) was developed by Fang and coworkers to synthesize hierarchical COF hollow nanofiber (COF HNFs)-coated stainless-steel fibers for the HS-SPME extraction of both six chlorophenols in water and thymol and carvacrol in milk samples [138]. The very high specific BET surface area (747 m^2 g^{-1}) and the hierarchical porosity of the coating allowed a dramatic improvement in the accessibility of the internal COF micropores, resulting in an overall superior performance of the TpBD-Me2HNFs-12 fiber with EFs in the 452–2632 range and a low matrix effect.

Table 2. COF-based coatings used for the SPME extraction of environmental samples.

Analyte	Material	Deposition Method	Extraction Mode	Matrix	Platform	LOD (ng/L)	EFs	References
5 PAHs	BTCH-PTA-COF	sol–gel deposition	HS	beverages and river water	GC-FID	30–50	767–1411	[101]
5 PAHs	TFPA-TAPP-COF	in situ growth	HS	river water	GC-FID	6–24	-	[105]
5 PAHs	porphyrin-COF	physical adhesion	HS	lake water and soil	GC-FID	250–5000	-	[106]
6 PAHs	TAPB-TMC-COF	physical adhesion	HS	river water, pond water, and industrial wastewater	GC-MS	0.29–0.94	819–2420	[107]
7 PAHs	imine- COF-SCU1	in situ deposition	HS	soil	NTD-GC-FID	0.01–0.05 [a]	-	[94]
6 PAHs	Zn-MOF/COF	physical adhesion	HS	soil	GC-FID	0.1–1c	-	[108]
6 PAHs	Cu-MOF/COF	in situ deposition	HS	soil	GC-FID	0.1–0.5c	-	[109]
6 PAHs and BTEX	2DTP/MIL-101-Cr	physical adhesion	HS	soil	VA-GC-FID	2.1–5 [a] (BTEX), 0.07–1.6 [a] (PAHs)	-	[110]
7 PAHs	TpBD-COF	in situ electrodeposition	DI	tap water and lake water	GC-FID	1000–5000	-	[111]
8 PAHs	triazine- COF	physical adhesion	IT	tap water, river water, rainwater, and beverages	LC-UV	4–10	1110–2763	[112]
8 PAHs	TpPa-1-1000	physical adhesion	DI	soil	GC-MS	3.1–8.6 [a]	-	[102]
8 PAHs	g-C₃N₄@TpBD	sol–gel deposition	DI	pond water, river water, lake water, well water, rainwater, and snow	GC-MS	20–50	-	[103]
8 PAHs, 4 estrogens and 4 bisphenols	TiO₂NARs-CFs	in situ growth	IT	tap water, rainwater, and river water	LC-UV	1–10	405–6784	[104]
5 nitroaromatic compounds	MA/PFC-1-HOF	physical adhesion	HS	lake water, river water, and domestic water	GC-MS	4.3–20.8	393–1708	[132]
5 CPhs	TPB-DMTP-COF	physical adhesion	HS	underground, reservoir, and drinking water	GC-MS/MS	0.0048–0.015	1741–4265	[113]
5 CPhs, 2 CPhs, 2-nitrophenol, 2 dimethylphenols	CuPc-MCOF	physical adhesion	EE-SPME	seawater and seafood	GC-MS/MS	0.8–5	339–988	[114]
	TpBD COF	in situ growth	HS	water and soil	GC-MS	0.39–0.72	11,080–58,762	[115]
6 CPhs	TpBD-Me2HNFs-12	in situ growth	HS	river water	GC-FID	-	452–2632	[138]
BPA	COF-GO	physical adhesion	DI	river water and seawater	CFDI-MS	22.2	-	[97]
5 PBDEs	TpPa-1	physical adhesion	DI	ground water, drinking water, and pond water	GC-NCI-MS	0.0058–0.022	2035–6859	[116]
6 PBDEs	PI@TPB-DMTP	in situ growth	DI	lake water, river water, and wastewater	GC-NCI-MS	0.0083–0.0190	1470–3555	[100]
6 PBDEs	COF-γ-PIL	physical adhesion	DI	lake water, river water, and seawater	GC-MS	0.0021–0.014	913–3625	[117]
6 PBBs	TAPB-DMTP-DB COF	physical adhesion	HS	river water	GC-MS	0.04–0.28	4400–11,360	[118]

Table 2. Cont.

Analyte	Material	Deposition Method	Extraction Mode	Matrix	Platform	LOD (ng/L)	EFs	References
TBBPA	TpBD-COF	in situ deposition	DI	tap water, river water, seawater, and beverages	CFDI-MS	0.92	185	[119]
4 TBBPA analoges	TpPaBD$_{50}$-COF	in situ deposition	DI	river water and seawater	CFDI-MS	0.5–12	-	[120]
4 TBBPA analoges	porous-TpBD	in situ deposition	DI	river water and seawater	CFDI-MS	0.1–1	-	[98]
6 PCBs	PAN-SiO$_2$@TpPa	physical adhesion	HS	river, lake, and seawater	GC-ECD	0.1–5	2602–5611	[121]
17 PCBs	chlorinated-TpPa-1	in situ growth	HS	seawater, river water, and reservoir water	GC-MS	0.0015–0.0088	699–4281	[122]
15 PCBs	3D TpTAM-COF	physical adhesion	HS	river water and soil	GC-MS	0.001–0.020	5308–10,305	[99]
5 OCPs	Tp-Azo-COF	physical adhesion	DI	tap and well water and beverages	GC-MS/MS	0.002–0.08	1061–3693	[123]
Trifluralin, chlorpyrifos	porous PTA/TAPPT COF	in situ deposition	DI	agriculture wastewater and vegetables	GC-CD-IMS	130, 150	1950, 2123	[124]
Triclosan, methyltriclosn	NiFe$_2$O$_4$@COF	physical adhesion	DI	tap water, river water, and barreled water	GC-ECD	1–7	279–334	[125]
14 OCPs	COF-CN	physical adhesion	HS	river water	GC-MS/MS	0.0010–13.54	540–5065	[126]
Benzoylurea insecticide	COF-(CF$_3$)$_2$	physical adhesion	DI	lake water, river water, pond water, wastewater and farmland water	UHPLC-MS/MS	0.06–0.50	44–105	[127]
11 Pyrethroid insecticides	COF$_{TDBA-TTL}$	physical adhesion	DI	river water	GC-MS	0.170–1.68	2584–7199	[128]
8 PFASs	TH-COF	physical adhesion	DI	drinking water, underground water, and river water	UPLC-MS/MS	0.0020–0.0045	-	[129]
14 PFASs	COF-NH-CO-F9	physical adhesion	DI	tap water, river water, lake water, pond water, wastewater, and farmland water	UHPLC-MS/MS	0.0035–0.18	66–160	[130]
14 PFASs	COF-F-1	physical adhesion	DI	lake water and blood	NanoESI-MS	0.02–0.8	105–4538	[131]
5 PAEs	TpTph-COF	in situ growth	DI	lake water and seawater	GC-MS/MS	0.02–0.08	1140–3720	[96]
6 PAEs	porphyrin-based COF	in situ growth	EE-SPME	beverages, lake water, industrial wastewater, and oysters	GC-MS/MS	50–2000	1329 (diethylhexyl phthalate)	[136]
6 Synthetic musks	TpPa-1	physical adhesion	DI	river water, tap water, and underground water	GC-MS/MS	0.04–0.31	1214–12,487	[137]

^{a} ng/g.

5. Supramolecular Macrocycles

Supramolecular macrocycles are molecular receptors that can be properly designed to achieve the desired selectivity towards target analytes. Host–guest interactions are based on the lock–key principle [139,140]: the receptors are properly designed to obtain specific interactions and host–guest shape complementarity. In this context, the adsorption of target analytes is obtained by the occurrence of simultaneous non-covalent interactions, such as hydrogen bonding, charge transfer, electrostatic, π-π, van der Waals, and hydrophilic/hydrophobic interactions, and size complementarity between the cavity of the macrocycle and the guest. Among the supramolecular macrocycles, cyclodextrins, calixarenes, and quinoxaline cavitands have been applied as SPME coatings [141] (Table 3).

5.1. Cyclodextrins as SPME Coatings

CDs are cyclic oligosaccharides containing D-glucopyranose units bound via α-(1,4)-glycosidic linkages. The shape is an asymmetrical toroid, with primary and secondary hydroxyl groups situated at the top (primary face, narrower) and at the bottom (secondary face, wider) of the system, respectively, resulting in hydrophilic outer faces and a hydrophobic inner cavity. CDs have been applied for the complexation of different compounds, including ions, volatile, and semi-volatile organic compounds [141,142]. These macrocyclic receptors are classified on the basis of the number of oligosaccharide units, with α-CD, β-CD, and γ-CD constituted by six, seven, and eight glucopyranose units, respectively. Considering that α-CDs present a distorted glucose and a smaller cavity and γ-CDs are characterized by higher flexibility and dimensions, resulting in a lower steric barrier, β-CDs are the most commonly used extraction sorbents [141]. CDs can be easily functionalized to improve the complexation capability towards specific classes of compounds, to obtain hybrid nanocomposite materials, or to increase the bonding stability onto the SPME surface. In this context, recent studies have focused on the development of both polymer-based CDs [143] and hybrid nanostructured materials [144–146].

A β-CD-crosslinked polymer was obtained via in situ functionalization of a glass fiber to be used for an IT-SPME extraction [143]. In particular, the glass fibers were benzylated and functionalized using β-CD via the Friedel–Crafts reaction; the coated fibers were then inserted in an SS tube to be used for the on-line IT-SPME-LC-UV analysis of low- and medium-size PAHs (Figure 10). The extraction capabilities of the developed coating were up to 208 times higher than those obtained using an unfunctionalized glass fiber, achieving EFs in the 2130–2670 range. Method validation provided LOQs in the 12–25 ng/L range and a good precision, with RSDs $\leq$ 7%. Finally, method reliability was tested for the extraction of bottled lake, soil, and tap water spiked with PAHs in the 3–5 µg/L range, obtaining extraction recoveries in the 80–107% range.

In recent studies, multiwalled carbon nanotubes (MWCNTs) were combined with CDs to develop nanostructured hybrid materials [144,145]. The new coatings featured the high selectivity of the supramolecular receptors combined with enhanced surface area/volume ratio and morphological structure of the MWCNTs. In a study by Ghorbani et al., β-CD-functionalized multiwalled carbon nanotubes and acyl chloride functionalized MWCNTs were synthesized and incorporated into a silica composite by the sol–gel technique [144]. A dip-coating process was performed to deposit the hybrid material onto a polypropylene hollow fiber to be used for the SPME extraction of fluoxetine and norfluoxetine, an antidepressant drug and its biologically active form, respectively. The β-CD-MWCNT coating provided a superior extraction capability than that obtained by using only the pristine and acyl-functionalized nanotubes, with EFs of 144 and 151 for fluoxetine and norfluoxetine, respectively. The DI-LC-UV method was validated under optimized conditions, obtaining LOQs of 400 and 300 ng/L and a good precision (RSDs < 10%). The method was applied for the analysis of well, tap, and river water samples, but none of the investigated analytes were detected in the analyzed samples. Similarly, Riboni et al. [145] developed a γ-CD functionalized MWCNT-based material to be used for the extraction of 16 US EPA priority pollutant PAHs from a water sample. In this study, both acidic and hydrogen peroxide pretreatment of the nanotubes were tested, and each of the pretreated MWCNT materials was functionalized using either β- or γ-CDs. Finally, the coatings were obtained by a dip-coating procedure using an epoxy resin. Among the nanocomposite-coated fibers, the highest enrichment capability was achieved by using the H_2O_2-treated MWCNTs with γ-CDs as macrocyclic receptors. The achieved results were explained considering both the lower degree of damage of the nanotube surface produced by the hydrogen peroxide, and the steric complementarity between the γ-CD and the analytes, maximizing the host–guest interactions. Under optimal conditions, the DI-SPME-GC-MS method exhibited LOQs in the 0.2–2.3 ng/L range, a linear range of two orders of magnitude, and RSDs ≤ 21%. Exceptionally high EFs, in the 3770(±260)–113,300(±3100) range, were observed. Snow samples collected in the Italian Alpine area were analyzed, detecting a ΣPAHs in the 11.5–34.0 ng/L range.

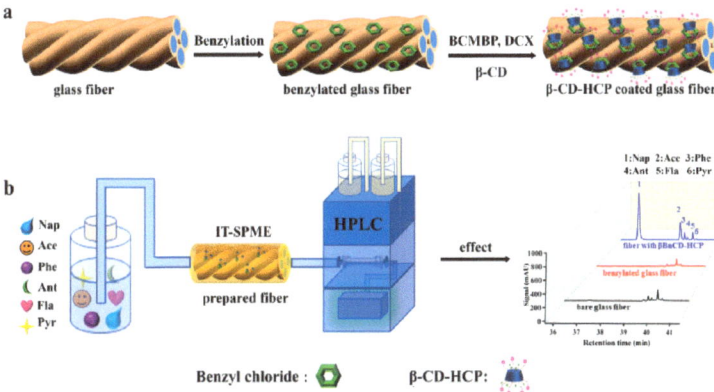

Figure 10. (a) Preparation of β-CD-coated glass fiber; (b) IT-SPME-LC-UV method for PAHs analysis. Reprinted with permission from [143].

Finally, Li et al. developed a novel coating using a γ-CD-MOF as an adsorbent material [146]. The MOF was obtained by coordination between γ-CD and potassium ions under alkaline conditions, even though a poor water stability was observed. The fibers were obtained by the dip-coating of SS wires with the pre-synthesized γ-CD-MOF using a neutral silicone sealant; finally, PDMS was deposited onto the surface by means of thermal vapor deposition to obtain a hydrophobic layer. An HS-SPME-GC-MS method was optimized and validated for the extraction of BTEX from aqueous samples. Owing to the high surface area and multifarious nanopores, the developed coating provided an extraction capability up to 11.6 times higher compared to commercial 30 µm PDMS fibers. The method resulted in LOQs in the 0.44–0.95 ng/L range and good precision, with RSDs ≤ 10%. When applied to the analysis of river and pond water, ΣBTEX of 25.6 and 24.6 ng/L were obtained, respectively.

5.2. Calixarenes and Cavitands as SPME Coatings

Calixarenes and cavitands are synthetic macrocycles able to host target guests inside their cavity. They are versatile multidentate receptors whose dimensions can be properly modified on the basis of number of monomeric units and whose surface properties can be tuned by the introduction of specific substituents surrounding the hydrophobic cavity.

Calixarenes are a class of synthetic macrocyclic receptors obtained by the condensation of para-substituted phenol and formaldehyde (Figure 11a). Functionalization can be performed on both faces of the macrocycle, exploiting both the hydroxyl group and the para-substituent of phenols to improve the selectivity of the receptor to anchor the macrocycle onto the substrate or to tune its properties, including the opening and flexibility of the cavity [147]. Najarzadekan et al. developed an SPME coating based on novel polyurethane–

polysulfone/calix[4]arene (PU-PSU/calix[4]arene) nanofibers [148] to be used for the HS extraction of 1,2,4-trichlorobenzene, 1,2,3-trichlorobenzene, and 1,2,3,4-tetrachlorobenzene. Four calixarene derivatives, namely, calix[4]arene, sulfonated calix[4]arene, p-tert-butyl-calix[4]arene, and calix[6]arene, were synthesized and electrospun with polyurethane and polysulfone directly onto the SPME fiber. The performance of the different coatings was compared in terms of extraction efficiency, showing the PU-PSU/calix[4]arene had the highest adsorption capability towards chlorobenzenes. After optimization, the HS-SPME-gas chromatography-micro electron capture detector (GC-μECD) method was validated, obtaining LOQs in the 0.4–4.0 ng/L range and RSDs always lower than 10%. Finally, the HS-SPME-GC-μECD method was applied for the quantitation of the target analytes in tap, river, sewage, and industrial water; since no analyte could be detected, the environmental matrices were spiked with the analytes at concentrations of 40 and 400 ng/L, obtaining recoveries in the 80–106% range.

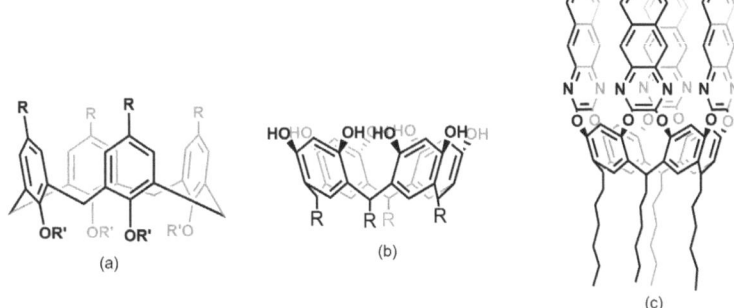

Figure 11. Representation of the structures of (**a**) calix[4]arene, (**b**) resorcinarene-based macrocycle, and (**c**) BenzoQxCav.

Other important synthetic macrocycles are the resorcinarene-based receptors obtained by the condensation of resorcinol and aldehydes (Figure 11b) [147,149]. As in the case of CDs and calixarenes, the upper rim of the receptors can be functionalized to obtain deeper cavities, tuned polarities, or different complexation capabilities, whereas the substituents at the lower rim are mostly used to tune the macrocycle solubility or to improve the anchoring on a substrate. Very recently, our research group developed and tested a novel cavitand characterized by four benzoquinoxaline walls at the upper rim (BenzoQxCav) as SPME coatings for the extraction of PAHs from water samples (Figure 11c) [150,151]. Compared to the tetraquinoxaline cavitand, the new receptor was characterized by a 2.5 Å deeper hydrophobic cavity, able to host large guests such as PAHs. The coating was obtained by the dip-coating of silica fibers into an epoxy resin followed by dipping into the powdered cavitand. The coating was tested for the extraction of the 16 US EPA priority pollutants PAHs and provided higher enrichment capabilities compared to commercially available PDMS fibers, owing to π-π, CH-π interactions, and analyte engulfment into the hydrophobic cavity. After optimization, the method was validated, obtaining LOQs in the 0.09–1.01 ng/L range, a four order of magnitude linear range, and RSDs $\leq$ 18%. Finally, the DI-SPME-GC-MS method was applied for the extraction of PAHs from snow samples from Antarctica and the Alps, obtaining ΣPAHs in the 32.2–49.5 ng/L and 10.19–59.13 ng/L ranges, respectively.

Table 3. Supramolecular receptor-based coatings used for the SPME extraction of environmental samples.

Analyte	Material	Deposition Method	Extraction Mode	Matrix	Platform	LOD (ng/L)	EFs	References
6 PAHs	β-CD-crosslinked polymer	in situ growth	bottled water, lake water, tap water, and soil water	IT	LC-UV	4–8	2130–2670	[143]
16 PAHs	γ-CD-MWCNTs-H$_2$O$_2$	physical adhesion	snow	DI	GC-MS	0.1–0.7	3770–113,300	[145]
Fluoxetine and norfluoxetine	β-CD-MWCNTs	physical adhesion	Tap water, river water, and well water	DI	LC-UV	300–400	144–151	[144]
BTEX	γ-CD-MOF	physical adhesion	River water and pond water	HS	LC-MS	0.13–0.29	-	[146]
1,2,4-trichlorobenzene, 1,2,3-trichlorobenzene, 1,2,3,4-tetrachlorobenzene	PU-PSU/Calix[4]arene nanofibers	in situ electrodeposition	Tap water, river water, sewage water, and wastewater	HS	GC-μECD	0.1–1.0	-	[148]
16 PAHs	BenzoQxCav	physical adhesion	snow	DI	GC-MS	0.03–0.30	10,260–125,500	[150]

PDMS	polydimethylsiloxane
PES	polyethersulfone
PFASs	poly- and perfluoroalkyl substances
PFOA	perfluorooctanoic acid
PIL	poly(ionic liquid)s
PPCPs	pharmaceutical and personal care products
PPy	polypyrrole
PPy@MIL-101(Cr)	PPy/chromium-based MOF nanocomposite
PU-PSU/calix[4]arene	polyurethane–polysulfone/calix[4]arene
RSD	relative standard deviation
SESI-IMS	secondary electrospray ionization-ion mobility spectrometry
SOM	single-crystal ordered macroporous
SPME	solid-phase microextraction
SS	stainless steel
TAM	(p-aminophenyl)methane
TAPB	1,3,5-tris(4-aminophenyl)benzene
TAPB-TMC-COF	1,3,5-tris(4-aminophenyl)benzene and trimesoyl chloride COF
TBBPA	tetrabromobisphenol A
TFPA-TAPP-COF	tris(4-formyl phenyl)amine-etra(4-aminophenyl)porphyrin COF
TH-COF	dioxin-linked COF
TMC	trimesoyl chloride
Tp	1,3,5-triformylphloroglucinol
Tp−Azo−COF	ketoenamine COF
TpPa-1–1000	1,3,5-triformylphloroglucinol—p-phenylenediamine N-doped porous carbons
TpTph-COF	1,3,5-triformylphloroglucinol—4,4′-diamino-p-terphenyl COF
UHPLC-MS/MS	ultra high performance liquid chromatography
US EPA	United States Environmental Protection Agency
ZIF	Zeolitic Imidazolate Frameworks
Zr/N-OMC	zirconium and nitrogen co-doped ordered mesoporous carbon
Zr-MOF@GO	zirconium-based MOF and graphene oxide coating
Σ	total concentration

References

1. Belardi, R.P.; Pawliszyn, J.B. The Application of Chemically Modified Fused Silica Fibers in the Extraction of Organics from Water Matrix Samples and Their Rapid Transfer to Capillary Columns. *Water Qual. Res. J.* **1989**, *24*, 179–191. [CrossRef]
2. Reyes-Garcés, N.; Gionfriddo, E.; Gómez-Ríos, G.A.; Alam, M.N.; Boyacl, E.; Bojko, B.; Singh, V.; Grandy, J.; Pawliszyn, J. Advances in Solid Phase Microextraction and Perspective on Future Directions. *Anal. Chem.* **2018**, *90*, 302–360. [CrossRef] [PubMed]
3. Gómez-Ríos, G.A.; Mirabelli, M.F. Solid Phase Microextraction-Mass Spectrometry: Metanoia. *TrAC-Trends Anal. Chem.* **2019**, *112*, 201–211. [CrossRef]
4. Riboni, N.; Fornari, F.; Bianchi, F.; Careri, M. Recent Advances in in Vivo Spme Sampling. *Separations* **2020**, *7*, 6. [CrossRef]
5. Zheng, J.; Kuang, Y.; Zhou, S.; Gong, X.; Ouyang, G. Latest Improvements and Expanding Applications of Solid-Phase Microextraction. *Anal. Chem.* **2023**, *95*, 218–237. [CrossRef] [PubMed]
6. Murtada, K. Trends in Nanomaterial-Based Solid-Phase Microextraction with a Focus on Environmental Applications—A Review. *Trends Environ. Anal. Chem.* **2020**, *25*, e00077. [CrossRef]
7. Lashgari, M.; Yamini, Y. An Overview of the Most Common Lab-Made Coating Materials in Solid Phase Microextraction. *Talanta* **2019**, *191*, 283–306. [CrossRef] [PubMed]
8. Souza-Silva, É.A.; Reyes-Garcés, N.; Gómez-Ríos, G.A.; Boyaci, E.; Bojko, B.; Pawliszyn, J. A Critical Review of the State of the Art of Solid-Phase Microextraction of Complex Matrices III. Bioanalytical and Clinical Applications. *TrAC-Trends Anal. Chem.* **2015**, *71*, 249–264. [CrossRef]
9. Gałuszka, A.; Migaszewski, Z.; Namieśnik, J. The 12 Principles of Green Analytical Chemistry and the SIGNIFICANCE Mnemonic of Green Analytical Practices. *TrAC-Trends Anal. Chem.* **2013**, *50*, 78–84. [CrossRef]
10. Zhou, W.; Wieczorek, M.N.; Javanmardi, H.; Pawliszyn, J. Direct Solid-Phase Microextraction-Mass Spectrometry Facilitates Rapid Analysis and Green Analytical Chemistry. *TrAC-Trends Anal. Chem.* **2023**, *166*, 117167. [CrossRef]
11. Llompart, M.; Celeiro, M.; García-Jares, C.; Dagnac, T. Environmental Applications of Solid-Phase Microextraction. *TrAC Trends Anal. Chem.* **2019**, *112*, 1–12. [CrossRef]
12. Qian, H.L.; Wang, Y.; Yan, X.P. Covalent Organic Frameworks for Environmental Analysis. *TrAC-Trends Anal. Chem.* **2022**, *147*, 116516. [CrossRef]

13. Xu, C.H.; Chen, G.S.; Xiong, Z.H.; Fan, Y.X.; Wang, X.C.; Liu, Y. Applications of Solid-Phase Microextraction in Food Analysis. *TrAC-Trends Anal. Chem.* **2016**, *80*, 12–29. [CrossRef]
14. Hu, K.; Chen, L.; Gao, S.; Liu, W.; Wei, B.; He, Q. Application Progress of Covalent Organic Frameworks (COFs) Materials in the Detection of Food Contaminants. *Food Control* **2024**, *155*, 110072. [CrossRef]
15. Leszczyńska, D.; Hallmann, A.; Treder, N.; Bączek, T.; Roszkowska, A. Recent Advances in the Use of SPME for Drug Analysis in Clinical, Toxicological, and Forensic Medicine Studies. *Talanta* **2024**, *270*, 125613. [CrossRef] [PubMed]
16. El-Deen, A.K.; Hussain, C.M. Magnetic Analytical Extractions of Forensic Samples: Latest Developments and Future Perspectives. *Trends Environ. Anal. Chem.* **2023**, *39*, e00209. [CrossRef]
17. Gao, Y.; Sheng, K.; Bao, T.; Wang, S. Recent Applications of Organic Molecule-Based Framework Porous Materials in Solid-Phase Microextraction for Pharmaceutical Analysis. *J. Pharm. Biomed. Anal.* **2022**, *221*, 115040. [CrossRef]
18. Hu, B.; Ouyang, G. In Situ Solid Phase Microextraction Sampling of Analytes from Living Human Objects for Mass Spectrometry Analysis. *TrAC-Trends Anal. Chem.* **2021**, *143*, 116368. [CrossRef]
19. Roszkowska, A.; Miękus, N.; Bączek, T. Application of Solid-Phase Microextraction in Current Biomedical Research. *J. Sep. Sci.* **2019**, *42*, 285–302. [CrossRef]
20. Kou, X.; Tong, L.; Huang, S.; Chen, G.; Zhu, F.; Ouyang, G. Recent Advances of Covalent Organic Frameworks and Their Application in Sample Preparation of Biological Analysis. *TrAC-Trends Anal. Chem.* **2021**, *136*, 116182. [CrossRef]
21. Piri-Moghadam, H.; Alam, M.N.; Pawliszyn, J. Review of Geometries and Coating Materials in Solid Phase Microextraction: Opportunities, Limitations, and Future Perspectives. *Anal. Chim. Acta* **2017**, *984*, 42–65. [CrossRef] [PubMed]
22. Delińska, K.; Rakowska, P.W.; Kloskowski, A. Porous Material-Based Sorbent Coatings in Solid-Phase Microextraction Technique: Recent Trends and Future Perspectives. *TrAC-Trends Anal. Chem.* **2021**, *143*, 116386. [CrossRef]
23. Godage, N.H.; Gionfriddo, E. A Critical Outlook on Recent Developments and Applications of Matrix Compatible Coatings for Solid Phase Microextraction. *TrAC-Trends Anal. Chem.* **2019**, *111*, 220–228. [CrossRef]
24. Zheng, J.; Huang, J.; Yang, Q.; Ni, C.; Xie, X.; Shi, Y.; Sun, J.; Zhu, F.; Ouyang, G. Fabrications of Novel Solid Phase Microextraction Fiber Coatings Based on New Materials for High Enrichment Capability. *TrAC Trends Anal. Chem.* **2018**, *108*, 135–153. [CrossRef]
25. Peng, S.; Huang, X.; Huang, Y.; Huang, Y.; Zheng, J.; Zhu, F.; Xu, J.; Ouyang, G. Novel Solid-Phase Microextraction Fiber Coatings: A Review. *J. Sep. Sci.* **2022**, *45*, 282–304. [CrossRef] [PubMed]
26. Fang, S.; Huang, Y.; Ruan, Q.; Chen, C.; Liu, S.; Ouyang, G. Recent Developments on Solid Phase Microextraction (SPME) Coatings for in Vivo Analysis. *Green Anal. Chem.* **2023**, *6*, 100069. [CrossRef]
27. Gong, X.; Lin, S.; Huang, X.; Peng, S.; Shen, M.; Ouyang, S.; Zheng, J.; Xu, J.; Ouyang, G. Applications of in Vivo SPME Based on Mass Spectrometry for Environmental Pollutants Analysis and Non-Target Metabolomics: A Review. *Green Anal. Chem.* **2022**, *1*, 100004. [CrossRef]
28. Feng, X.; Kuang, Y.; Gan, L.; Zhou, S.; Zheng, J.; Ouyang, G. Solid Phase Microextraction for the Bioanalysis of Emerging Organic Pollutants. *TrAC Trends Anal. Chem.* **2024**, *177*, 117786. [CrossRef]
29. Omarova, A.; Bakaikina, N.V.; Muratuly, A.; Kazemian, H.; Baimatova, N. A Review on Preparation Methods and Applications of Metal–Organic Framework-Based Solid-Phase Microextraction Coatings. *Microchem. J.* **2022**, *175*, 107147. [CrossRef]
30. Guo, W.; Tao, H.; Tao, H.; Shuai, Q.; Huang, L. Recent Progress of Covalent Organic Frameworks as Attractive Materials for Solid-Phase Microextraction: A Review. *Anal. Chim. Acta* **2023**, *1287*, 341953. [CrossRef]
31. Rocío-Bautista, P.; Pacheco-Fernández, I.; Pasán, J.; Pino, V. Are Metal-Organic Frameworks Able to Provide a New Generation of Solid-Phase Microextraction Coatings?—A Review. *Anal. Chim. Acta* **2016**, *939*, 26–41. [CrossRef] [PubMed]
32. Torabi, E.; Mirzaei, M.; Bazargan, M.; Amiri, A. A Critical Review of Covalent Organic Frameworks-Based Sorbents in Extraction Methods. *Anal. Chim. Acta* **2022**, *1224*, 340207. [CrossRef]
33. Choy, K.L. Chemical Vapour Deposition of Coatings. *Prog. Mater. Sci.* **2003**, *48*, 57–170. [CrossRef]
34. Zhou, Z.; Xu, L.; Ding, Y.; Xiao, H.; Shi, Q.; Li, X.; Li, A.; Fang, G. Atomic Layer Deposition Meets Metal–Organic Frameworks. *Prog. Mater. Sci.* **2023**, *138*, 101159. [CrossRef]
35. Bianchi, F.; Mattarozzi, M.; Careri, M.; Mangia, A.; Musci, M.; Grasselli, F.; Bussolati, S.; Basini, G. An SPME-GC-MS Method Using an Octadecyl Silica Fibre for the Determination of the Potential Angiogenesis Modulators 17β-Estradiol and 2-Methoxyestradiol in Culture Media. *Anal. Bioanal. Chem.* **2010**, *396*, 2639–2645. [CrossRef]
36. Amiri, A. Solid-Phase Microextraction-Based Sol–Gel Technique. *TrAC Trends Anal. Chem.* **2016**, *75*, 57–74. [CrossRef]
37. Aziz-Zanjani, M.O.; Mehdinia, A. Electrochemically Prepared Solid-Phase Microextraction Coatings-A Review. *Anal. Chim. Acta* **2013**, *781*, 1–13. [CrossRef] [PubMed]
38. Zhang, Q.; Yan, S.; Yan, X.; Lv, Y. Recent Advances in Metal-Organic Frameworks: Synthesis, Application and Toxicity. *Sci. Total Environ.* **2023**, *902*, 165944. [CrossRef] [PubMed]
39. Główniak, S.; Szczęśniak, B.; Choma, J.; Jaroniec, M. Mechanochemistry: Toward Green Synthesis of Metal–Organic Frameworks. *Mater. Today* **2021**, *46*, 109–124. [CrossRef]
40. Cui, X.Y.; Gu, Z.Y.; Jiang, D.Q.; Li, Y.; Wang, H.F.; Yan, X.P. In Situ Hydrothermal Growth of Metal-Organic Framework 199 Films on Stainless Steel Fibers for Solid-Phase Microextraction of Gaseous Benzene Homologues. *Anal. Chem.* **2009**, *81*, 9771–9777. [CrossRef]
41. Jiang, H.; Li, J.; Hu, X.; Shen, J.; Sun, X.; Han, W.; Wang, L. Ordered Mesoporous Silica Film as a Novel Fiber Coating for Solid-Phase Microextraction. *Talanta* **2017**, *174*, 307–313. [CrossRef]

42. Shamsipur, M.; Gholivand, M.B.; Shamizadeh, M.; Hashemi, P. Preparation and Evaluation of a Novel Solid-Phase Microextraction Fiber Based on Functionalized Nanoporous Silica Coating for Extraction of Polycyclic Aromatic Hydrocarbons From Water Samples Followed by GC–MS Detection. *Chromatographia* **2015**, *78*, 795–803. [CrossRef]
43. Shirey, R.E. SPME Commercial Devices and Fibre Coatings. In *Handbook of Solid Phase Microextraction*; Pawliszyn, J., Ed.; Elsevier: Amsterdam, The Netherlands, 2012; pp. 99–133. ISBN 9780124160170.
44. Liu, X.; Qian, B.; Zhang, D.; Yu, M.; Chang, Z.; Bu, X. Recent Progress in Host–Guest Metal–Organic Frameworks: Construction and Emergent Properties. *Coord. Chem. Rev.* **2023**, *476*, 214921. [CrossRef]
45. Trzciński, J.W.; Pinalli, R.; Riboni, N.; Pedrini, J.; Bianchi, F.; Zampolli, S.; Elmi, I.; Massera, C.; Ugozzoli, F.; Dalcanale, E. In Search of the Ultimate Benzene Sensor: The EtQxBox Solution. *ACS Sens.* **2017**, *2*, 590–598. [CrossRef]
46. Riboni, N.; Trzcinski, J.W.; Bianchi, F.; Massera, C.; Pinalli, R.; Sidisky, L.; Dalcanale, E.; Careri, M. Conformationally Blocked Quinoxaline Cavitand as Solid-Phase Microextraction Coating for the Selective Detection of BTEX in Air. *Anal. Chim. Acta* **2016**, *905*, 79–84. [CrossRef] [PubMed]
47. Chang, N.; Gu, Z.Y.; Wang, H.F.; Yan, X.P. Metal-Organic-Framework-Based Tandem Molecular Sieves as a Dual Platform for Selective Microextraction and High-Resolution Gas Chromatographic Separation of n-Alkanes in Complex Matrixes. *Anal. Chem.* **2011**, *83*, 7094–7101. [CrossRef] [PubMed]
48. Maya, F.; Ghani, M. Ordered Macro/Micro-Porous Metal-Organic Framework of Type ZIF-8 in a Steel Fiber as a Sorbent for Solid-Phase Microextraction of BTEX. *Microchim. Acta* **2019**, *186*, 425. [CrossRef]
49. Wei, S.; Kou, X.; Liu, Y.; Zhu, F.; Xu, J.; Ouyang, G. Facile Construction of Superhydrophobic Hybrids of Metal-Organic Framework Grown on Nanosheet for High-Performance Extraction of Benzene Homologues. *Talanta* **2020**, *211*, 120706. [CrossRef]
50. Hu, Q.; Liu, S.; Chen, X.; Xu, J.; Zhu, F.; Ouyang, G. Enhancing Enrichment Ability of a Nanoporous Carbon Based Solid-Phase Microextraction Device by a Morphological Modulation Strategy. *Anal. Chim. Acta* **2019**, *1047*, 1–8. [CrossRef]
51. Kardani, F.; Mirzajani, R. Electrospun Polyacrylonitrile/MIL-53(Al) MOF@ SBA-15/4, 4′-Bipyridine Nanofibers for Headspace Solid-Phase Microextraction of Benzene Homologues in Environmental Water Samples with GC-FID Detection. *Microchem. J.* **2022**, *180*, 107591. [CrossRef]
52. Omarova, A.; Baimatova, N.; Kazemian, H. MOF-199-Based Coatings as SPME Fiber for Measurement of Volatile Organic Compounds in Air Samples: Optimization of in Situ Deposition Parameters. *Microchem. J.* **2023**, *185*, 108263. [CrossRef]
53. United States Environmental Protection Agency. *Priority Pollutant List*; United States Environmental Protection Agency: Washington, DC, USA, 2014.
54. Kong, J.; Zhu, F.; Huang, W.; He, H.; Hu, J.; Sun, C.; Xian, Q.; Yang, S. Sol–Gel Based Metal-Organic Framework Zeolite Imidazolate Framework-8 Fibers for Solid-Phase Microextraction of Nitro Polycyclic Aromatic Hydrocarbons and Polycyclic Aromatic Hydrocarbons in Water Samples. *J. Chromatogr. A* **2019**, *1603*, 92–101. [CrossRef]
55. Rocío-Bautista, P.; Gutiérrez-Serpa, A.; Cruz, A.J.; Ameloot, R.; Ayala, J.H.; Afonso, A.M.; Pasán, J.; Rodríguez-Hermida, S.; Pino, V. Solid-Phase Microextraction Coatings Based on the Metal-Organic Framework ZIF-8: Ensuring Stable and Reusable Fibers. *Talanta* **2020**, *215*, 120910. [CrossRef]
56. Lian, C.; Feng, X.; Tian, M.; Tian, Y.; Zhang, Y. Electrodeposition of Zeolitic Imidazolate Framework Coating on Stainless Steel Wire for Solid-Phase Microextraction of Polycyclic Aromatic Hydrocarbons in Water Samples. *Microchem. J.* **2022**, *175*, 107146. [CrossRef]
57. Zeng, J.; Li, Y.; Zheng, X.; Li, Z.; Zeng, T.; Duan, W.; Li, Q.; Shang, X.; Dong, B. Controllable Transformation of Aligned ZnO Nanorods to ZIF-8 as Solid-Phase Microextraction Coatings with Tunable Porosity, Polarity, and Conductivity. *Anal. Chem.* **2019**, *91*, 5091–5097. [CrossRef]
58. Du, J.; Zhang, R.; Wang, F.; Wang, X.; Du, X. Template-Directed Fabrication of Zeolitic Imidazolate Framework-67-Derived Coating Materials on Nickel/Titanium Alloy Fiber Substrate for Selective Solid-Phase Microextraction. *J. Chromatogr. A* **2020**, *1618*, 460855. [CrossRef]
59. Liu, M.; Liu, J.; Guo, C.; Li, Y. Metal Azolate Framework-66-Coated Fiber for Headspace Solid-Phase Microextraction of Polycyclic Aromatic Hydrocarbons. *J. Chromatogr. A* **2019**, *1584*, 57–63. [CrossRef] [PubMed]
60. Wang, X.; Wang, J.; Du, T.; Kou, H.; Du, X.; Lu, X. Zn(II)-Imidazole Derived Metal Azolate Framework as an Effective Adsorbent for Double Coated Solid-Phase Microextraction of Sixteen Polycyclic Aromatic Hydrocarbons. *Talanta* **2020**, *214*, 120866. [CrossRef] [PubMed]
61. Sun, S.; Huang, L.; Xiao, H.; Shuai, Q.; Hu, S. In Situ Self-Transformation Metal into Metal-Organic Framework Membrane for Solid-Phase Microextraction of Polycyclic Aromatic Hydrocarbons. *Talanta* **2019**, *202*, 145–151. [CrossRef]
62. Bianchi, F.; Pankajakshan, A.; Fornari, F.; Mandal, S.; Pelagatti, P.; Bacchi, A.; Mazzeo, P.P.; Careri, M. A Zinc Mixed-Ligand Microporous Metal-Organic Framework as Solid-Phase Microextraction Coating for Priority Polycyclic Aromatic Hydrocarbons from Water Samples. *Microchem. J.* **2020**, *154*, 104646. [CrossRef]
63. Qiu, J.; Zhang, T.; Wang, F.; Zhu, F.; Ouyang, G. Sheathed in Situ Heteroepitaxial Growth Metal-Organic Framework Probe for Detection of Polycyclic Aromatic Hydrocarbons in River Water and Living Fish. *Sci. Total Environ.* **2020**, *729*, 138971. [CrossRef] [PubMed]
64. Campanale, C.; Massarelli, C.; Losacco, D.; Bisaccia, D.; Triozzi, M.; Uricchio, V.F. The Monitoring of Pesticides in Water Matrices and the Analytical Criticalities: A Review. *TrAC-Trends Anal. Chem.* **2021**, *144*, 116423. [CrossRef]

65. Amini, S.; Ebrahimzadeh, H.; Seidi, S.; Jalilian, N. Preparation of Polyacrylonitrile/Ni-MOF Electrospun Nanofiber as an Efficient Fiber Coating Material for Headspace Solid-Phase Microextraction of Diazinon and Chlorpyrifos Followed by CD-IMS Analysis. *Food Chem.* **2021**, *350*, 129242. [CrossRef]
66. Mohammadi, V.; Jafari, M.T.; Saraji, M. Flexible/Self-Supported Zeolitic Imidazolate Framework-67 Film as an Adsorbent for Thin-Film Microextraction. *Microchem. J.* **2019**, *146*, 98–105. [CrossRef]
67. Gong, X.; Xu, L.; Huang, S.; Kou, X.; Lin, S.; Chen, G.; Ouyang, G. Application of the NU-1000 Coated SPME Fiber on Analysis of Trace Organochlorine Pesticides in Water. *Anal. Chim. Acta* **2022**, *1218*, 339982. [CrossRef] [PubMed]
68. Xu, S.; Dong, P.; Liu, H.; Li, H.; Chen, C.; Feng, S.; Fan, J. Lotus-like Ni@NiO Nanoparticles Embedded Porous Carbon Derived from MOF-74/Cellulose Nanocrystal Hybrids as Solid Phase Microextraction Coating for Ultrasensitive Determination of Chlorobenzenes from Water. *J. Hazard. Mater.* **2022**, *429*, 128384. [CrossRef]
69. Zarębska, M.; Bajkacz, S. Poly– and Perfluoroalkyl Substances (PFAS)—Recent Advances in the Aquatic Environment Analysis. *TrAC Trends Anal. Chem.* **2023**, *163*, 117062. [CrossRef]
70. Suwannakot, P.; Lisi, F.; Ahmed, E.; Liang, K.; Babarao, R.; Gooding, J.J.; Donald, W.A. Metal-Organic Framework-Enhanced Solid-Phase Microextraction Mass Spectrometry for the Direct and Rapid Detection of Perfluorooctanoic Acid in Environmental Water Samples. *Anal. Chem.* **2020**, *92*, 6900–6908. [CrossRef] [PubMed]
71. Gong, X.; Xu, L.; Kou, X.; Zheng, J.; Kuang, Y.; Zhou, S.; Huang, S.; Zheng, Y.; Ke, W.; Chen, G.; et al. Amino-Functionalized Metal–Organic Frameworks for Efficient Solid-Phase Microextraction of Perfluoroalkyl Acids in Environmental Water. *Microchem. J.* **2022**, *179*, 107661. [CrossRef]
72. Ouyang, S.; Liu, G.; Peng, S.; Zheng, J.; Ye, Y.X.; Zheng, J.; Tong, Y.; Hu, Y.; Zhou, N.; Gong, X.; et al. Superficially Capped Amino Metal-Organic Framework for Efficient Solid-Phase Microextraction of Perfluorinated Alkyl Substances. *J. Chromatogr. A* **2022**, *1669*, 462959. [CrossRef]
73. Jia, Y.; Qian, J.; Pan, B. Dual-Functionalized MIL-101(Cr) for the Selective Enrichment and Ultrasensitive Analysis of Trace Per- And Poly-Fluoroalkyl Substances. *Anal. Chem.* **2021**, *93*, 11116–11122. [CrossRef] [PubMed]
74. Chakraborty, A.; Adhikary, S.; Bhattacharya, S.; Dutta, S.; Chatterjee, S.; Banerjee, D.; Ganguly, A.; Rajak, P. Pharmaceuticals and Personal Care Products as Emerging Environmental Contaminants: Prevalence, Toxicity, and Remedial Approaches. *ACS Chem. Health Saf.* **2023**, *30*, 362–388. [CrossRef]
75. Liu, H.; Fan, H.; Dang, S.; Li, M.; A, G.; Yu, H. A Zr-MOF@GO-Coated Fiber with High Specific Surface Areas for Efficient, Green, Long-Life Solid-Phase Microextraction of Nonsteroidal Anti-Inflammatory Drugs in Water. *Chromatographia* **2020**, *83*, 1065–1073. [CrossRef]
76. Khodayari, P.; Jalilian, N.; Ebrahimzadeh, H.; Amini, S. Trace-Level Monitoring of Anti-Cancer Drug Residues in Wastewater and Biological Samples by Thin-Film Solid-Phase Micro-Extraction Using Electrospun Polyfam/Co-MOF-74 Composite Nanofibers Prior to Liquid Chromatography Analysis. *J. Chromatogr. A* **2021**, *1655*, 462484. [CrossRef]
77. Zhang, L.; Jiang, R.; Li, W.; Muir, D.C.G.; Zeng, E.Y. Development of a Solid-Phase Microextraction Method for Fast Analysis of Cyclic Volatile Methylsiloxanes in Water. *Chemosphere* **2020**, *250*, 126304. [CrossRef] [PubMed]
78. González-Hernández, P.; Pacheco-Fernández, I.; Bernardo, F.; Homem, V.; Pasán, J.; Ayala, J.H.; Ratola, N.; Pino, V. Headspace Solid-Phase Microextraction Based on the Metal-Organic Framework CIM-80(Al) Coating to Determine Volatile Methylsiloxanes and Musk Fragrances in Water Samples Using Gas Chromatography and Mass Spectrometry. *Talanta* **2021**, *232*, 122440. [CrossRef] [PubMed]
79. Liu, G.; Liu, H.; Tong, Y.; Xu, L.; Ye, Y.X.; Wen, C.; Zhou, N.; Xu, J.; Ouyang, G. Headspace Solid-Phase Microextraction of Semi-Volatile Ultraviolet Filters Based on a Superhydrophobic Metal-Organic Framework Stable in High-Temperature Steam. *Talanta* **2020**, *219*, 121175. [CrossRef] [PubMed]
80. Pang, J.; Liao, Y.; Huang, X.; Ye, Z.; Yuan, D. Metal-Organic Framework-Monolith Composite-Based in-Tube Solid Phase Microextraction on-Line Coupled to High-Performance Liquid Chromatography-Fluorescence Detection for the Highly Sensitive Monitoring of Fluoroquinolones in Water and Food Samples. *Talanta* **2019**, *199*, 499–506. [CrossRef]
81. Jalili, V.; Ghanbari Kakavandi, M.; Ghiasvand, A.; Barkhordari, A. Microextraction Techniques for Sampling and Determination of Polychlorinated Biphenyls: A Comprehensive Review. *Microchem. J.* **2022**, *179*, 107442. [CrossRef]
82. Zhang, N.; Huang, C.; Feng, Z.; Chen, H.; Tong, P.; Wu, X.; Zhang, L. Metal Organic Framework-Coated Stainless Steel Fiber for Solid-Phase Microextraction of Polychlorinated Biphenyls. *J. Chromatogr. A* **2018**, *1570*, 10–18. [CrossRef]
83. Guo, Y.; He, X.; Huang, C.; Chen, H.; Lu, Q.; Zhang, L. Metal–Organic Framework-Derived Nitrogen-Doped Carbon Nanotube Cages as Efficient Adsorbents for Solid-Phase Microextraction of Polychlorinated Biphenyls. *Anal. Chim. Acta* **2020**, *1095*, 99–108. [CrossRef] [PubMed]
84. Xu, H.; Huang, S.; Liu, Y.; Wei, S.; Chen, G.; Gong, Z.; Ouyang, G. Hollow Carbon Nanobubbles-Coated Solid-Phase Microextraction Fibers for the Sensitive Detection of Organic Pollutants. *Anal. Chim. Acta* **2020**, *1097*, 85–93. [CrossRef] [PubMed]
85. Chen, Z.; Wang, J.; Li, Q.; Wu, Y.; Liu, Y.; Ding, Q.; Chen, H.; Zhang, W.; Zhang, L. Hollow Zirconium-Porphyrin-Based Metal-Organic Framework for Efficient Solid-Phase Microextraction of Naphthols. *Anal. Chim. Acta* **2022**, *1200*, 339586. [CrossRef] [PubMed]
86. Darabi, J.; Ghiasvand, A. Chromium-Based Polypyrrole/MIL-101 Nanocomposite as an Effective Sorbent for Headspace Microextraction of Methyl Tert-Butyl Ether in Soil Samples. *Molecules* **2020**, *25*, 644. [CrossRef] [PubMed]

87. Bianchi, F.; Careri, M.; Marengo, E.; Musci, M. Use of Experimental Design for the Purge-and-Trap-Gas Chromatography-Mass Spectrometry Determination of Methyl Tert.-Butyl Ether, Tert.-Butyl Alcohol and BTEX in Groundwater at Trace Level. *J. Chromatogr. A* **2002**, *975*, 113–121. [CrossRef] [PubMed]
88. Wei, F.; He, Y.; Qu, X.; Xu, Z.; Zheng, S.; Zhu, D.; Fu, H. In Situ Fabricated Porous Carbon Coating Derived from Metal-Organic Frameworks for Highly Selective Solid-Phase Microextraction. *Anal. Chim. Acta* **2019**, *1078*, 70–77. [CrossRef] [PubMed]
89. Ni, C.; Huang, J.; Xie, X.; Shi, Y.; Zheng, J.; Ouyang, G. Simple Fabrication of Zirconium and Nitrogen Co-Doped Ordered Mesoporous Carbon for Enhanced Adsorption Performance towards Polar Pollutants. *Anal. Chim. Acta* **2019**, *1070*, 43–50. [CrossRef] [PubMed]
90. Zhang, N.; Mu, M.; Qin, M.; Zhu, J.; Tian, X.; Lou, X.; Zhou, Q.; Lu, M. Confinement Effect of Ionic Liquid: Improve of the Extraction Performance of Parent Metal Organic Framework for Phthalates. *J. Chromatogr. A* **2023**, *1703*, 464101. [CrossRef]
91. Xu, Z.; Zhang, Z.; She, Z.; Lin, C.; Lin, X.; Xie, Z. Aptamer-Functionalized Metal-Organic Framework-Coated Nanofibers with Multi-Affinity Sites for Highly Sensitive, Selective Recognition of Ultra-Trace Microcystin-LR. *Talanta* **2022**, *236*, 122880. [CrossRef]
92. Cóté, A.P.; Benin, A.I.; Ockwig, N.W.; O'Keeffe, M.; Matzger, A.J.; Yaghi, O.M. Porous, Crystalline, Covalent Organic Frameworks. *Science* **2005**, *310*, 1166–1170. [CrossRef]
93. Feng, J.; Feng, J.; Ji, X.; Li, C.; Han, S.; Sun, H.; Sun, M. Recent Advances of Covalent Organic Frameworks for Solid-Phase Microextraction. *TrAC-Trends Anal. Chem.* **2021**, *137*, 116208. [CrossRef]
94. Chegeni, S.; Nouriasl, K.; Ghiasvand, A. A Novel Needle Trap Device Containing an Imine-Based Covalent Organic Framework for Sampling of PAHs in Soil. *Microchem. J.* **2023**, *193*, 109034. [CrossRef]
95. Jagirani, M.S.; Gumus, Z.P.; Soylak, M. Covalent Organic Frameworks, a Renewable and Emergent Source for the Separation and Pre-Concentration of the Traces of Targeted Species. *Microchem. J.* **2023**, *191*, 108820. [CrossRef]
96. Yu, Q.; Wu, Y.; Zhang, W.; Ma, W.; Wang, J.; Chen, H.; Ding, Q.; Zhang, L. Rapidly Covalent Immobilization of β-Ketoenamine-Linked Covalent Organic Framework on Fibers for Efficient Solid-Phase Microextraction of Phthalic Acid Esters. *Talanta* **2022**, *243*, 123380. [CrossRef]
97. Gao, W.; Cheng, J.; Yuan, X.; Tian, Y. Covalent Organic Framework-Graphene Oxide Composite: A Superior Adsorption Material for Solid Phase Microextraction of Bisphenol A. *Talanta* **2021**, *222*, 121501. [CrossRef]
98. Gao, W.; Li, M.; Fa, Y.; Zhao, Z.; Cai, Y.; Liang, X.; Yu, Y.; Jiang, G. Porous Covalent Organic Frameworks-Improved Solid Phase Microextraction Ambient Mass Spectrometry for Ultrasensitive Analysis of Tetrabromobisphenol-A Analogs. *Chin. Chem. Lett.* **2022**, *33*, 3849–3852. [CrossRef]
99. Lu, F.; Wu, M.; Lin, C.; Lin, X.; Xie, Z. Efficient and Selective Solid-Phase Microextraction of Polychlorinated Biphenyls by Using a Three-Dimensional Covalent Organic Framework as Functional Coating. *J. Chromatogr. A* **2022**, *1681*, 463419. [CrossRef] [PubMed]
100. Song, C.; Shao, Y.; Yue, Z.; Hu, Q.; Zheng, J.; Yuan, H.; Yu, A.; Zhang, W.; Zhang, S.; Ouyang, G. Sheathed In-Situ Room-Temperature Growth Covalent Organic Framework Solid-Phase Microextraction Fiber for Detecting Ultratrace Polybrominated Diphenyl Ethers from Environmental Samples. *Anal. Chim. Acta* **2021**, *1176*, 338772. [CrossRef] [PubMed]
101. Huo, S.; Deng, Y.; Yang, N.; Qin, M.; Zhang, X.; Yao, X.; An, C.; Zhou, P.; Lu, X. A Durable Hydrophobicity Hydrazone Covalent Organic Framework Coating for Solid Phase Microextraction of Polycyclic Aromatic Hydrocarbons in Food and Environmental Sample. *Chem. Eng. J.* **2024**, *481*, 148562. [CrossRef]
102. Yan, Q.; Huang, L.; Mao, N.; Shuai, Q. Covalent Organic Framework Derived Porous Carbon as Effective Coating for Solid Phase Microextraction of Polycyclic Aromatic Hydrocarbons Prior to Gas-Chromatography Mass Spectrometry Analysis. *Talanta Open* **2021**, *4*, 100060. [CrossRef]
103. Zang, X.; Pang, Y.; Li, H.; Chang, Q.; Zhang, S.; Wang, C.; Wang, Z. Solid Phase Microextraction of Polycyclic Aromatic Hydrocarbons from Water Samples by a Fiber Coated with Covalent Organic Framework Modified Graphitic Carbon Nitride. *J. Chromatogr. A* **2020**, *1628*, 461428. [CrossRef]
104. Sun, M.; Feng, J.; Feng, J.; Sun, H.; Feng, Y.; Ji, X.; Li, C.; Han, S.; Sun, M. Biochar Nanosphere- and Covalent Organic Framework Nanosphere-Functionalized Titanium Dioxide Nanorod Arrays on Carbon Fibers for Solid-Phase Microextraction of Organic Pollutants. *Chem. Eng. J.* **2022**, *433*, 133645. [CrossRef]
105. Tian, Y.; Hou, Y.; Yu, Q.; Wang, X.; Tian, M. Layer-by-Layer Self-Assembly of a Novel Covalent Organic Frameworks Microextraction Coating for Analyzing Polycyclic Aromatic Hydrocarbons from Aqueous Solutions via Gas Chromatography. *J. Sep. Sci.* **2020**, *43*, 896–904. [CrossRef] [PubMed]
106. Yu, C.; Wu, F.; Luo, X.; Zhang, J. Porphyrin-Based Covalent Organic Framework Coated Stainless Steel Fiber for Solid-Phase Microextraction of Polycyclic Aromatic Hydrocarbons in Water and Soil Samples. *Microchem. J.* **2021**, *168*, 106364. [CrossRef]
107. Yang, X.; Wang, J.; Wang, W.; Zhang, S.; Wang, C.; Zhou, J.; Wang, Z. Solid Phase Microextraction of Polycyclic Aromatic Hydrocarbons by Using an Etched Stainless-Steel Fiber Coated with a Covalent Organic Framework. *Microchim. Acta* **2019**, *186*, 145. [CrossRef]
108. Koonani, S.; Ghiasvand, A. A Highly Porous Fiber Coating Based on a Zn-MOF/COF Hybrid Material for Solid-Phase Microextraction of PAHs in Soil. *Talanta* **2024**, *267*, 125236. [CrossRef] [PubMed]
109. Nouriasl, K.; Ghiasvand, A. A Copper-Based MOF/COF Hybrid as an Innovative Fiber Coating for SPME Sampling of Polycyclic Aromatic Hydrocarbons from Environmental Matrices. *Talanta Open* **2023**, *8*, 100262. [CrossRef]

110. Maleki, S.; Hashemi, P.; Adeli, M. A Simple and Portable Vacuum Assisted Headspace Solid Phase Microextraction Device Coupled to Gas Chromatography Based on Covalent Organic Framework/Metal Organic Framework Hybrid for Simultaneous Analysis of Volatile and Semi-Volatile Compounds in Soil. *J. Chromatogr. A* **2023**, *1705*, 464195. [CrossRef] [PubMed]
111. Li, Z.; Yang, M.; Shen, X.; Zhu, H.; Li, B. The Preparation of Covalent Bonding COF-TpBD Coating in Arrayed Nanopores of Stainless Steel Fiber for Solid-Phase Microextraction of Polycyclic Aromatic Hydrocarbons in Water. *Int. J. Environ. Res. Public Health* **2023**, *20*, 1393. [CrossRef]
112. Feng, J.; Feng, J.; Han, S.; Ji, X.; Li, C.; Sun, M. Triazine-Based Covalent Porous Organic Polymer for the Online in-Tube Solid-Phase Microextraction of Polycyclic Aromatic Hydrocarbons Prior to High-Performance Liquid Chromatography-Diode Array Detection. *J. Chromatogr. A* **2021**, *1641*, 462004. [CrossRef]
113. Liu, L.; Meng, W.K.; Li, L.; Xu, G.J.; Wang, X.; Chen, L.Z.; Wang, M.L.; Lin, J.M.; Zhao, R.S. Facile Room-Temperature Synthesis of a Spherical Mesoporous Covalent Organic Framework for Ultrasensitive Solid-Phase Microextraction of Phenols Prior to Gas Chromatography-Tandem Mass Spectrometry. *Chem. Eng. J.* **2019**, *369*, 920–927. [CrossRef]
114. Wang, J.; Zhang, W.; Chen, H.; Ding, Q.; Xu, J.; Yu, Q.; Fang, M.; Zhang, L. Piperazine-Linked Metal Covalent Organic Framework-Coated Fibers for Efficient Electro-Enhanced Solid-Phase Microextraction of Chlorophenols. *J. Chromatogr. A* **2023**, *1692*, 463847. [CrossRef] [PubMed]
115. Guo, W.; Tao, H.; Shuai, Q.; Huang, L. Architectural Engineering Inspired in Situ Growth of Covalent Organic Frameworks as Outstanding Fiber Coating for Solid-Phase Microextraction of Phenols. *Microchem. J.* **2023**, *189*, 108564. [CrossRef]
116. Liu, L.; Meng, W.K.; Zhou, Y.S.; Wang, X.; Xu, G.J.; Wang, M.L.; Lin, J.M.; Zhao, R.S. B-Ketoenamine-Linked Covalent Organic Framework Coating for Ultra-High-Performance Solid-Phase Microextraction of Polybrominated Diphenyl Ethers from Environmental Samples. *Chem. Eng. J.* **2019**, *356*, 926–933. [CrossRef]
117. Su, L.; Zheng, X.; Tang, J.; Wang, Q.; Zhang, L.; Wu, X. Poly(Ionic Liquid)s Threaded into Covalent Organic Framework for Synergistic Capture of Polybrominated Diphenyl Ethers. *J. Hazard. Mater.* **2024**, *461*, 132657. [CrossRef] [PubMed]
118. Zhou, S.; Kuang, Y.; Shi, Y.; Hu, Y.; Chen, L.; Zheng, J.; Ouyang, G. Modulated Covalent Organic Frameworks with Higher Specific Surface Area for the Ultrasensitive Detection of Polybrominated Biphenyls. *Chem. Eng. J.* **2023**, *453*, 139743. [CrossRef]
119. Gao, W.; Tian, Y.; Liu, H.; Cai, Y.; Liu, A.; Yu, Y.L.; Zhao, Z.; Jiang, G. Ultrasensitive Determination of Tetrabromobisphenol A by Covalent Organic Framework Based Solid Phase Microextraction Coupled with Constant Flow Desorption Ionization Mass Spectrometry. *Anal. Chem.* **2019**, *91*, 772–775. [CrossRef]
120. Gao, W.; Li, G.; Liu, H.; Tian, Y.; Li, W.T.; Fa, Y.; Cai, Y.; Zhao, Z.; Yu, Y.L.; Qu, G.; et al. Covalent Organic Frameworks with Tunable Pore Sizes Enhanced Solid-Phase Microextraction Direct Ionization Mass Spectrometry for Ultrasensitive and Rapid Analysis of Tetrabromobisphenol A Derivatives. *Sci. Total Environ.* **2021**, *764*, 144388. [CrossRef] [PubMed]
121. Lv, Y.; Ma, J.; Yu, Z.; Liu, S.; Yang, G.; Liu, Y.; Lin, C.; Ye, X.; Shi, Y.; Liu, M. Fabrication of Covalent Organic Frameworks Modified Nanofibrous Membrane for Efficiently Enriching and Detecting the Trace Polychlorinated Biphenyls in Water. *Water Res.* **2023**, *235*, 119892. [CrossRef]
122. Su, L.; Zhang, N.; Tang, J.; Zhang, L.; Wu, X. In-Situ Fabrication of a Chlorine-Functionalized Covalent Organic Framework Coating for Solid-Phase Microextraction of Polychlorinated Biphenyls in Surface Water. *Anal. Chim. Acta* **2021**, *1186*, 339120. [CrossRef]
123. Xin, J.; Xu, G.; Zhou, Y.; Wang, X.; Wang, M.; Lian, Y.; Zhao, R.S. Ketoenamine Covalent Organic Framework Coating for Efficient Solid-Phase Microextraction of Trace Organochlorine Pesticides. *J. Agric. Food Chem.* **2021**, *69*, 8008–8016. [CrossRef] [PubMed]
124. Tabibi, A.; Jafari, M.T. High Efficient Solid-Phase Microextraction Based on a Covalent Organic Framework for Determination of Trifluralin and Chlorpyrifos in Water and Food Samples by GC-CD-IMS. *Food Chem.* **2022**, *373*, 131527. [CrossRef]
125. Li, Y.; Dong, G.; Li, J.; Xiang, J.; Yuan, J.; Wang, H.; Wang, X. A Solid-Phase Microextraction Fiber Coating Based on Magnetic Covalent Organic Framework for Highly Efficient Extraction of Triclosan and Methyltriclosan in Environmental Water and Human Urine Samples. *Ecotoxicol. Environ. Saf.* **2021**, *219*, 112319. [CrossRef] [PubMed]
126. Yang, Y.; Guo, Y.; Jia, X.; Zhang, Q.; Mao, J.; Feng, Y.; Yin, D.; Zhao, W.; Zhang, Y.; Ouyang, G.; et al. An Ultrastable 2D Covalent Organic Framework Coating for Headspace Solid-Phase Microextraction of Organochlorine Pesticides in Environmental Water. *J. Hazard. Mater.* **2023**, *452*, 131228. [CrossRef]
127. Song, C.; Luo, Y.; Zheng, J.; Zhang, W.; Yu, A.; Zhang, S.; Ouyang, G. Facile and Large-Scale Synthesis of Trifluoromethyl-Grafted Covalent Organic Framework for Efficient Microextraction and Ultrasensitive Determination of Benzoylurea Insecticides. *Chem. Eng. J.* **2023**, *462*, 142220. [CrossRef]
128. Han, J.; Yu, Y.; Wen, H.; Chen, T.; Chen, Y.; Chen, G.; Qiu, J.; Zhu, F.; Ouyang, G. Sea-Urchin-like Covalent Organic Framework as Solid-Phase Microextraction Fiber Coating for Sensitive Detection of Trace Pyrethroid Insecticides in Water. *Sci. Total Environ.* **2024**, *912*, 169129. [CrossRef]
129. Ji, W.; Guo, Y.S.; Xie, H.M.; Wang, X.; Jiang, X.; Guo, D.S. Rapid Microwave Synthesis of Dioxin-Linked Covalent Organic Framework for Efficient Micro-Extraction of Perfluorinated Alkyl Substances from Water. *J. Hazard. Mater.* **2020**, *397*, 122793. [CrossRef] [PubMed]
130. Song, C.; Zheng, J.; Zhang, Q.; Yuan, H.; Yu, A.; Zhang, W.; Zhang, S.; Ouyang, G. Multifunctionalized Covalent Organic Frameworks for Broad-Spectrum Extraction and Ultrasensitive Analysis of Per- and Polyfluoroalkyl Substances. *Anal. Chem.* **2023**, *95*, 7770–7778. [CrossRef]

131. Hou, Y.J.; Deng, J.; He, K.; Chen, C.; Yang, Y. Covalent Organic Frameworks-Based Solid-Phase Microextraction Probe for Rapid and Ultrasensitive Analysis of Trace Per- A Nd Polyfluoroalkyl Substances Using Mass Spectrometry. *Anal. Chem.* **2020**, *92*, 10213–10217. [CrossRef]
132. Hu, Y.; Liu, Y.; Kuang, Y.; Zhou, S.; Chen, L.; Zhou, N.; Zheng, J.; Ouyang, G. Melamine-Participant Hydrogen-Bonded Organic Frameworks with Strong Hydrogen Bonds and Hierarchical Micropores Driving Extraction of Nitroaromatic Compounds. *Anal. Chim. Acta* **2023**, *1277*, 341652. [CrossRef]
133. Śmiełowska, M.; Zabiegała, B. Current Trends in Analytical Strategies for the Determination of Polybrominated Diphenyl Ethers (PBDEs) in Samples with Different Matrix Compositions–Part 2: New Approaches to PBDEs Determination. *TrAC-Trends Anal. Chem.* **2020**, *132*, 115889. [CrossRef]
134. Qureshi, M.S.; Yusoff, A.R.b.M.; Wirzal, M.D.H.; Barek, J.; Afridi, H.I.; Üstündag, Z. Methods for the Determination of Endocrine-Disrupting Phthalate Esters. *Crit. Rev. Anal. Chem.* **2016**, *46*, 146–159. [CrossRef]
135. Lorre, E.; Riboni, N.; Bianchi, F.; Orlandini, S.; Furlanetto, S.; Careri, M.; Zilius, M. Quality by Design in the Optimization of the Ultrasonic Assisted Solvent Extraction for the GC-MS Determination of Plasticizers in Sediments and Shells. *Talanta Open* **2023**, *8*, 100258. [CrossRef]
136. Chen, H.; Wang, J.; Zhang, W.; Guo, Y.; Ding, Q.; Zhang, L. In Situ Rapid Electrochemical Fabrication of Porphyrin-Based Covalent Organic Frameworks: Novel Fibers for Electro-Enhanced Solid-Phase Microextraction. *ACS Appl. Mater. Interfaces* **2023**, *15*, 12453–12461. [CrossRef]
137. Wen, L.; Wu, P.; Wang, L.-L.; Chen, L.-Z.; Wang, M.-L.; Wang, X.; Lin, J.-M.; Zhao, R.-S. Solid-Phase Microextraction Using a β-Ketoenamine-Linked Covalent Organic Framework Coating for Efficient Enrichment of Synthetic Musks in Water Samples. *Anal. Methods* **2020**, *12*, 2434–2442. [CrossRef]
138. Fang, Y.; Zhou, F.; Zhang, Q.; Deng, C.; Wu, M.; Shen, H.; Tang, Y.; Wang, Y. Hierarchical Covalent Organic Framework Hollow Nanofibers-Bonded Stainless Steel Fiber for Efficient Solid Phase Microextraction. *Talanta* **2024**, *267*, 125223. [CrossRef] [PubMed]
139. Bertani, F.; Riboni, N.; Bianchi, F.; Brancatelli, G.; Sterner, E.S.; Pinalli, R.; Geremia, S.; Swager, T.M.; Dalcanale, E. Triptycene-Roofed Quinoxaline Cavitands for the Supramolecular Detection of BTEX in Air. *Chem.-A Eur. J.* **2016**, *22*, 3312–3319. [CrossRef]
140. Bianchi, F.; Mattarozzi, M.; Betti, P.; Bisceglie, F.; Careri, M.; Mangia, A.; Sidisky, L.; Ongarato, S.; Dalcanale, E. Innovative Cavitand-Based Sol-Gel Coatings for the Environmental Monitoring of Benzene and Chlorobenzenes via Solid-Phase Microextraction. *Anal. Chem.* **2008**, *80*, 6423–6430. [CrossRef] [PubMed]
141. Ma, J.; Zhang, Y.; Zhao, B.; Jia, Q. Supramolecular Adsorbents in Extraction and Separation Techniques—A Review. *Anal. Chim. Acta* **2020**, *1122*, 97–113. [CrossRef]
142. Gentili, A. Cyclodextrin-Based Sorbents for Solid Phase Extraction. *J. Chromatogr. A* **2020**, *1609*, 460654. [CrossRef]
143. Zhang, Y.; Wang, Y.; Li, C.X.; Chen, X.; Lu, Z.; Zhang, M.; Zhu, X.; Yu, Y. In-Situ Preparation of β-Cyclodextrin-Crosslinked Polymer Coated Glass Fiber for in-Tube Solid-Phase Microextraction of Polycyclic Aromatic Hydrocarbons from Water Samples. *Microchem. J.* **2023**, *194*, 109285. [CrossRef]
144. Ghorbani, M.; Esmaelnia, M.; Aghamohammadhasan, M.; Akhlaghi, H.; Seyedin, O.; Azari, Z.A. Preconcentration and Determination Of Fluoxetine and Norfluoxetine in Biological and Water Samples with β-Cyclodextrin Multi-Walled Carbon Nanotubes as a Suitable Hollow Fiber Solid Phase Microextraction Sorbent and High Performance Liquid Chromatography. *J. Anal. Chem.* **2019**, *74*, 540–549. [CrossRef]
145. Riboni, N.; Bianchi, F.; Scaccaglia, M.; Bisceglie, F.; Secchi, A.; Massera, C.; Luches, P.; Careri, M. A Novel Multiwalled Carbon Nanotube–Cyclodextrin Nanocomposite for Solid-Phase Microextraction–Gas Chromatography–Mass Spectrometry Determination of Polycyclic Aromatic Hydrocarbons in Snow Samples. *Microchim. Acta* **2023**, *190*, 212. [CrossRef] [PubMed]
146. Li, N.; Pu, W.; Yu, L.D.; Tong, Y.J.; Liu, X.; Wang, S.; Fu, Q.; Yang, H.; Chen, G.; Zhu, F.; et al. PDMS-Coated ΓCD-MOF Solid-Phase Microextraction Fiber for BTEX Analysis with Boosted Performances. *Anal. Chim. Acta* **2022**, *1189*, 339259. [CrossRef] [PubMed]
147. Vincenti, M.; Irico, A. Gas-Phase Interactions of Calixarene- and Resorcinarene-Cavitands with Molecular Guests Studied by Mass Spectrometry. *Int. J. Mass Spectrom.* **2002**, *214*, 23–36. [CrossRef]
148. Najarzadekan, H.; Kamboh, M.A.; Sereshti, H.; Ahmad, I.; Sridewi, N.; Shahabuddin, S.; Rashidi Nodeh, H. Headspace Extraction of Chlorobenzenes from Water Using Electrospun Nanofibers Fabricated with Calix[4]Arene-Doped Polyurethane–Polysulfone. *Polymers* **2022**, *14*, 3760. [CrossRef] [PubMed]
149. Wang, K.; Liu, Q.; Zhou, L.; Sun, H.; Yao, X.; Hu, X. State-of-the-Art and Recent Progress in Resorcinarene-Based Cavitand R. *Chin. Chem. Lett.* **2023**, *34*, 108559. [CrossRef]

150. Riboni, N.; Amorini, M.; Bianchi, F.; Pedrini, A.; Pinalli, R.; Dalcanale, E.; Careri, M. Ultra-Sensitive Solid-Phase Microextraction–Gas Chromatography–Mass Spectrometry Determination of Polycyclic Aromatic Hydrocarbons in Snow Samples Using a Deep Cavity BenzoQxCavitand. *Chemosphere* **2022**, *303*, 135144. [CrossRef] [PubMed]
151. Amorini, M.; Riboni, N.; Pesenti, L.; Dini, V.A.; Pedrini, A.; Massera, C.; Gualandi, C.; Bianchi, F.; Pinalli, R.; Dalcanale, E. Reusable Cavitand-Based Electrospun Membranes for the Removal of Polycyclic Aromatic Hydrocarbons from Water. *Small* **2021**, *18*, 2104946. [CrossRef]

Disclaimer/Publisher's Note: The statements, opinions and data contained in all publications are solely those of the individual author(s) and contributor(s) and not of MDPI and/or the editor(s). MDPI and/or the editor(s) disclaim responsibility for any injury to people or property resulting from any ideas, methods, instructions or products referred to in the content.

Review

Advances and Applications of Hybrid Graphene-Based Materials as Sorbents for Solid Phase Microextraction Techniques

Alessandra Timóteo Cardoso, Rafael Oliveira Martins and Fernando Mauro Lanças *

Laboratory of Chromatography, Institute of Chemistry at Sao Carlos, University of Sao Paulo, P.O. Box 780, São Carlos 13566590, Brazil
* Correspondence: flancas@iqsc.usp.br

Citation: Cardoso, A.T.; Martins, R.O.; Lanças, F.M. Advances and Applications of Hybrid Graphene-Based Materials as Sorbents for Solid Phase Microextraction Techniques. *Molecules* **2024**, *29*, 3661. https://doi.org/10.3390/molecules29153661

Academic Editor: Hiroyuki Kataoka

Received: 2 July 2024
Revised: 29 July 2024
Accepted: 30 July 2024
Published: 2 August 2024

Copyright: © 2024 by the authors. Licensee MDPI, Basel, Switzerland. This article is an open access article distributed under the terms and conditions of the Creative Commons Attribution (CC BY) license (https://creativecommons.org/licenses/by/4.0/).

Abstract: The advancement of traditional sample preparation techniques has brought about miniaturization systems designed to scale down conventional methods and advocate for environmentally friendly analytical approaches. Although often referred to as green analytical strategies, the effectiveness of these methods is intricately linked to the properties of the sorbent utilized. Moreover, to fully embrace implementing these methods, it is crucial to innovate and develop new sorbent or solid phases that enhance the adaptability of miniaturized techniques across various matrices and analytes. Graphene-based materials exhibit remarkable versatility and modification potential, making them ideal sorbents for miniaturized strategies due to their high surface area and functional groups. Their notable adsorption capability and alignment with green synthesis approaches, such as bio-based graphene materials, enable the use of less sorbent and the creation of biodegradable materials, enhancing their eco-friendly aspects towards green analytical practices. Therefore, this study provides an overview of different types of hybrid graphene-based materials as well as their applications in crucial miniaturized techniques, focusing on offline methodologies such as stir bar sorptive extraction (SBSE), microextraction by packed sorbent (MEPS), pipette-tip solid-phase extraction (PT-SPE), disposable pipette extraction (DPX), dispersive micro-solid-phase extraction (d-µ-SPE), and magnetic solid-phase extraction (MSPE).

Keywords: graphene-based materials; miniaturized sample preparation; stir bar sorptive extraction; microextraction by packed sorbent; pipette-tip solid-phase extraction; disposable pipette extraction; magnetic solid-phase extraction; green analytical chemistry

1. Introduction

Since the confirmation of the existence of graphene (G), a single-layer carbon allotrope, in 2004 [1,2], this material has stood out in various areas, such as energy generation, electronics, sensors, and other material science fields, due to its unique properties, such as flexibility, lightness, thermal and electrical conductivity, and mechanical resistance to high pressures [3,4]. In addition to these, some peculiar characteristics of this nanomaterial, such as its extensive surface area, its binding sites arranged in a "honeycomb" pattern, the presence of delocalized π electrons, and its single-layer structure, among others, have prompted investigation of its use as a sorbent in sample preparation techniques, with the earliest records dating back to the early 2010s [5–7].

Among the various methods for obtaining graphene, such as mechanical or chemical exfoliation from graphite, pyrolysis, and chemical vapor deposition (CVD), the most widely used method in the field of sorbent preparation is chemical synthesis or chemical reduction from graphene oxide (GO) [8]. GO is the precursor to graphene obtained from the oxidation of graphite, a step in which the Hummers methodology is often used and has undergone modifications over the years to improve the process. The single-layer structure of GO is very similar to that of graphene (Figure 1), except for the presence of oxygen-containing functional groups, such as epoxy, hydroxyl, and carboxyl, protruding from the nanosheet obtained through a change in the hybridization of carbon atoms from sp^2 to sp^3, imparting this derivative with a hydrophilic characteristic and enabling interaction with compounds containing more polar groups. Finally, after obtaining GO, a reducing agent is used to reduce the oxygenated groups of GO, resulting in reduced graphene oxide (rGO), which exhibits properties very similar to graphene, such as high hydrophobicity, albeit with different chemical structures [9].

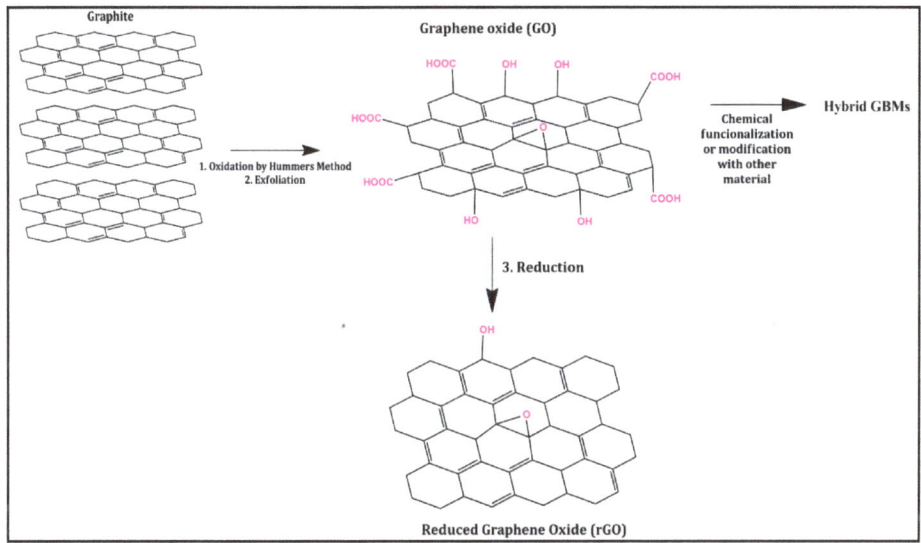

Figure 1. Scheme of obtaining GBMs from graphite oxidation.

In this context, especially in solid phase-based extraction techniques, there has been an increasing demand for sorbents with enhanced selectivity, i.e., those that promote efficient interactions with various analytes in the same sample, providing a less costly method aligned with the principles of green analytical chemistry (GAC) [10,11]. Thus, GO, in particular, allows for chemical functionalization with other components, such as ionic liquids (ILs) [12,13], silica derivatives [14,15], magnetic materials [16,17], molecularly imprinted polymers (MIPs) [18,19], resins [20], deep eutectic solvents (DESs) [21,22], and carbon-based biosorbents [23–25], among other materials. These materials are typically obtained through covalent bonding with oxygenated groups, resulting in hybrid materials promoting multiple interactions with the analytes and improving method performance, even in complex matrices [26].

Notably, modified graphene-based materials (GBMs) have been widely and successfully utilized in solid phase-based extraction techniques in conventional and miniaturized formats, with applications across various matrices, including biological, food, beverage, and environmental samples [27]. Their versatile applications underscore the importance of evaluating the methods' efficiency, the phases' selectivity, and the ecological effects they may cause. Several aspects, such as the toxicity of reagents and solvents used in synthesis and extraction methods, the quantity of sorbent and its potential for reuse, energy consumption, waste generation, sample consumption, and the volume of organic solvents used, need to be considered to minimize the environmental impact of each extraction process.

Following this brief introduction, this study aims to provide an overview of the literature on using modified GBMs in miniaturized solid-phase sample preparation techniques over the past seven years (from 2018 to May 2024). The first part of the article focuses on presenting the advances, types, and properties of hybrid GBMs, accompanied by a discussion of the environmental aspects of each material covered and their advantages and disadvantages. Some recent reviews have addressed the use of GBMs in extraction techniques, such as Jiang, Zhang, and Sun [27], who provided a comprehensive review of the use of graphene and graphene oxide in solid-phase extraction (SPE) and solid-phase microextraction (SPME) techniques, and Ibukun et al. [21], who compiled studies on the synthesis and characterization of graphene oxide modified with DESs for dispersive and magnetic solid-phase extractions. However, the existence of limited or outdated reviews on using hybrid GBMs in miniaturized extraction techniques suggests the need for additional studies to explore their potential in these approaches. Therefore, the second part of this review will focus on delineating recent applications specifically in miniaturized solid-phase sample preparation techniques, such as microextraction by packed sorbent (MEPS), pipette-tip solid-phase extraction (PT-SPE), disposable pipette extraction (DPX), dispersive micro-solid-phase extraction (d-μ-SPE), and stir bar sorptive extraction (SBSE), for the evaluation of different analyte(s) across a wide range of matrices. Finally, we present some of the current challenges and prospects related to the development and application of hybrid GBMs, providing a critique from the authors' perspective.

2. Hybrid Graphene-Based Materials (Hybrid GBMs)

As discussed in the introduction, the presence of oxygenated groups in graphene oxide makes this material highly modifiable, driving the development of novel sorbent phases for various miniaturized sample preparation applications. Moreover, these materials have been proposed as low-cost and environmentally friendly alternatives to traditional sorbent phases for different sample preparation strategies [8]. This attribute stems from their biomass waste production, such as camphor leaves, wheat straw, and rice husks. These and other agricultural or forestry residues are rich in organic compounds like cellulose, hemicellulose, and lignin, which possess a high carbon content, thus making them suitable as precursors for the synthesis of GBMs via processes such as pyrolysis, CVD, and high-temperature carbonization (HTC), among others. Furthermore, due to their abundance and renewability, these materials typically incur lower costs than graphene synthesis from graphite and, depending on the production process, are also less detrimental to the environment [3,4]. Consequently, the following sections will provide a comprehensive literature overview detailing the main trends of materials anchored onto graphene and their utilization in diverse sample preparation methodologies. Table 1 shows some applications of hybrid GBMs that will be discussed in the following topics.

Table 1. Studies on the application of hybrid graphene-based materials in microextraction techniques.

Modifier Material	Type of GBM	Analyte	Matrix	Microextraction Technique [1]	Recovery (%)	Reusability Rate (cycles)	Ref.
Silica Derivatives	GO@SiO$_2$	Benzylpenicillin, Cefalexin, Cefoperazone, and Ceftiofur	Wastewater	Column-switching	>60	100	[28]
	GO@SiO$_2$	Carbamazepine, Citalopram, Desipramine, Sertraline and Clomipramine	Urine	Multidimensional LC-MS/MS	-	250	[29]
	SiGO-C18	Simazine, Atrazine, Carbofuran, Tebuthiuron, Diuron, Ametryn, Clomazone, Thiacloprid, Hexazinone, and Imidacloprid	Sugarcane spirits	In-tube SPME	>80	-	[30]
	SiGO-C18	Aflatoxins G2, G1, B2 and B1	Food	PT-SPE	>70	10	[31]
	Si@GO@βCD	Daidzein, Genistein, Formononetin and Biochanin A	Urine	Online SPME-LC	-	200	[23]

Table 1. Cont.

Modifier Material	Type of GBM	Analyte	Matrix	Microextraction Technique [1]	Recovery (%)	Reusability Rate (cycles)	Ref.
Ionic Liquids	IL-TGO	Fipronil	Chicken eggs	PT-SPE and DSPE	>90	15	[13]
	IL–CS–GOA	3-bromocarbazole, 2,7-dibromocarbazole and 1,3,6,8-tetrabromocarbazole	Sediment	Glass dropper Extraction	>80	6	[25]
	GO@MIL	Inorganic antimony (Sb III and Sb IV)	Water, Tea And Honey	d-μ-SPE	>97	-	[32]
Magnetic Materials	MGO	Triazine Herbicides	Fruit and vegetable	MSPE	>70	8	[33]
	MGO@UIO-66	Food colorants	Soft drinks, candies, and pastilles	UA-DSPE	>95	6	[34]
	rGO@MNS	Copper(II)	Environmental waters	SPME	>95	-	[35]
Molecularly imprinted polymers	PDESs-MIP/GO	Anti-adipogenic drugs	*Solidago decurrens*	CPT-MSPD	>94	-	[36]
	N-GQDs/Fe_3O_4@SiO_2/IRMOF-1/MIP	Phenylureas	Cucumber, tomato, and radish	d-MSPE	>80	4	[37]
	TPhP-MIPs/GO	Triphenyl phosphate	Environmental water	DI-SPME	>70	110	[38]
	GO/MIP-FA	Organophosphate flame retardants	Environmental water	SPME	>70	110	[39]
Carbon-based biosorbents	GO/CS	Organic pollutants	Water	SPME	>90	100	[40]
	ACGO	Chlorophenols	Food and environmental samples	TFME	>80	48	[41]
	$NiFe_2O_4$@GO@β-CD	Bisphenols	Milk and Milk packing	MPSE	>78	12	[42]
	PDMS/GO/β-CD sponge	Lavander essential oil	Lavender	HS-SPME	-	6	[43]
	β-CD@GO@Si	Isoflavones	Soy-based juice	MEPS	>90	50	[24]
Deep eutectic solvents	GO@DES	Chlorpyrifos, diazinon, Tebuconazole, Deltamethrin, Permethrin, Haloxyfop-methyl, Penconazole, and Cyhalothrin	Zucchini	d-μ-SPE	>70%	-	[22]
	DFG	Hippuric acid and Methylhippuric acid	Urine	PT-SPE	>90%	-	[44]
	MGO@DES	Estrone, 17β-estradiol and 17α-ethinylestradiol	Milk	MSPE	>90%	7	[45]

[1] **Abbreviations:** **GO@SiO_2**: graphene oxide supported on aminopropyl silica; **LC-MS/MS**: liquid chromatography-tandem mass spectrometry; **SiGO-C18**: graphene supported on aminopropyl silica with octadecylsilane; **In-tube SPME**: in-tube solid-phase microextraction; **PT-SPE**: pipette-tip solid-phase extraction; **Si@GO@βCD**: graphene oxide supported on aminopropyl silica particles and modified with β-cyclodextrin; **Online SPME-LC**: Online solid-phase microextraction coupled with liquid chromatography; **IL-TGO**: ionic liquid-thiol-graphene oxide composite; DSPE: dispersive solid-phase extraction; **IL–CS–GOA**: ionic liquid-chitosan-graphene oxide aerogel; **GO@MIL**: magnetized graphene oxide nanoparticles; **d-μ-SPE**: dispersive micro-solid-phase extraction; **MGO**: magnetic graphene oxide; **MSPE**: magnetic solid-phase extraction; **MGO@UIO-66**: magnetic graphene oxide@UIO-66; **UA-DSPE**: ultrasound-assisted dispersive solid-phase micro-extraction; **rGO@MNS**: metal-doped graphene nanostructure; **SPME**: solid-phase microextraction; **PDESs-MIP/GO**: poly(deep eutectic solvents) surface imprinted graphene oxide composite; **CPT-MSPD**: centrifugation-accelerated pipette-tip matrix solid-phase dispersion method; **N-GQDs/Fe_3O_4@SiO_2/IRMOF-1/MIP**: Fe_3O_4/SiO_2/molecularly imprinted polymer with N-GQDs and IRMOF-1; **d-MSPE**: dispersive magnetic solid-phase extraction; **TPhP-MIPs/GO**: triphenyl phosphate molecularly imprinted polymers immobilized on graphene oxide; **DI-SPME**: direct immersion solid-phase microextraction; **GO/MIP-FA**: GO-based surface molecularly imprinted polymeric fiber array; **GO/CS**: graphene oxide-chitosan; **ACGO**: agarose/chitosan/graphene oxide; TFME: thin film microextraction; **β-CD@GO@Si**: β-Cyclodextrin, coupled to graphene oxide supported on aminopropyl silica; **MEPS**: microextraction by packed sorbent; **$NiFe_2O_4$@GO@β-CD**: β-cyclodextrin-functionalized magnetic graphene oxide; **HS-SPME**: headspace solid-phase microextraction; **PDMS/GO/β-CD sponge**: polydimethylsiloxane/graphene oxide/β-cyclodextrin sponge; **GO@DES**: GO nanoparticles modified with a deep eutectic solvent; **DFG**: deep eutectic solvent functionalized graphene oxide; **MGO@DES**: magnetic graphene oxide modified with deep eutectic solvent.

2.1. Silica Derivatives

The use of graphene derivatives in microextraction techniques is mainly due to their advantages, such as their high surface area, which allow these materials to be successfully employed in dispersive techniques. However, the irregular morphology of GBM particles is a challenge in terms of their use in techniques involving the packing of sorbents, such as automated in-tube solid-phase microextraction (IT-SPME) or online SPE, as well as offline techniques such as MEPS and PT-SPE, due to issues with high system pressure, device clogging, or particle aggregation. Functionalizing the oxygenated groups of GO or rGO with ionic liquids (ILs), deep eutectic solvents (DESs), carbon-based biosorbents, or magnetic particles can help resolve particle aggregation challenges, especially in dispersive techniques. However, in packed techniques, grafting GBMs onto the surface of the support material with precise and well-defined particles ensures better method performance, avoiding backpressure problems, and even improving the sorbent's extraction capacity [46,47].

As a result, silica-derived materials, such as aminopropyl silica particles, are widely used to modify GBMs [8]. Maciel et al. [28] synthesized a sorbent based on graphene oxide supported on aminopropyl silica (GO@SiO$_2$). They used it in a miniaturized extraction column within a column-switching method coupled with LC-MS/MS to extract and determine some β-lactam antibiotics from wastewater samples. The coupling between the aminopropyl silica and the graphene oxide occurred through covalent bonds between the carboxyl groups of the GO and the amino groups of the silica, using a water-based synthetic route, which is consequently more sustainable. The results showed that the method achieved recoveries above 60% for the analytes, demonstrating high performance due to using the prepared sorbent, which showed good stability and reusability for up to 100 cycles without significant losses in the extraction process. Previously, the same research group used GO@SiO$_2$ to extract antidepressants and antiepileptics from urine samples in a capillary column in the first dimension of a multidimensional liquid chromatography system. In this study, low detection limits (LOD) for the analytes (0.5–20 µg L^{-1}) were obtained, and the samples were analyzed without any prior treatment, such as precipitation and/or dilution. Additionally, the extraction column containing the modified graphene phase could be used more than 250 times, attesting to the robustness and stability of the sorbent [29].

To enhance the selectivity of graphene particles supported on silica for a variety of compounds, in addition to the interactions promoted by the structure of GO and rGO (such as π-π and n-π interactions), hydrophobic interactions are further encouraged by functionalizing the graphene supported on aminopropyl silica with octadecylsilane (C18), thereby forming hybrid particles known as SiGO-C18 [47,48]. In this context, dos Santos et al. [30] employed this sorbent in an automated in-tube SPME method coupled with LC-MS/MS to quantify ten multi-class pesticides in Brazilian sugarcane spirits called "cachaças". As a result, the authors demonstrated excellent analytical performance of the method, with accuracy percentages exceeding 80%, attributed to the interactions of the sorbent even with analytes of different physicochemical characteristics. More recently, Peng et al. [31] reported a sorbent with similar properties, consisting of GO anchored to silica through amino-terminated groups and doped with C18 applied in PT-SPE for the pre-concentration of aflatoxins in food, and analysis by liquid chromatography with fluorescence detector (LC-FLD) analysis. As the main advantage, the authors noted the appropriate polarity of the sorbent for the efficient enrichment of analytes compared to matrix compounds, achieving recovery rates greater than 70% for the four compounds studied and high precision between extractions. Figure 2 shows the process of obtaining GO particles anchored to silica.

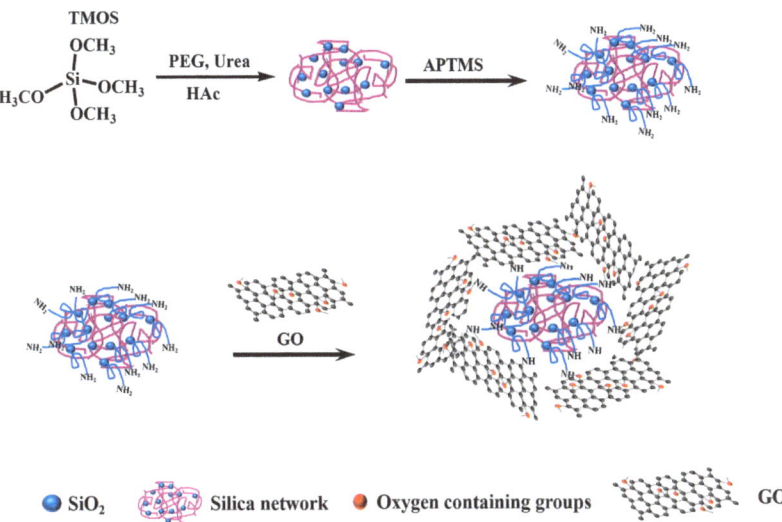

Figure 2. Schematic of preparation procedure of GO-anchored silica composite. Reproduced with permission [31] Copyright © 2024 Elsevier.

Nonetheless, the support of graphene sheets on the surface of silica derivatives still allows for their functionalization with other compounds to further enhance these materials' extraction capacity. The modification of Si@GO particles with β-cyclodextrin was addressed by da Silva et al. [23] to extract isoflavones from human urine. In their work, the authors reported using the sorbent as a coating in a "needle-sleeve" device in a fully automated solid-phase microextraction coupled with liquid chromatography (SPME-LC) system. The hybrid sorbent showed excellent performance for the intended purpose, attributed to the high stability and selectivity of GO and silica particles and both hydrophobic and hydrophilic characteristics of cyclodextrins. The authors highlighted the versatility of the sorbent by providing enhanced affinity with enrichment factors better than 3.8 times. The reported device could also be used more than 200 times while demonstrating adequate precision [23].

The studies mentioned in this topic demonstrate that incorporating graphene derivatives onto the surface of silica expands the application possibilities of these sorbents in various microextraction devices, especially in packing techniques (offline or online). Such a remarkable achievement is attributed to the improved mechanical resistance and morphology of the GBMs. Furthermore, the structural changes in these materials caused by these modifications result in the high stability and adequate reusability rates of these devices. Furthermore, the reusability of these materials significantly impacts green analytical practices since it reduces waste generation and contributes to the method's sustainability, promoting a more environmentally friendly approach to sample preparation.

2.2. Ionic Liquids

ILs, composed of organic cations and inorganic anions, have been highlighted in sample preparation due to their unique properties, such as thermal and chemical stability, low volatility, and non-flammability [26,49,50]. Adjustable according to the choice of cations and anions, these liquids enhance the affinity and pre-concentration for various classes of analytes, improving sample cleanup through multiple interactions, including π-systems, hydrogen bonding, dispersive, electrostatic, and dipolar interactions [12,51].

In this regard, several studies have employed different classes of ILs in combination with other sorbent materials in microextraction techniques. They have been serving various roles, such as functional monomers, cross-linkers, porogenic solvents, coating materials, and, in some cases, even magnetic components, depending on their class [52]. Previously, ILs were used as alternatives to volatile organic solvents in liquid-liquid extractions (LLE) and elution solvents in SPE [49,50]. However, due to the toxicological potential of certain IL combinations and their persistence in aquatic and terrestrial environments, the use of small amounts of ILs as sorbent modifiers in solid-based sample preparation techniques appears to be a more environmentally viable option, as they are stable and highly reusable [50].

Nevertheless, functionalizing graphene derivatives with ILs makes the sorbent more stable. It helps to overcome problems such as the aggregation of its nanosheets, especially in dispersive extraction techniques in aqueous media. Furthermore, it increases the absorption capacity of the sorbent and reduces its loss during extractions [22,25,51]. Li et al. [13] developed an ionic liquid-thiol-graphene oxide composite (IL-TGO) that exhibited high extraction capacity for the pesticide fipronil, surpassing other commercial sorbents. The covalent functionalization of graphene oxide with 1-Allyl-3-pentafluorobenzyl imidazole bromide (PFBr) through intermediate thiol groups allowed hydrophobic and hydrogen interactions as well as the creation of new interaction sites through the IL. The stability of the sorbent enabled its reuse in over fifteen extraction cycles in chicken egg samples using a combined solid-phase microextraction technique.

In addition to particulate sorbents, the use of ILs for monolith preparation has also been recently reported. Xiao et al. [25] prepared an aerogel using natural chitosan supported on GO layers through covalent bonds (IL–CS–GOA). Additionally, they added the ionic liquid—1-allyl-3-methylimidazolium tetrafluoroborate ([AMIm]BF$_4$)—as a porogen and modifier, increasing the surface area of the sorbent, resulting in a porous three-dimensional aerogel for extracting polyhalogenated carbazoles from sediment samples. The authors reported that this combination of materials resulted in a high specific surface area of 173 m^2 g^{-1}, attributed to the addition of the IL. The sorbent was employed in a glass dropper extraction system, with just 8 mg showing a reusability rate of six cycles and recovery values above 80%, demonstrating its high adsorption capacity and efficient mass transfer for the analytes. In the field of magnetic ionic liquids (MILs), Oviedo and collaborators [32] described the synthesis of magnetized graphene oxide nanoparticles with trihexyl(tetradecyl)phosphonium hexachlorodysprosiate(III) ([P6,6,6,14]3DyCl6) (GO@MIL) sorbent, using the dispersive solid-phase microextraction technique (DμSPE) to extract inorganic antimony from water and food samples. The extraction efficiency for Sb(III) was over 99%, using only 3 mg of graphene oxide and 40 μL of MIL, resulting in a greenness metric for the method of 0.61, as confirmed by the Analytical GREEnness calculator (AGREE).

In this context, while the benefits of using ILs as modifiers and their associated adsorptive properties are increasingly recognized, evaluating sustainability metrics in both the preparation and use of these sorbents is essential. It is known that the toxic nature of an IL depends solely on its synthetic route and structure, steps such as single-stage synthesis reactions to avoid waste generation, careful selection of ions based on their toxicity to living organisms, and assessing sorbent reusability before method optimization can reduce environmental impact [26]. Additionally, new generations of ionic liquids, such as protic ionic liquids (PIL) formed by simple acid-base reactions and zwitterionic ionic liquids (ZILs), which feature functional ionic groups conferring a neutral overall charge, have been suggested as compounds with lower ecotoxicity compared to aprotic ionic liquids, thereby reinforcing their potential in more sustainable sample preparation practices [50,53,54].

2.3. Magnetic Materials

The continual advancements in sample preparation within the scientific realm have paved the way for the introduction of magnetic nanoparticles (MNPs) or magnetic beads (MBs) as sorbent materials in countless miniaturized extraction strategies [55,56]. According to Vállez-Gomis [57], the magnetism inherent in these materials primarily stems from spinel ferrites (MFe$_2$O$_4$, where M represents Fe, Co, Mn, Ni, or Cu). Moreover, the authors assert that magnetic materials offer the advantage of being straightforward and cost-effectively synthesized, presenting remarkable application versatility. On the other hand, common drawbacks include their tendency to form agglomerates and undergo oxidation [56] quickly. Consequently, conventional strategies use inorganic protection or organic coatings to enhance their chemical stability [21,57]. Furthermore, given their active sites containing hydroxy, amino, or vinyl groups, the literature has increasingly demonstrated the potential functionalization of magnetic materials, particularly GBMs, for their application as sorbent materials in miniaturized extraction approaches [26,57].

The utilization of G in the synthesis of MNPs and MBs has showcased a remarkable synergy, leveraging the inherent properties of graphene. This synergy is particularly evident due to graphene's robust π-π stacking interactions, mechanical durability, and thermal resilience [27]. Moreover, graphene-based magnetic materials are mainly fabricated by physical attachment or covalent binding [58]. The first approach is often regarded as the simplest, involving the straightforward mixing of graphene-based materials and MNPs either in solution or through the in-situ growth of MNPs. Conversely, covalent bonding entails an amidation reaction with crosslinking reagents, such as N-hydroxisuccinimide (NHS). Furthermore, the exploration of GO and rGO for their magnetization and application as sorbents in sample preparation techniques shows significant potential for their use as MNPs [57].

Most recently, Carvalho et al. [59] reported an ex-situ magnetization method for rGO material. The technique mainly involves the mixing and interaction of two previously prepared phases. The authors highlighted that, while the in-situ approach is commonly employed for producing graphene-based magnetic materials due to its one-step execution, it still suffers from limited control over reaction products and by-products. For the experimental part, the start rGO nanoparticles were immersed in a suspension of 1 mg mL^{-1} of Fe_2O_4-NPs for four days with later filtration and drying (50 °C). While the study highlights the significant potential for scalability, its drawbacks primarily stem from challenges related to the dispersion of nanostructures within the matrix. Therefore, future studies are needed to address this challenge effectively. Apart from this, an intriguing comparative study was proposed by Thy et al. [60], in which the authors conducted a comparison between iron magnetic nanoparticles-doped graphene oxide (Fe_3O_4-GO) synthesized via both in-situ and ex-situ approaches. The results primarily highlighted the larger specific area of Fe_3O_4-GO obtained by the in-situ approach compared to those from the ex-situ method. However, each approach's advantages and drawbacks must be thoroughly evaluated before determining the optimal synthesis method for obtaining Fe_3O_4-GO sorbent materials.

Although in-situ approaches offer simplicity, the physical adsorption between magnetic materials and graphene-based nanoparticles may not ensure sufficient stability for repetitive use. In contrast, covalent bonding offers intriguing versatility in fabricating carbon-based MNPs. Li et al. [33] reported the development of graphene oxide-based magnetic covalent organic framework composites (GO@MCOFs). In their study, the MNPs were synthesized by grafting two monomers, tris(4-aminophenyl)amine, and 2,4,6-triformylphloroglucinol, onto Fe_3O_4 nanoparticles anchored on graphene oxide scaffolds. The reaction proceeded under continuous stirring for 2 h until the formation of the final composite material. Comprehensive characterization studies have revealed the resultant material exhibited significant paramagnetism and effective adsorption properties for the target analytes. While the covalent bonding approach may appear more laborious with more steps performed than the in-situ method, it holds immense potential for yielding more stable carbon-based MNPs. Hence, there is a crucial need for concerted efforts within the scientific community to explore and expand the applications of this synthesis approach.

The fabrication of graphene-based magnetic materials offers significant advantages over traditional materials, including high surface area, water solubility, and exceptional adsorption capabilities for target analytes. However, current challenges include the detachment of MNPs from the graphene surface and their tendency to agglomerate. Developing new synthesis strategies, such as functionalizing graphene magnetic nanoparticles and improving them with covalent bounding protocols, is crucial to addressing these issues.

Moreover, from an environmental perspective, there is a pressing need to propose new green synthesis routes for producing these materials. Many existing synthesis protocols still rely heavily on large volumes of organic solvents. Achieving a balance between improving graphene-based magnetic materials and adopting eco-friendly synthesis protocols is essential to unlocking a new era of advanced materials.

2.4. Molecularly Imprinted Polymers

MIPs represent a distinct category of materials primarily utilized as sorbent phases in various conventional and miniaturized sample preparation methodologies. These materials are designed to facilitate selective interactions between the sorbent and the target analyte, thereby potentially enhancing the efficiency and specificity of the method. The synthesis of MIPs predominantly involves three common approaches: (I) non-covalent technique, (II) covalent pre-organized approach, and (III) semi-covalent strategy [61]. Across these synthesis methodologies, the fundamental principle remains consistent: employing a template-based approach wherein a template molecule, typically representing the target analyte, is utilized to construct a three-dimensional polymeric structure. The obtained structure contains specific cavities molded by the presence of the template molecule. Subsequent template removal through a washing procedure renders these cavities available for particular interactions with the target analyte, which shares analogous physicochemical properties with the template. This enables effective recognition and separation within complex matrices [52,61]. Given the inherent advantages of utilizing MIPs as sorbent phases, there has been a notable surge of interest in their combination with GBMs.

Yuan et al. [36] demonstrated the synthesis of a novel poly(deep eutectic solvents) surface-imprinted graphene oxide composite (PDESs-MIP/GO). The synthesis involved several distinct steps. Initially, sulfhydryl-functionalized GO (SH-GO) was prepared by adding (3-mercaptopropyl)trimethoxysilane to a GO dispersion. Subsequently, two DESs were synthesized. The first DES (DES 1), intended as the monomer for the final MIP, was created by mixing allyl trimethylammonium chloride and urea. The second DES (DES 2), used as the crosslinker, was synthesized by combining allyl trimethylammonium chloride and itaconic acid. The final PDESs-MIP/GO sorbent was formed using SH-GO as the carrier, DES 1 as the functional monomer, DES 2 as the crosslinker, and epinephrine as the template molecule. The resultant polymer exhibited a remarkable imprinting factor (~2.0), primarily attributed to the choice of starting materials. Moreover, this study highlights the potential of DESs in providing an eco-friendly alternative for synthesizing MIPs based on graphene oxide, effectively circumventing the use of traditional and harmful organic solvents typically employed in conventional MIP synthesis procedures.

Another modification of GO with MIP was proposed by Jian et al. [38]. In this study, the authors first modified GO using a polymerizable silane group to create reactive sites for polymer grafting. For the MIP synthesis, the molecularly imprinted polymer-graphene oxide (MIP-GO) was immersed in a solution containing triphenyl phosphate (TPhP) as the template molecule, acrylamide (AM), ethylene glycol dimethacrylate (EGDMA), and azobisisobutyronitrile (AIBN). Fibers were also prepared without the addition of GO. The main results demonstrated that the incorporation of GO significantly improved the physicochemical properties of the polymer, such as specific surface area and adsorption capability, highlighting the importance of including this graphene-based material to improve analytical aspects of MIP applications.

Chen et al. [39] proposed a similar MIP-GO using different organophosphate flame retardants (OPFRs) as the template molecules to evaluate their analytical performance. Among the synthesized polymers, the one with TPhP exhibited the highest imprinting factor value (10.3). Despite this, the authors introduced an innovative strategy by assembling the three synthesized fibers into a 250 µm o.d. array called the GO-based surface molecularly imprinted polymeric fiber array (GO/MIP-FA). This strategy demonstrated superior analytical performance compared to traditional fibers such as polydimethylsiloxane (PDMS) (100 µm o.d.) and polyacrylate (PA) (80 µm o.d.). The authors highlighted that traditional phases primarily cover the polarity scale but lack the specificity provided by the MIP-GO, underscoring the enhanced performance of their novel fiber array.

A nitrogen-doped graphene quantum dots adsorbent was recently synthesized, integrating a zinc metal-organic framework and magnetic nanoparticles into a MIP framework (N-GQDs/Fe$_3$O$_4$ @SiO$_2$/IRMOF-1/MIP) [37]. The hybrid material was crafted via the sol-gel method, employing motoneuron, chlorotoluron, and monolinuron as template molecules. Additionally, (3-aminopropyl)triethoxysilane (APTES) served as the functional monomer, while tetraethyl orthosilicate (TEOS) acted as the crosslinker for the MIP, N-GQDs, MNPs, and MOF components. The primary synthesis pathway of the hybrid material is depicted in Figure 3. Notably, the carbon-based sorbent exhibited a surface area featuring a mesoporous structure of 255.67 m^2 g^{-1}, surpassing the non-imprinted polymer (NIP) at 104.36 m^2 g^{-1}.

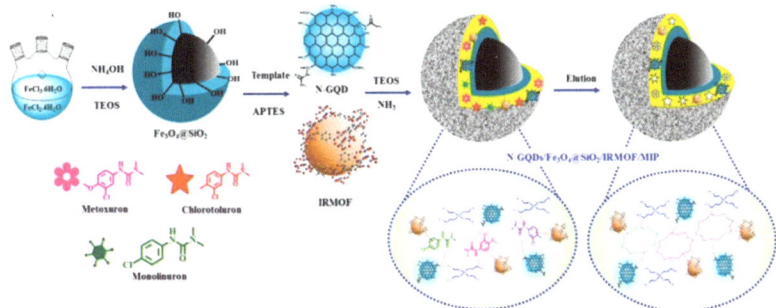

Figure 3. Schematic illustration of the N-GQDs/Fe$_3$O$_4$ synthesis approach @SiO$_2$/IRMOF-1/MIP adsorbent. Reproduced with permission [37] Copyright © 2024 Elsevier.

One of the notable advantages of MIPs, beyond their well-known selectivity, is their versatility in incorporating various sorbent materials for sample preparation methods. This section highlights that combining graphene-based materials with MIPs offers several analytical benefits, including significantly enhanced adsorption capabilities. Despite their unique features, a critical aspect of synthesizing graphene-based MIPs must be addressed. As for graphene-based magnetic materials, the synthesis of graphene-based MIPs often relies on traditional methods. This is particularly important because these protocols typically include a washing step to remove the template, which, depending on the affinity of the template molecule with the specific binding sites, can require a substantial amount of organic solvent. Various strategies have been explored in the literature to develop greener synthesis methods, such as using water and DESs [62]. These green strategies for graphene-based MIP synthesis can significantly reduce the environmental impact.

2.5. Graphene-Based Biosorbents

Aligned with green analytical chemistry (GAC) principles, utilizing biosorbents emphasizes employing renewable and biodegradable materials. This concept not only opens new avenues for advancing eco-analytical strategies but also underscores the significance of sustainability in analytical methodologies [63,64]. Furthermore, the integration of biomaterials, such as chitosan (CS) and cyclodextrins (CDs), offers additional benefits to GBMs, enhancing their versatility and efficacy across various complex matrices [65].

2.5.1. Graphene-Based Materials Immobilized with Chitosan

A significant challenge in utilizing GO, especially in water matrices, is the substantial aggregation stemming from π-π interactions among GO sheets and their inadequate dispersion in aqueous environments [66]. These drawbacks result in a considerable decline in adsorption efficiency by constricting the accessible surface area for molecular interaction. To address these challenges, using CS is a promising strategy for immobilizing graphene-derived materials [66,67]. CS, a biopolymer endowed with notable biodegradability, non-toxicity, and intriguing physicochemical attributes, presents a compelling solution for graphene-based materials. Its functional groups, including -OH and -NH$_2$, facilitate the activation of adsorption sites while enhancing electrostatic interactions and hydrogen bonding with functional groups inherent in GO, for example [66,68]. Despite the CS's intrinsic limitations in thermal stability and mechanical properties, these shortcomings are mitigated through synergistic interactions upon integration with graphene-based materials. Consequently, CS serves as a stabilizing agent for GO sheets, effectively mitigating aggregation issues [66].

Graphene oxide/chitosan (GO/CS) immobilized sorbents are synthesized through various methods, including sol-gel [69] processing and chemical crosslinking [70]. However, the conventional approach typically entails dissolving chitosan in an acetic acid aqueous solution (1–3% v/v). This process yields a yellowish-colored homogeneous solution obtained from the solubilization of chitosan. Moreover, the solution of graphene-based material is combined with the CS solution, followed by sonication until a homogeneous mixture is achieved. This amalgamation forms the basis of the GO/CS material, which can be utilized directly or subjected to subsequent processes, such as freeze drying [66]. Due to its inherent simplicity, this synthesis procedure has been widely used to produce various sorbent materials, primarily for application in microextraction protocols, as discussed in this section.

Peng et al. [40] reported the fabrication of a GO/CS aerogel for the extraction of hydrophobic pollutants. The authors employed a blend of ethanol and the cross-linking agent terephthalaldehyde in their synthesis approach, followed by vacuum freeze-drying for 24 h. Key findings underscored the superior analytical performance of the resultant GO/CS sorbent compared to conventional sorbent phases, such as polydimethylsiloxane/divinylbenzene (PDMS/DVB). This enhancement was primarily attributed to the robust hydrophobic, π-π, halogen bond, and hydrogen bond interactions between the coating and the analytes. While using cross-linkers in the synthesis of GO/CS has garnered considerable attention, conventional procedures outlined in the literature remain relevant. Ayazi, Saei, and Sarnaghi [71] demonstrated the synthesis of GO/CS material via the traditional method, as previously described, followed by filtration and drying steps.

In pollutant detection, incorporating cross-linking agents into GO/CS sorbent materials has been extensively documented in the literature for their significant impact on the abundance of functional groups available for binding with these compounds. As has been previously reported by Alves et al. [66], the authors stated that the hydrolysis reaction of borax leads to the formation of tetrahydroxyborate ions B(OH)$_4^-$ upon interaction with the -OH groups inherent in CS and GO. This reaction results in the formation of orthoborate chemical bonds, thereby exerting a notable influence on the ultimate adsorption characteristics of the GO/CS material. As previously mentioned, Peng et al. [40] the use of terephthalaldehyde as a cross-linking agent. According to the authors, this cross-linking agent was crucial for ensuring the stability of the GO/CS biosorbent in the SPME fibers. Cross-linking agents can be a valuable strategy to enhance the stability of sorbents in microextraction procedures, as these protocols often involve fragile systems. By incorporating cross-linkers, the durability and performance of the sorbents in microextraction applications can be significantly improved.

The cross-linking process has garnered significant attention in the literature, aiming to decipher its true implications for the ultimate efficacy of graphene-based chitosan biosorbents [72]. Given its pivotal role in determining the abundance of functional groups primed for binding with target analytes, the timing of this step holds considerable sway over the resultant polymer composition [66]. Despite this, different protocols that do not incorporate cross-linking in their synthesis have been reported. An interesting example is provided by Ghani et al. [41] who demonstrated the synthesis of a graphene oxide-coated agarose/chitosan (ACGO) biosorbent. The biosorbent was obtained through a simple freeze-drying procedure involving a mixture of chitosan and agarose with graphene oxide.

Remarkably, the resulting polymer exhibited adequate mechanical structure and stability without cross-linking treatment.

2.5.2. Graphene-Based Materials Immobilized with Cyclodextrin

CDs are oligopolysacchrides typically derived from starch through the enzymatic action of cyclodextrin glucanosyl transferase [24]. These molecules boast a distinctive structure characterized by a hydrophobic inner cavity and a hydrophilic surface, imparting remarkable physicochemical properties [73,74]. Considering their sample preparation application, one of the main advantages of CDs is their unique shape and functionalization potential. Werner et al. [65] CDs exhibit a cone-like structure reminiscent of hollow truncated cones, facilitating the encapsulation of guest molecules within their framework. Furthermore, the abundance of hydroxyl groups renders CDs ideal candidates for functionalization with various materials, enabling the creation of tailored sorbent phases [75].

When considering their role as sorbent phases, covalent bonding to solid supports bolsters the stability of cyclodextrins against water solubility challenges. Consequently, strategies for developing CD-based sorbents predominantly involve (I) immobilization onto inert supports, (II) integration with diverse nanomaterials, and (III) formation of nanosponges (NSs) through polymerization of cyclodextrins with an appropriate cross-linking agent [75,76]. Regarding item II, coupling cyclodextrins with graphene-based materials has showcased significant promise in forging novel biosorbent phases. This combination amplifies the utility of graphene-based phases and expands their applicability across a spectrum of analytical strategies and complex sample evaluations [8].

The literature reports many synthesis approaches for obtention cyclodextrins with carbon-based materials. Ning et al. [42] detailed the synthesis of a magnetic graphene oxide sorbent functionalized with β-cyclodextrin (NiFe$_2$O$_4$@GO@β-CD), employed for the magnetic solid-phase extraction of bisphenols. In their method, the previously fabricated NiFe$_2$O$_4$@GO phase was combined with β-CD to create the ultimate sorbent phase. This resultant product showcased a substantial increase in specific surface area, attributed to the incorporation of β-CD, which enhanced the adsorption of the target analyte by host-guest interactions, such as hydrogen bonding and π-π interactions. Hybrid materials based on β-CD and GO have also been introduced to enhance the applicability of these materials and propose different synthesis strategies for their obtention. For example, Silva and Lanças [24] reported the synthesis of β-cyclodextrin coupled to graphene oxide supported on aminopropyl silica (β-CD@GO@Sil). This synthesis involved using APTES combined with β-cyclodextrin for a subsequent reaction with graphene oxide and aminopropyl silica. The results showed that the final β-CD@GO@Sil material exhibited the adsorption characteristics typical of graphene-based materials. Furthermore, the selectivity of the material was attributed to the inclusion of β-cyclodextrin. The authors highlighted that supporting the material on aminopropyl silica opens new and unexplored horizons for applying graphene biosorbents in microextraction protocols.

Wang et al. [43] introduced an NS composed of polydimethylsiloxane/graphene oxide/β-cyclodextrin (PDMS/GO/β-CD) as a solid sorbent phase for detecting lavender essential oil. The fabrication of this NS involved three primary steps: preparing the PDMS sponge, modifying it with GO, and subsequently functionalizing it with β-CD. In the penultimate steps, the polydimethylsiloxane/graphene oxide (PDMS/GO) sponge was derived through the amidation process of graphene oxide with polydimethylsiloxane/polydopamine/3-aminopropyltriethoxysilane (PDMS/PDA/APT), followed by solvent washing. Subsequently, the PDMS/GO/β-CD composite was obtained by functionalizing the PDMS/GO sponge with β-cyclodextrin, after which it was allowed to dry at 60 °C to achieve the final NS sorbent. The results revealed a hierarchical porous interconnected morphology featuring continuous macro-sized cavities, enabling the material to be utilized in six sorption-desorption cycles while maintaining satisfactory analytical performance.

While CS and CD are widely recognized as two of the most commonly utilized biopolymers for graphene derivative modification, the literature also highlights alternative methods for crafting graphene-based biosorbents, particularly for application in microextraction procedures. For instance, cellulose has proven to be an excellent material for modification with graphene-based sorbents. Recently, Wang et al. [77] described the synthesis of a reduced graphene oxide/cellulose nanocrystal (rGO/CNC) composite for microextraction applications. The biosorbent was produced by mixing GO and cellulose in a reaction medium, followed by a hydrothermal reaction and a freeze-drying process. Furthermore, agarose [41] and alginate [78] biopolymers have emerged as viable candidates for crafting graphene-based materials, showcasing the adaptability of graphene in manifesting diverse modification potentials with these natural materials.

CS and CD have found widespread application in conjunction with GBMs, leveraging their unique attributes such as remarkable supramolecular recognition and facile functionalization. Indeed, the exploration extends beyond CS and CD; various other natural materials have been investigated, broadening the spectrum of synthesis approaches and material choices for diverse applications. Furthermore, it is equally important to note that utilizing biosorbents based on GBMs represents a significant advancement towards environmentally conscious practices in analytical chemistry. Due to their inherent biodegradability, synthesis procedures employing GBMs with biomaterials are environmentally friendlier than traditional methods. As a result, this approach protects the environment from chemical waste production and ensures the well-being of researchers interested in applying such materials.

2.6. Deep Eutectic Solvents

Since the discovery of the adsorptive properties of GBMs and their utilization in sample preparation techniques, various materials with unique properties have been reported as functionalizers aimed at enhancing the extraction capacity of these sorbents, as discussed in the sections of this topic. However, some of these materials have a significant drawback: the use of toxic organic reagents or solvents in their preparation and an additional purification step. Therefore, the quest for more sustainable materials has brought DESs into the spotlight for modifying GBMs [21]. DESs, such as ILs, are designer solvents, meaning their physicochemical properties can be tailored according to their compounds, which are formed from eutectic mixtures of two or more components, one acting as a hydrogen bond donor (HBD) and the other as a hydrogen bond acceptor (HBA) through non-covalent interactions, resulting in compounds with melting points < 100 °C. Additionally, their preparation is carried out in just one step and does not require a purification step [79,80]. Their properties include thermal stability, low vapor pressure, low flammability, low volatility, adjustable composition, high solvation capacity, and higher biodegradability rates.

Additionally, they are easy to obtain and synthesize, low-cost, and highly reusable, and their constituents are derived from renewable sources such as alcohols, amines, carboxylic acids, sugars, glycols, phenols, quaternary ammonium salts, and phosphonium salts. These qualities make DESs a greener and safer alternative than ILs [79,81]. Various applications have been reported in modifying GBMs, with investigations in diverse areas such as food, beverages, environmental, and biological samples [22,44,45,82].

Miyardan et al. [22] developed a sorbent based on GO nanoparticles modified with a DES consisting of phosphocholine chloride/1-naphthol for pesticide residue extraction in zucchini using an approach combining d-μ-SPE combined with dispersive liquid-liquid microextraction (DLLME). The authors reported that GO was modified by simple ultrasonication, and its adsorption capacity was enhanced due to the presence of DES compared to unmodified GO, with recovery values above 70% for the eight pesticides investigated. In the field of hydrophobic DESs, Yuan et al. [44] modified GO with allyltriethylammonium bromide/ethylene glycol to extract toluene and xylene exposure biomarkers in urine samples by PT-SPE. The selectivity of the prepared sorbent, deep eutectic solvent functionalized graphene oxide (DFG), proved superior to some evaluated commercial phases and even pure GO for extracting the same analytes. This fact was attributed to functional groups in the DES, which promote multiple adsorption mechanisms with the biomarkers, including hydrophobic interactions. Additionally, 2 mg of the sorbent in the extraction device recovered percentages above 90% of the analytes.

Hao et al. [45] developed a magnetic graphene oxide based on Fe_3O_4 modified with choline chloride/citric acid, resulting in a hybrid material (MGO@DES) used in magnetic solid phase extraction (MSPE) for the extraction of three estrogens in milk samples (Figure 4). In addition to the π-π and hydrogen interactions already promoted by GO, the hydrophobicity of the sorbent was enhanced by the addition of DES, which showed excellent selectivity for more non-polar analytes through the increase in hydrophobic interactions, as well as electrostatic interactions promoted by the opposite charges of the analyte and the sorbent in the extraction medium. The method showed recovery rates above 90% for the investigated estrogens using only 3 mg of sorbent and the possibility of reuse for seven cycles without considerable loss in extraction efficiency.

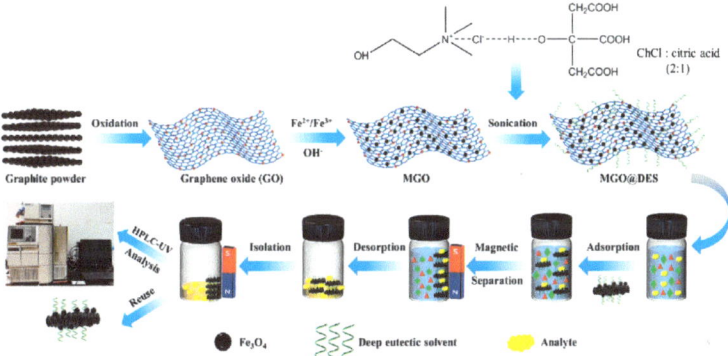

Figure 4. Schematic diagram of the MGO@DES and MSPE procedure. Reproduced with permission [45] Copyright © 2024 Elsevier.

Therefore, as observed in this topic, DESs can be tailored according to the desired interaction mechanism. When coupled with GBMs, they can extract various types of analytes, even in complex samples. Moreover, the modification of these sorbents also addresses the issue of aggregation, particularly in dispersive techniques, where the sorbent needs to be homogeneously dispersed to achieve extraction efficiency [44,82]. Consequently, their ecological advantages and ease of obtainment make methods more environmentally friendly, rendering them promising in sample preparation, specifically in functionalizing GBMs.

The synthesis of GBMs has demonstrated versatile approaches for obtaining enhanced sorbent materials. These materials possess remarkable physicochemical properties, showcasing the great potential of GBMs in creating hybrid and superior sorbents compared to traditional ones. However, significant effort is still needed to propose green synthesis approaches for these materials, as many current methods are not environmentally friendly. The use of biodegradable materials and reagents can overcome the main drawbacks of non-eco-friendly synthesis. This is particularly important given the literature highlighting the risks of toxicity to living organisms and the environment. These materials have been noted for their potential to penetrate cellular structures through various exposure routes [83]. According to Ghulam et al. [84], factors such as lateral size, surface structure, functional groups, purity, dosage, and exposure time influence the toxicity of these materials. Therefore, green analytical practices must include the conscious use and production of these materials, along with adequate synthesis conditions, to avoid human and environmental risks. The application of green alternatives for obtaining GBMs has the potential to expand their applications across different fields, enhancing their practical uses.

3. Selected Applications of GBMs in Key Miniaturized Techniques

As demonstrated earlier, combining GBMs with various materials can produce a multitude of new sorbents with unique physicochemical properties and enhanced selectivity, thereby expanding their utility for diverse applications. Figure 5 illustrates the miniaturized sample preparation approaches in which graphene compounds have been a growing trend in hybrid GBMs. Therefore, this section will discuss some applications of graphene-based sorbents in different matrices for SBSE, MEPS, PT-SPE and DPX, d-µSPE, and MSPE, miniaturized methods.

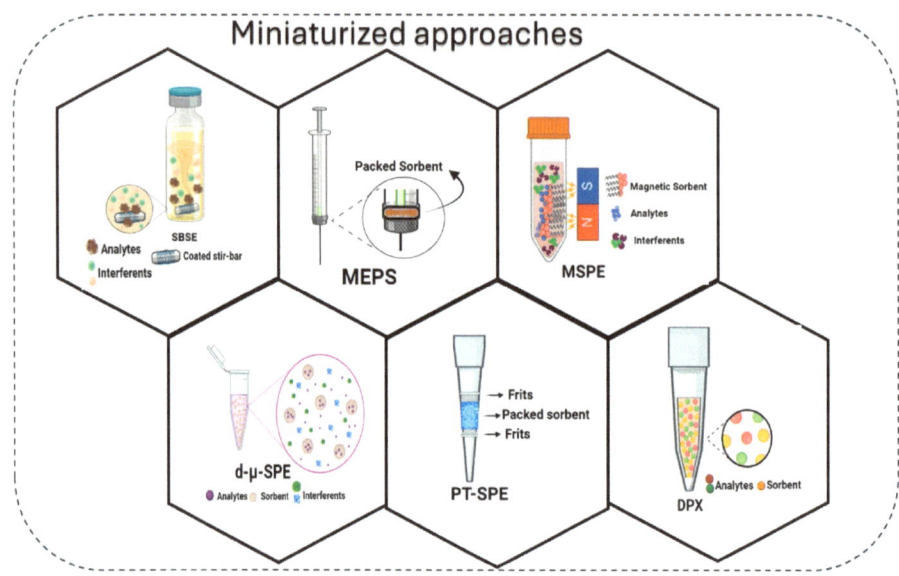

Figure 5. Miniaturized offline sample preparation techniques employing hybrid GBMs.

3.1. Stir Bar Sorptive Extraction

SBSE is a miniaturized technique that emerged in 1999. It is based on using a solid phase employed on a magnetic stirring bar as a device for compound extraction. This technique operates on the principle of partitioning, making it a non-exhaustive technique, in contrast to SPE. The extraction occurs with the exposition of the device containing the extracting phase to the sample medium for a certain period under constant agitation until the partition equilibrium of analytes between the solid and liquid phases is reached. Subsequently, this magnetic bar is exposed to an organic solvent that has an affinity for the compounds adsorbed on the bar, allowing the migration and subsequent injection of this extract into the analytical system [85]. Various coating materials for the bars have been investigated and reported in the literature since its inception, and in the field of graphene derivatives, their applications are diverse [86].

In the environmental field, Zhang et al. [87] used a porous nickel foam modified with reduced graphene oxide on its surface (rGO-NF) as a stirring bar for the extraction of six benzotriazole ultraviolet absorbents (BZTs) and their determination by liquid chromatography with a diode array detector (LC-DAD). One of this study's advantages was the extraction device's cost-effective preparation, as the authors reported acquiring the nickel foam commercially, and the hydrothermal reduction of GO occurred in situ on the substrate surface. Additionally, the method demonstrated enrichment factors exceeding 30% for the proposed method, with LODs in the range of 0.33–0.50 µg L^{-1} for the studied compounds. The device could be reused approximately 25 times without a loss in extraction efficiency, showcasing the significant stability of the rGO coating. Another study focused on developing a new coating for stirring bars based on a zirconium dioxide-reduced graphene oxide nanocomposite (ZrO$_2$-rGO). The authors used the sol-gel methodology to coat the extraction bar made of a magnet encapsulated in glass and its application for the extraction of the organophosphate pesticide ethion in agricultural and river wastewater samples, followed by analysis using negative corona discharge ion mobility spectrometry (NCD-IMS). The method demonstrated a LOD of 1.5 μg L^{-1} and recoveries above 90% for the pesticide in actual samples. Furthermore, the ZrO2-rGO-coated bar showed higher sorption efficiency for ethion compared to a bar coated with polydimethylsiloxane (PDMS), attributed to the greater polarity of ZrO$_2$-rGO, which enhances affinity interactions with the pesticide [88].

Another approach to this method is stir bar sorptive-dispersive microextraction (SBSDME), which combines the advantages of SBSE and d-μSPE. This method utilizes sorbents based on magnetic nanoparticles that dynamically coat an apparatus containing a bar-shaped magnet. When the device is introduced into the sample at a low stirring speed, the particles remain attached to the magnet, operating like SBSE. As the speed increases, the coating material disperses in the sample due to rotational force. At the end of the procedure, when the rotation stops, the magnetic particles reassemble on the magnet [89]. This methodology is promising for enhancing the pre-concentration process of analytes and reducing factors such as extraction time and the recovery of the sorbent at the end of each extraction.

In the context of combining magnetic nanoparticles with GBMs for this method, Madej et al. [90] prepared a magnetic composite consisting of Fe$_3$O$_4$ aggregates grafted onto the surface of rGO for the isolation of seven multiclass pesticides in water samples, followed by HPLC-DAD analysis. The SBSDME method was compared to MSPE. Although SBSDME showed lower recovery values (22–82%) compared to MSPE (20–75%) for the analytes, significant parameters such as good precision, shorter extraction time, ease of coating the bar, and ease of separating the sorbent at the end of extraction/desorption make the SBSDME approach preferable for its practicality.

A study conducted by Vállez-Gomis et al. [91] utilized the same analytical strategy for the determination of residues of ten polycyclic aromatic hydrocarbons (PAHs) in cosmetic samples, followed by analysis using gas chromatography-mass spectrometry (GC-MS). Magnetic cobalt ferrite nanoparticles were grafted onto rGO sheets (CoFe$_2$O$_4$-rGO) and used as the extraction sorbent after being magnetically deposited onto a neodymium stirring bar. The performance of the method was attributed to the selectivity of the extraction phase, which promoted π-π interactions with the aromatic rings of the analytes and hydrophobic interactions, thus resulting in LOQs ranging from 0.15 to 24.22 ng g^{-1}, excellent precision, and enrichment factors between 0.80 and 5.73 for the studied PAHs. Consequently, the method was successfully applied for trace-level investigations in actual cosmetic samples. Approaches such as these have the potential to emerge as promising trends in the functionalization of GBMs with magnetic compounds for SBSDME methodologies.

A notable example is the use of MILs, previously discussed in Section 2.2 of this review. MILs can impart excellent selectivity to GBMs, expanding the range of possible interactions with analytes. This results in a hybrid sorbent with enhanced efficiency, combining the unique properties of graphene compounds with the characteristics of MILs. Such a hybrid sorbent can be applied to isolate a wide range of compounds with diverse traits, showcasing its versatility in analytical applications.

3.2. Microextraction by Packed Sorbent

MEPS stands as another pivotal analytical technique within the realm of solid-based microextraction methods. It was pioneered in 2004 by Abdel-Rehim, drawing upon conventional SPE principles [92]. In MEPS, a packed sorbent (±2 mg) is housed within a syringe, either as a plug or as a cartridge between the barrel and the needle [93,94]. Moreover, the system's format primarily comprises two key components: the MEPS syringe and the MEPS cartridge, also referred to as the barrel insert and needle (BIN). The packed sorbent within the BIN section is pivotal in interacting with the target, significantly influencing the method's analytical performance [95]. This analytical approach offers significant advantages by integrating sample extraction, pre-concentration, and clean-up into a single methodology. Therefore, this consolidation enhances the sample throughput for large-scale analyses, streamlining the analytical process and improving efficiency [93,96]. Since the analytical performance of the method is intricately linked to the sorbent phase, the evolution of GMBs has greatly enhanced MEPS applications, ushering in new horizons for this miniaturized approach.

Jordan-Sinisterra and Lanças [54] reported the synthesis of an ionic liquid supported on silica, functionalized with graphene oxide (ILz/Si@GO) as a sorbent phase for MEPS extraction of pesticides in coffee samples, followed by gas-chromatography-tandem mass spectrometry (GC-MS/MS). The authors demonstrated the application of the covalent bonding method, employing N, N'-dicyclohexylcarbodiimide (DCC) as the coupling agent to anchor the IL onto the modified silica. Characterization assays of the synthesized material revealed irregularly shaped particles of uniform size, affirming the successful deposition of IL particles onto GO leaves. Moreover, adsorption experiments indicated a rise in adsorption at 100 μg mL^{-1}, reaching saturation at 200 μg mL^{-1}. Additionally, the authors stated a substantial enhancement in the adsorption properties of the final sorbent with the application of the IL in the ILz/Si@GO composite. The final application of the sorbent in the MEPS procedure achieved recoveries ranging from 35 to 97%. Although the pesticides aldrin and endosulfan sulfate weren't detected at quantifiable levels, their identification through the proposed method underscores their persistence in the environment. Such a finding highlights the significance of the developed MEPS sorbent featuring the ILz/Si@GO sorbent phase as a valuable analytical tool for pesticide environmental monitoring.

The rGO sorbent phase was reported by Ahmadi et al. [97] for the MEPS extraction of local anesthetics in plasma and saliva samples with later LC-MS/MS analysis. The study highlighted the exceptional performance of the rGO phase in mitigating matrix effects, eliminating interference peaks at the retention times of target analytes. Notably, the reported method achieved recovery values ranging from 97.26% to 106.83% in plasma and 95.21% to 105.83% in saliva, with LOQ values within the nanomolar range (nmol L^{-1}). An outstanding feature of the reported rGO sorbent was its reusability in MEPS extraction. The material demonstrated remarkable durability, enduring over 100 extractions without compromising the analytical performance of the method.

Karimiyan et al. [98] detailed the synthesis and application of a polyacrylonitrile/graphene oxide (PAN/GO) sorbent for microextraction by MEPS, followed by LC-MS/MS analysis of anesthetic drugs (lidocaine, prilocaine) and their metabolites (2,6-xylidine, o-toluidine). The optimization of MEPS revealed that pH significantly influences the interaction of target analytes with the sorbent phase. A basic pH enhances the distribution of the analytes in the sorbent phase, as the target compounds contain amino groups attached to aromatic rings, which are more interactive under alkaline conditions. The obtained LOQ ranged from 2.0 to 10 nmol L^{-1}, while the recovery values ranged from 91% to 111%. The authors emphasized that the proposed method was suitable for monitoring drugs in biological samples at low concentration levels, which could be crucial for forensic analysis.

One common challenge encountered when GBMs are applied in MEPS applications is the generation of overpressure within the MEPS syringe due to material obstruction [15]. Addressing this issue, Maciel et al. [99] introduced graphene oxide supported on silica (GO-Sil) to extract tetracyclines from milk samples, followed by LC-MS/MS analysis. Optimization of the method revealed that sampling and elution cycles were the most influential parameters affecting the analytical performance of MEPS. The analytical assessment of the method yielded LOQs ranging from 0.05 to 0.9 μg L^{-1}. This exceptionally low LOQ value was instrumental in analyzing 11 commercial samples, detecting traces of tetracyclines in only two. Despite the method's capability to detect tetracycline traces, the levels observed were within regulatory limits, affirming the efficacy of the GBM sorbent in MEPS applications.

GMBs have found wide application in MEPS across various matrices and analytes. However, there remains a noticeable gap in recent studies that thoroughly explore the potential of these sorbents in MEPS. From the author's critical perspective, scientific endeavors still need to introduce new graphene-based sorbents into MEPS, particularly with recent applications. Furthermore, there is a pressing need to explore the automation and semi-automation potential of MEPS applications to develop new green analytical strategies and transcend the boundaries of traditional offline methods. By integrating GBMs with automated or semi-automated MEPS alternatives, this miniaturized technique can be elevated to new heights, opening up numerous potential applications.

3.3. Pipette-Tip Solid-Phase Extraction and Disposable Pipette Extraction

Over the years, various forms of miniaturization of conventional SPE have been developed using different devices. PT-SPE is a miniaturized format of SPE, employing a small amount of sorbent packed between two frits or cotton filters within a pipette tip. This tip is then connected to a suitable pipettor or a syringe in more cost-effective setups [13]. Analyte sorption is conducted through sample aspiration and dispensing cycles until partition equilibrium is reached. Generally, the execution of this technique is divided into several steps similar to conventional SPE: (I) conditioning of the sorbent; (II) sorption of analytes (occurring through cycles of aspiration/dispensing); (III) washing of the sorbent to remove sample interferents; (IV) desorption of analytes using an appropriate organic solvent. Advantages over conventional SPE include shorter extraction times, ease of execution, the possibility of on-site extraction, a reduced extraction phase and organic solvent use, and a lower sample volume required [100]. However, since this technique is non-exhaustive, unlike conventional SPE, its performance depends directly on the characteristics of the sorbent used and its specificity for analytes [101].

The unique characteristics of GBMs have led to their successful application in various fields using this technique. For instance, Shen et al. [102] grafted GO nanosheets onto the surface of fibrous silica nanospheres through a water-vapor-induced internal hydrolysis method, aiming at isolating protease inhibitors in yellow catfish, followed by analysis by liquid chromatography-tandem mass spectrometry (LC-MS/MS). Using only 220 mg of the sorbent prepared in a 1 mL pipette tip, the authors achieved LOQs of 0.8–1.6 ng mL^{-1}, with analyte recoveries above 70% and good reproducibility. These parameters were attributed to the performance of the sorbent produced, which exhibited a highly rough surface after modification with GO, leading to an increased adsorption capacity of the investigated analytes. Furthermore, the sorbent preparation involved a low amount of organic solvents, with aqueous reaction media and renewable washing solvents (water and ethanol).

Recently, Zhang et al. [103] modified GO sheets with polydopamine for the extraction of alectinib and its metabolite (M4) in plasma samples using PT-SPE and determination by liquid chromatography with ultraviolet detection (LC-UV). The resulting material exhibited high porosity, increased specific surface area, and was functionalized with functional groups such as amino, hydroxyl, and benzene rings derived from polydopamine. As a consequence, low LOQ values (13.1–16.1 ng mL^{-1}), excellent method reproducibility (RSDs $\leq$ 2.8%), and high analyte recovery rates (>80%) were achieved with just 1 mg of the sorbent packed into the extraction device. The porosity of the material helped to mitigate backpressure issues caused by packing, and the performance of the method results indicated high sorbent selectivity for the target compounds, facilitated by its multiple interaction mechanisms, even in complex biological samples.

In a semi-automated approach, Tsai et al. [104] packed 10 mg of microwave-assisted synthesized rGO nanosheets into a 200 µL pipette tip, connecting it to a commercial plastic syringe attached to a dual syringe pump for the extraction of triclosan from environmental water samples and analysis by HPLC-UV. The authors reported that the primary analyte-sorbent interaction mechanism was the π bonds of rGO with the aromatic rings of the analyte, resulting in a reproducible method with recoveries exceeding 90%. This method was described as simple, cost-effective, and environmentally friendly, as it only required 1 mL of conditioning solvent and the same amount for desorption and sorbent regeneration, achieving reusability for up to 20 extraction cycles. Automated or semi-automated PT-SPE strategies like this hold promise in pursuing greener sample preparation methods, as they promote more sustainable laboratory practices while improving operational efficiency and analyst safety, reducing exposure to toxic substances and the risks associated with manual procedures.

Another approach that uses sorbents in pipette tips is DPX, which emerged in 2003. This method employs an extraction device similar to PT-SPE but with a crucial difference: in DPX, the sorbent is not fixed between two filters (frits). Instead, the sorbent remains free inside the pipette tip, allowing more efficient interaction with the sample and analytes. Only a bottom filter, made of polyethylene or glass wool, retains the sorbent phase within the tip. This configuration facilitates handling and enhances the extraction process efficiency, which follows principles similar to those of dispersive solid-phase extraction (DSPE), based on extraction driven by the contact surface of the sorbent with the sample, achieving a partition equilibrium of analytes between phases. The extraction/desorption steps are the same as those used in PT-SPE, with the washing step not mandatory [105]. Once again, as it is a non-exhaustive technique, the affinity of the sorbent for analytes will play a crucial role in the performance of DPX methods.

When it comes to graphene compounds, Oliveira and Lanças [14] recently anchored GO nanosheets onto silica, which was subsequently functionalized with octadecyl silane endcapped to produce SiGOC18end sorbent employed in DPX for the extraction of eleven multi-class herbicides in sugar cane-derived food and beverage samples, followed by determination by LC-MS/MS. The anchoring of GO onto aminopropyl silica was performed to minimize aggregation issues of GO sheets when dispersed in the sample matrix, and the functionalization of this material with C18 was carried out to enhance the analytes' surface area and adsorption capacity. The end-capping procedure was conducted to replace surface hydroxyl groups with trimethylsilane groups, rendering the sorbent more apolar and capable of providing hydrophobic interactions with the analytes. Using SiGOC18end in this technique could offer LOQs in the 1–25 ng g^{-1} range for the analytes among the investigated samples and satisfactory linearity ($r^2 > 0.99$) for all matrices. The extraction was completed in six minutes, using 0.68 mL of organic solvents, 0.8 mL of sample, and 10 mg of extraction phase.

Until today (May 2024), no further records involving graphene derivatives in DPX methods have been found. One contributing factor may be attributed to the two-dimensional structure of its nanosheets, which could potentially interfere with the dispersive adsorption/desorption process within the pipette tip. However, the advantages of this technique, when combined with the possibility of anchoring and functionalizing these phases with various materials (such as silica derivatives, ILs, and DESs), offer significant potential. These combinations can improve the material structure, address device clogging and particle aggregation issues, and improve selectivity. Consequently, they present many opportunities for future research, providing more efficient and sustainable methods for sample preparation.

3.4. Dispersive Micro Solid-Phase Extraction

Aligned with introducing alternative technologies to conventional SPE, d-μ-SPE emerged as a potential analytical approach for sample preparation [106]. This miniaturization strategy involves dispersing the sorbent into the sample matrix to enhance the kinetic interaction between the sorbent particles and the target analyte(s) [107,108]. The close contact between the sorbent particles and the analytes creates an advantageous condition reflecting improved extraction achievement. The enhanced adsorption capacity of this method relies on retaining matrix components while the analyte remains in the liquid phase, achieved through the introduction of the dispersed sorbent [109]. Given the critical role of the dispersion sorbent in enhancing analytical performance, evaluating the assisted dispersion approach primarily involves studying the utilization of external energy sources (such as mechanical stirring) or chemicals [110]. Employing GBMs as dispersion sorbents in d-μ-SPE has proven to be a versatile strategy for tackling the analysis of complex matrices [21].

Recently, Feist [111] outlined the utilization of GO nanosheets in conjunction with complexing reagents such as neocuproine or batocuproine as the dispersion sorbent for assessing trace-level metal ions in food samples, followed by inductively coupled plasma-optical emission spectrometry (ICP-OES). According to the results, the utilization of both neocuproine and batocuproine reagents enhanced the adsorption of GO by forming cationic complexes with metal ions. Furthermore, a significant achievement was made regarding the preconcentration factor of the GO/neocuproine and GO/batocuproine systems. Both sorbents exhibited preconcentration factors ranging from 10 to 100 and 20 to 200, respectively. Moreover, LOQ values ranged from 0.035 to 0.84, with recovery values over 90% for evaluating heavy metals in food. The developed method was demonstrated as a promising analytical tool for assessing heavy metals in food matrices.

Another heavy metal detection was proposed by Greda et al. [112] focusing on Cadmium (Cd), using a d-μ-SPE protocol with a GO dispersive sorbent in rose dry wine samples, analyzed by solution anode glow discharge optical emission spectrometry (SAGD-OES). The authors thoroughly investigated the sorption of Cd in comparison to common ions found in the matrix, such as Na^+, K^+, Mg^{2+}, and Ca^{2+}. The results revealed the remarkable adsorption capability of the GO phase, attributed to its functional groups (e.g., –COOH, –OH, =O, and epoxy) acting as electron pair donors, forming strong covalent bonds with metal ions. While most ions were adsorbed onto the GO phase, the analysis emphasized the superior affinity of the GO sorbent towards Cd ions over alkali and alkaline metals. This observation suggests that the strength of covalent bonds with d-block metals like Cd surpassed that of other ions, indicating the GO sorbent's pronounced selectivity towards Cd ions.

While adopting GBMs as dispersion sorbents has yielded significant advantages for d-μ-SPE, a notable drawback persists: their tendency to aggregate in such applications [110]. This aggregation tendency can diminish analyte adsorption, reducing the available contact area between the sorbent and analyte. As demonstrated in this review, using biosorbents like chitosan can mitigate this aggregation trend, offering a promising solution to enhance the efficacy of d-μ-SPE methodologies. As evidenced by literature examples, biosorbents have already found application in d-μSPE methodologies. For instance, Nakhonchai et al. [113] utilized green hairy basil seed mucilage biosorbent for d-μ-SPE extraction of tetracyclines in bovine milk. This study showcased the biosorbent's ability to enhance contact between sorbent and analytes, resulting in remarkable recovery values ranging from 83.1% to 109.9%. Moreover, there's a growing emphasis on developing environmentally friendly biosorbents based on GBMs. This innovation holds the promise of opening new avenues for d-μ-SPE applications, leveraging graphene-based sorbents' unique properties to enhance extraction efficiencies further.

3.5. Magnetic Solid-Phase Extraction

MSPE is a dispersive extraction technique that resembles the principles of DSPE (and d-μSPE), differing in the type of sorbent used, which in this case consists of magnetic particles. The sorbent comes into contact with the aqueous sample medium, and through an appropriate agitation rate, the analytes are retained in the magnetic phase. Subsequently, the sorbent is separated by applying an external magnetic force to the vial (usually a magnet) and placed in contact with an organic solvent for the desorption process of the compounds, which is performed by ultrasound or agitation. After this step, the sorbent is separated again, and the extract is collected for instrumental analysis [21]. Since this technique also represents a miniaturization of SPE, several advantages are attributed to MSPE, such as eliminating the need for extraction devices like cartridges or syringes. This elimination helps prevent issues like backpressure and clogging, allowing for the use of particles with different physical characteristics [114].

GBMs in this technique have proven very effective due to their relatively high surface area and capacity for extracting organic molecules. The magnetic nanoparticles most commonly attached to GO/rGO sheets in various studies in the literature are magnetite (Fe_3O_4), which are deposited on the surface of these materials through electrostatic interactions [114]. Thus, the applications are diverse, as demonstrated in the study by Akamine, Medina, and Lanças [17], where a GO-Fe_3O_4 nanocomposite was synthesized for the extraction of gingerols from foods, supplements, and ginger-derived beverages using MSPE and analysis by UHPLC-MS/MS. When compared to other sorbents, such as pure GO, GO@SiO_2, or commercial C18 particles, the prepared GO-Fe_3O_4 sorbent showed equal or superior adsorption capacity for the investigated compounds, standing out for its practicality in being used in the MSPE technique. Consequently, 8 mL of sample was extracted with only 10 mg of the sorbent, resulting in enrichment factors between 9.9 and 30.8 with LOQs of 5 µg L^{-1} for the investigated analytes. The reusability of GO-Fe_3O_4 was determined to be 10 cycles without any loss in extraction efficiency. In another study, Kalaboka and Sakkas [115] evaluated the performance of two magnetic sorbents: graphene oxide modified with magnetite particles (Fe_3O_4@GO) and magnetite anchored on aminopropyl silica particles functionalized with C18 (Fe_3O_4@SiO_2@C18) for the extraction of 33 emerging multiclass contaminants from various groups in wastewater samples, with determination by LC-Orbitrap MS. The authors found that the GO-based sorbent exhibited a better affinity for more polar compounds due to its hydrophilic functional groups, such as hydroxyl and carbonyl. For 19 of the investigated analytes, using 15 mg of Fe_3O_4@GO, the method achieved recoveries above 50% for effluent and tap water samples, with LOQs $\geq$ 1.2 ng L^{-1} and a reusability of 10 cycles. Both studies demonstrate the high applicability of magnetically modified GBMs for the clean-up and pre-concentration of organic analytes in complex samples.

Mohammadi et al. [116] recently introduced a novel separable sorbent composed of magnetic calcined layered double hydroxide onto graphene oxide (MgO/$MgFe_2O_4$/GO) for efficient extraction of anionic food dyes from water samples, coupled with a straightforward ultraviolet-visible detection (UV-vis) method. In their synthesis, GO and MNPs were amalgamated in methanol as the reaction medium, stirred for 4 h, and dried in an oven at 60 °C. Characterization assays demonstrated a porous and high-surface material essential for the application in actual samples. The LOD and LOQ for the anionic food dyes ranged from 0.167 to 0.14 mg L^{-1} and 0.55 to 0.47 mg L^{-1}, respectively. The recovery values for water, saffron, and soft drink samples ranged from 93% to 107%. The proposed graphene-based magnetic sorbent demonstrated exceptional efficacy in extracting anionic food dyes from water samples, offering a versatile and cost-effective analytical approach.

However, to enhance the affinity of graphene sorbents modified with magnetite particles, other materials can anchor and/or functionalize these sorbents, creating hybrid materials. This process improves material performance during the MSPE extraction step, resulting in a more homogeneous dispersion in the aqueous medium and enhanced selectivity for the compounds of interest, achieving a quick and efficient equilibrium. In this line of research, Cao et al. [117] anchored magnetite particles modified with TEOS and APTES (Fe_3O_4@SiO_2-NH_2) onto graphene oxide, followed by successive in-situ functionalization with β-cyclodextrin and an ionic liquid composed of a cation with an imidazole ring and an anion anthraquinone ($VOIm^+AQSO_3^-$), resulting in Fe_3O_4@SiO_2/GO/β-CD/IL. This composite was employed in an MSPE method for extracting seven plant regulators from vegetable samples with determination by LC-MS/MS. Utilizing 60 mg of sorbent, equilibrium extraction was achieved within 5 min, leveraging a range of interaction mechanisms inherent in its structure, including π-π stacking, hydrogen bonding, stronger hydrophobic interactions, host-guest inclusion complex formation, and electrostatic interactions. Furthermore, recovery percentages exceeding 80% were attained, along with excellent precision and low relative standard deviation values ($\leq$10.4). The method exhibited outstanding performance, with LOQs ranging from 0.03 to 0.58 µg kg^{-1}.

Recently, Gong and Liu [118] conducted a multi-step synthesis using glutaraldehyde as a crosslinker to modify silica-coated Fe_3O_4 with GO. They then functionalized it with chitosan and β-CD for extracting four bisphenols from environmental water and food samples, followed by analysis by LC-FLD. This process resulted in a high surface area (91.83 m^2 g^{-1}), facilitating compound extraction. Adsorption mechanisms involved hydrogen bonding, electrostatic interactions, π-π stacking, and hydrophobic interactions mediated by β-CD. Employing 50 mg of the adsorbent, the method achieved low LOQ levels (0.03 μg L^{-1}), with recovery percentages above 80% for the investigated bisphenols and possibly recycling the sorbent up to five times. As discussed, combining MSPE methods with magnetic sorbents containing GBMs, when combined with other materials, enables obtaining more selective phases and more efficient extraction methods. This integration enhances the selectivity and efficiency of extraction processes, reducing the need for large amounts of extraction phase and, consequently, eluent, significantly improving the quality of the results.

4. Concluding Remarks and Future Trends

This review offers a comprehensive discussion of the trends in the preparation and application of GBMs, providing an extensive overview of the literature from 2018 to May 2024. As demonstrated, GBMs, including GO and rGO, boast modification possibilities owing to their unique physicochemical properties. Notably, the presence of functional oxygen groups renders them ideal starting materials for developing novel and enhanced sorbent phases. Novel graphene-based sorbents encompass modifications with silica derivatives, ionic liquids, magnetic materials, molecularly imprinted polymers, biomaterials, and deep eutectic solvents. Sorbent materials obtained through these synthesis approaches hold significant promise for different applications. Furthermore, leveraging graphene's unique physicochemical properties, such as its high surface area and π-π interactions, in combination with these anchored materials, introduces distinct advantages over traditional sorbent materials, thereby enhancing the utility of these sorbents.

While synthesizing graphene-based sorbents offers limitless possibilities, the continued reliance on non-green synthesis methods remains a significant drawback. Most reported methods still involve large solvent volumes, generating substantial chemical residues, posing environmental harm, and compromising human safety. Another critical challenge in graphene-based synthesis is the tendency of 2D structures, such as GO, to form irregular aggregations or self-stacks. Such a drawback directly impacts the specific surface area, consequently affecting the adsorption capability of these materials. To overcome this limitation, it is crucial to suggest using anchoring materials capable of effectively infiltrating between graphene layers to mitigate the aggregation effect. As demonstrated in this study, various materials have been reported in the literature, such as CS, ILs, DESs solvents, and silica derivatives. These materials are primarily applied to mitigate the aggregation drawback of graphene-based materials in microextraction protocols. Incorporating these materials also shows promise for increasing the specific surface area of graphene-based materials and enhancing their recognition capacity.

Moreover, some critical green aspects of GBMs primarily hinge on three main points: (I) biodegradability, (II) reusability, and (III) potential for automation in application methods. Regarding the first aspect, there is a recent trend toward fabricating biodegradable sorbents, representing an eco-friendly concern, especially when these phases are discarded after their useful lifetime. The synthesis of graphene-biosorbents has predominantly centered on utilizing CS, CDs, and DESs as anchoring materials to graphene phases. Besides enhancing environmental consciousness, applying these materials has also bolstered the physicochemical properties of GBMs, opening up new horizons for their application. On the other hand, there is a pressing need to explore new materials, such as agarose, alginate, and cellulose, in synthesizing graphene-biosorbents. Introducing new graphene biosorbents can expand the scope of graphene towards greener analytical approaches, paving the way for novel advancements in the scientific field of sample preparation.

Furthermore, regarding item II, the reusability of the sorbent is crucial for ensuring sustainability. Although the most common protocols for synthesizing GBMs typically involve large volumes of organic solvents and, in some cases, toxic reagents, exploring greener methodologies for obtaining these materials—such as deriving them from biomass or growing graphene from living organisms—appears to be an excellent initiative. Additionally, since miniaturized sample preparation techniques use small amounts of sorbent, investigating their recycling rates is advantageous whenever possible. This reduces the need for repeated syntheses, minimizes waste generation, and enhances the method's sustainability.

Finally, concerning item III, many applications still rely on conventional offline methodologies despite the significant advantages of using graphene-based sorbents in miniaturized methods, such as minimal organic solvent use and small sample sizes. Focused efforts to propose and explore the automation potential of classical miniaturization methods can significantly advance green analytical practices. Key benefits include higher analytical throughput and reduced human error, thereby enhancing the analytical performance of the process. From the author's perspective, integrating graphene-based sorbents with miniaturization strategies holds excellent promise for future applications. Furthermore, these enhanced analytical strategies can open new avenues for evaluating diverse analytes and addressing the challenges of analyzing complex matrices, paving the way for a greener analytical future with graphene applications.

Author Contributions: Conceptualization, A.T.C.; writing—original draft preparation, A.T.C. and R.O.M.; writing—review and editing, A.T.C., R.O.M. and F.M.L.; supervision, F.M.L.; funding acquisition, F.M.L. All authors have read and agreed to the published version of the manuscript.

Funding: This work was partially funded by the Fundação de Amparo à Pesquisa do Estado de São Paulo (Grants: 2023/06258-1, 2023/07159-7, 2023/15675-5) and Conselho Nacional de Desenvolvimento Científico e Tecnológico, (CNPq—Grant 308843/2019-3; Instituto Nacional de Ciência e Tecnologia de Alimentos INCT-ALIM Grant 406760/2022-5).

Conflicts of Interest: The authors declare no conflicts of interest.

References

1. Novoselov, K.S.; Geim, A.K.; Morozov, S.V.; Jiang, D.; Zhang, Y.; Dubonos, S.V.; Grigorieva, I.V.; Firsov, A.A. Electric Field Effect in Atomically Thin Carbon Films. *Science* **2004**, *306*, 666–668. [CrossRef] [PubMed]
2. Novoselov, K.S.; Jiang, D.; Schedin, F.; Booth, T.J.; Khotkevich, V.V.; Morozov, S.V.; Geim, A.K. Two-Dimensional Atomic Crystals. *Proc. Natl. Acad. Sci. USA* **2005**, *102*, 10451–10453. [CrossRef]
3. Saha, J.K.; Dutta, A. A Review of Graphene: Material Synthesis from Biomass Sources. *Waste Biomass Valor.* **2022**, *13*, 1385–1429. [CrossRef] [PubMed]
4. Wu, Y.; Li, Y.; Zhang, X. The Future of Graphene: Preparation from Biomass Waste and Sports Applications. *Molecules* **2024**, *29*, 1825. [CrossRef] [PubMed]
5. Zhang, H.; Lee, H.K. Plunger-in-Needle Solid-Phase Microextraction with Graphene-Based Sol-Gel Coating as Sorbent for Determination of Polybrominated Diphenyl Ethers. *J. Chromatogr. A* **2011**, *1218*, 4509–4516. [CrossRef] [PubMed]
6. Luo, Y.B.; Shi, Z.G.; Gao, Q.; Feng, Y.Q. Magnetic Retrieval of Graphene: Extraction of Sulfonamide Antibiotics from Environmental Water Samples. *J. Chromatogr. A* **2011**, *1218*, 1353–1358. [CrossRef] [PubMed]
7. Luo, Y.B.; Cheng, J.S.; Ma, Q.; Feng, Y.Q.; Li, J.H. Graphene-Polymer Composite: Extraction of Polycyclic Aromatic Hydrocarbons from Water Samples by Stir Rod Sorptive Extraction. *Anal. Methods* **2011**, *3*, 92–98. [CrossRef] [PubMed]
8. Maciel, E.V.S.; Mejía-Carmona, K.; Jordan-Sinisterra, M.; da Silva, L.F.; Vargas Medina, D.A.; Lanças, F.M. The Current Role of Graphene-Based Nanomaterials in the Sample Preparation Arena. *Front. Chem.* **2020**, *8*. [CrossRef] [PubMed]
9. Gupta, T.; Ratandeep; Dutt, M.; Kaur, B.; Punia, S.; Sharma, S.; Sahu, P.K.; Pooja; Saya, L. Graphene-Based Nanomaterials as Potential Candidates for Environmental Mitigation of Pesticides. *Talanta* **2024**, *272*, 125748. [CrossRef] [PubMed]
10. Wojnowski, W.; Tobiszewski, M.; Pena-Pereira, F.; Psillakis, E. AGREEprep—Analytical Greenness Metric for Sample Preparation. *TrAC-Trends Anal. Chem.* **2022**, *149*, 116553. [CrossRef]
11. Pena-Pereira, F.; Wojnowski, W.; Tobiszewski, M. AGREE—Analytical GREEnness Metric Approach and Software. *Anal. Chem.* **2020**, *92*, 10076–10082. [CrossRef] [PubMed]
12. Hou, X.; Lu, X.; Tang, S.; Wang, L.; Guo, Y. Graphene Oxide Reinforced Ionic Liquid-Functionalized Adsorbent for Solid-Phase Extraction of Phenolic Acids. *J. Chromatogr. B Anal. Technol. Biomed. Life Sci.* **2018**, *1072*, 123–129. [CrossRef] [PubMed]
13. Li, M.; Yang, C.; Yan, H.; Han, Y.; Han, D. An Integrated Solid Phase Extraction with Ionic Liquid-Thiol-Graphene Oxide as Adsorbent for Rapid Isolation of Fipronil Residual in Chicken Eggs. *J. Chromatogr. A* **2020**, *1631*, 461568. [CrossRef] [PubMed]
14. Oliveira, T.C.; Lanças, F.M. Determination of Selected Herbicides in Sugarcane-Derived Foods by Graphene-Oxide Based Disposable Pipette Extraction Followed by Liquid Chromatography-Tandem Mass Spectrometry. *J. Chromatogr. A* **2023**, *1687*, 463690. [CrossRef] [PubMed]

15. Fumes, B.H.; Lanças, F.M. Use of Graphene Supported on Aminopropyl Silica for Microextraction of Parabens from Water Samples. *J. Chromatogr. A* **2017**, *1487*, 64–71. [CrossRef]
16. Shah, J.; Jan, M.R.; Rahman, I. Dispersive Solid Phase Microextraction of Fenoxaprop-p-Ethyl Herbicide from Water and Food Samples Using Magnetic Graphene Composite. *J. Inorg. Organomet. Polym. Mater.* **2020**, *30*, 1716–1725. [CrossRef]
17. Akamine, L.A.; Vargas Medina, D.A.; Lanças, F.M. Magnetic Solid-Phase Extraction of Gingerols in Ginger Containing Products. *Talanta* **2021**, *222*, 121683. [CrossRef] [PubMed]
18. Luo, J.; Gao, Y.; Tan, K.; Wei, W.; Liu, X. Preparation of a Magnetic Molecularly Imprinted Graphene Composite Highly Adsorbent for 4-Nitrophenol in Aqueous Medium. *ACS Sustain. Chem. Eng.* **2016**, *4*, 3316–3326. [CrossRef]
19. Cheng, L.; Pan, S.; Ding, C.; He, J.; Wang, C. Dispersive Solid-Phase Microextraction with Graphene Oxide Based Molecularly Imprinted Polymers for Determining Bis(2-Ethylhexyl) Phthalate in Environmental Water. *J. Chromatogr. A* **2017**, *1511*, 85–91. [CrossRef] [PubMed]
20. Yuan, Y.; Zhang, Y.; Wang, M.; Cao, J.; Yan, H. Green Synthesis of Superhydrophilic Resin/Graphene Oxide for Efficient Analysis of Multiple Pesticide Residues in Fruits and Vegetables. *Food Chem.* **2024**, *450*, 139341. [CrossRef] [PubMed]
21. Emmanuel Ibukun, A.; Yahaya, N.; Husaini Mohamed, A.; Semail, N.F.; Abd Hamid, M.A.; Nadhirah Mohamad Zain, N.; Anuar Kamaruddin, M.; Hong Loh, S.; Kamaruzaman, S. Recent Developments in Synthesis and Characterisation of Graphene Oxide Modified with Deep Eutectic Solvents for Dispersive and Magnetic Solid-Phase Extractions. *Microchem. J.* **2024**, *199*, 110111. [CrossRef]
22. Miyardan, F.N.; Afshar Mogaddam, M.R.; Farajzadeh, M.A.; Nemati, M. Combining Modified Graphene Oxide-Based Dispersive Micro Solid Phase Extraction with Dispersive Liquid–Liquid Microextraction in the Extraction of Some Pesticides from Zucchini Samples. *Microchem. J.* **2022**, *182*, 107884. [CrossRef]
23. Da Silva, L.F.; Vargas Medina, D.A.; Lanças, F.M. Automated Needle-Sleeve Based Online Hyphenation of Solid-Phase Microextraction and Liquid Chromatography. *Talanta* **2021**, *221*, 121608. [CrossRef] [PubMed]
24. Da Silva, L.F.; Lanças, F.M. β-Cyclodextrin Coupled to Graphene Oxide Supported on Aminopropyl Silica as a Sorbent Material for Determination of Isoflavones. *J. Sep. Sci.* **2020**, *43*, 4347–4355. [CrossRef] [PubMed]
25. Xiao, M.; Li, P.; Lu, Y.; Cao, J.; Yan, H. Development of a Three-Dimensional Porous Ionic Liquid-Chitosan-Graphene Oxide Aerogel for Efficient Extraction and Detection of Polyhalogenated Carbazoles in Sediment Samples. *Talanta* **2024**, *271*, 125711. [CrossRef] [PubMed]
26. Vargas Medina, D.A.; Cardoso, A.T.; Maciel, E.V.S.; Lanças, F.M. Current Materials for Miniaturized Sample Preparation: Recent Advances and Future Trends. *TrAC-Trends Anal. Chem.* **2023**, *165*, 117120. [CrossRef]
27. Jiang, Q.; Zhang, S.; Sun, M. Recent Advances on Graphene and Graphene Oxide as Extraction Materials in Solid-Phase (Micro)Extraction. *TrAC-Trends Anal. Chem.* **2023**, *168*, 117283. [CrossRef]
28. Maciel, E.V.S.; Vargas-Medina, D.A.; Lancas, F.M. Analyzes of β-Lactam Antibiotics by Direct Injection of Environmental Water Samples into a Functionalized Graphene Oxide-Silica Packed Capillary Extraction Column Online Coupled to Liquid Chromatography Tandem Mass Spectrometry. *Talanta Open* **2023**, *7*, 100185. [CrossRef]
29. Soares Maciel, E.V.; de Toffoli, A.L.; da Silva Alves, J.; Lanças, F.M. Multidimensional Liquid Chromatography Employing a Graphene Oxide Capillary Column as the First Dimension: Determination of Antidepressant and Antiepileptic Drugs in Urine. *Molecules* **2020**, *25*, 1092. [CrossRef] [PubMed]
30. Pereira, N.G.; Santos, D.; Vasconcelos, E.; Maciel, S.; Mejía-Carmona, K.; Lanças, F.M. Multidimensional Capillary Liquid Chromatography-Tandem Mass Spectrometry for the Determination of Multiclass Pesticides in "Sugarcane Spirits" (Cachaças). *Anal. Bioanal. Chem.* **2020**, *412*, 7789–7797. [CrossRef]
31. Peng, C.; Zhang, S.; Huang, J.; Wu, C.; Zhao, X.; Feng, Y.; Gao, Y. Adaptive Polarity of Graphene Oxide Anchored Silica Doped with C18 for Effective Enrichment of Aflatoxins from Foodstuff. *Microchem. J.* **2024**, *197*, 109728. [CrossRef]
32. Oviedo, M.N.; Botella, M.B.; Fiorentini, E.F.; Pacheco, P.; Wuilloud, R.G. A Simple and Green Dispersive Micro-Solid Phase Extraction Method by Combined Application of Graphene Oxide and a Magnetic Ionic Liquid for Selective Determination of Inorganic Antimony Species in Water, Tea and Honey Samples. *Spectrochim. Acta Part B At. Spectrosc.* **2023**, *199*, 106591. [CrossRef]
33. Li, Y.; Xu, X.; Guo, H.; Bian, Y.; Li, J.; Zhang, F. Magnetic Graphene Oxide–based Covalent Organic Frameworks as Novel Adsorbent for Extraction and Separation of Triazine Herbicides from Fruit and Vegetable Samples. *Anal. Chim. Acta* **2022**, *1219*, 339984. [CrossRef] [PubMed]
34. Arabkhani, P.; Sadegh, N.; Asfaram, A. Nanostructured Magnetic Graphene Oxide/UIO-66 Sorbent for Ultrasound-Assisted Dispersive Solid-Phase Microextraction of Food Colorants in Soft Drinks, Candies, and Pastilles Prior to HPLC Analysis. *Microchem. J.* **2023**, *184*, 108149. [CrossRef]
35. Soylak, M.; Jagirani, M.S.; Uzcan, F. Metal-Doped Magnetic Graphene Oxide Nanohybrid for Solid-Phase Microextraction of Copper from Environmental Samples. *Iran. J. Sci. Technol. Trans. A Sci.* **2022**, *46*, 807–817. [CrossRef]
36. Yuan, Y.; Wang, Y.; Zhang, Y.; Yin, J.; Han, Y.; Han, D.; Yan, H. Miniaturized Centrifugation Accelerated Pipette-Tip Matrix Solid-Phase Dispersion Based on Poly(Deep Eutectic Solvents) Surface Imprinted Graphene Oxide Composite Adsorbent for Rapid Extraction of Anti-Adipogenesis Markers from *Solidago decurrens* Lour. *J. Chromatogr. A* **2024**, *1715*, 464599. [CrossRef] [PubMed]

37. Sa-nguanprang, S.; Phuruangrat, A.; Bunkoed, O. A Magnetic Adsorbent of Nitrogen-Doped Graphene Quantum Dots, Zinc Metal-Organic Framework and Molecularly Imprinted Polymer to Extract Phenylureas. *J. Food Compos. Anal.* **2024**, *126*, 105911. [CrossRef]
38. Jian, Y.; Chen, L.; Cheng, J.; Huang, X.; Yan, L.; Li, H. Molecularly Imprinted Polymers Immobilized on Graphene Oxide Film for Monolithic Fiber Solid Phase Microextraction and Ultrasensitive Determination of Triphenyl Phosphate. *Anal. Chim. Acta* **2020**, *1133*, 1–10. [CrossRef] [PubMed]
39. Chen, L.; Jian, Y.; Cheng, J.; Yan, L.; Huang, X. Preparation and Application of Graphene Oxide-Based Surface Molecularly Imprinted Polymer for Monolithic Fiber Array Solid Phase Microextraction of Organophosphate Flame Retardants in Environmental Water. *J. Chromatogr. A* **2020**, *1623*, 461200. [CrossRef] [PubMed]
40. Peng, S.; Huang, Y.; Ouyang, S.; Huang, J.; Shi, Y.; Tong, Y.J.; Zhao, X.; Li, N.; Zheng, J.; Zheng, J.; et al. Efficient Solid Phase Microextraction of Organic Pollutants Based on Graphene Oxide/Chitosan Aerogel. *Anal. Chim. Acta* **2022**, *1195*, 339462. [CrossRef] [PubMed]
41. Ghani, M.; Jafari, Z.; Raoof, J.B. Porous Agarose/Chitosan/Graphene Oxide Composite Coupled with Deep Eutectic Solvent for Thin Film Microextraction of Chlorophenols. *J. Chromatogr. A* **2023**, *1694*, 463899. [CrossRef] [PubMed]
42. Ning, Y.; Xu, Y.; Bao, J.; Wang, W.; Wang, A.J. β-Cyclodextrin-Functionalized Magnetic Graphene Oxide for the Efficient Enrichment of Bisphenols in Milk and Milk Packaging. *J. Chromatogr. A* **2023**, *1692*, 463854. [CrossRef] [PubMed]
43. Wang, L.; Li, D.; Jiang, X.; Fu, J. Polydimethylsiloxane/Graphene Oxide/β-Cyclodextrin Sponge as a Solid-Phase Extraction Sorbent Coupled with Gas Chromatography-Mass Spectrometry for Rapid Adsorption and Sensitive Determination of Lavender Essential Oil. *J. Sep. Sci.* **2022**, *45*, 1904–1917. [CrossRef] [PubMed]
44. Yuan, Y.; Han, Y.; Yang, C.; Han, D.; Yan, H. Deep Eutectic Solvent Functionalized Graphene Oxide Composite Adsorbent for Miniaturized Pipette-Tip Solid-Phase Extraction of Toluene and Xylene Exposure Biomarkers in Urine Prior to Their Determination with HPLC-UV. *Microchim. Acta* **2020**, *187*, 387. [CrossRef] [PubMed]
45. Hao, Y.; Zhou, W.; Wang, X.; Liu, Y.; Di, X. Carboxyl-Based Deep Eutectic Solvent Modified Magnetic Graphene Oxide as a Novel Adsorbent for Fast Enrichment and Extraction of Estrogens in Milk Prior to HPLC-UV Analysis. *Microchem. J.* **2023**, *193*, 109050. [CrossRef]
46. De Toffoli, A.L.; Maciel, E.V.S.; Fumes, B.H.; Lanças, F.M. The Role of Graphene-Based Sorbents in Modern Sample Preparation Techniques. *J. Sep. Sci.* **2018**, *41*, 288–302. [CrossRef] [PubMed]
47. Maciel, E.V.S.; Borsatto, J.V.B.; Mejia-Carmona, K.; Lanças, F.M. Application of an In-House Packed Octadecylsilica-Functionalized Graphene Oxide Column for Capillary Liquid Chromatography Analysis of Hormones in Urine Samples. *Anal. Chim. Acta* **2023**, *1239*, 340718. [CrossRef] [PubMed]
48. Mejía-Carmona, K.; Lanças, F.M. Modified Graphene-Silica as a Sorbent for in-Tube Solid-Phase Microextraction Coupled to Liquid Chromatography-Tandem Mass Spectrometry. Determination of Xanthines in Coffee Beverages. *J. Chromatogr. A* **2020**, *1621*, 461089. [CrossRef] [PubMed]
49. Gong, J.; Liang, C.; Majeed, Z.; Tian, M.; Zhao, C.; Luo, M.; Li, C. Advances of Imidazolium Ionic Liquids for the Extraction of Phytochemicals from Plants. *Separations* **2023**, *10*, 151. [CrossRef]
50. Ražić, S.; Gadžurić, S.; Trtić-Petrović, T. Ionic Liquids in Green Analytical Chemistry—Are They That Good and Green Enough? *Anal. Bioanal. Chem.* **2024**, *416*, 2023–2029. [CrossRef]
51. Zhang, H.; Wu, X.; Yuan, Y.; Han, D.; Qiao, F.; Yan, H. An Ionic Liquid Functionalized Graphene Adsorbent with Multiple Adsorption Mechanisms for Pipette-Tip Solid-Phase Extraction of Auxins in Soybean Sprouts. *Food Chem.* **2018**, *265*, 290–297. [CrossRef] [PubMed]
52. Cardoso, A.T.; Martins, R.O.; Lanças, F.M.; Chaves, A.R. Molecularly Imprinted Polymers in Online Extraction Liquid Chromatography Methods: Current Advances and Recent Applications. *Anal. Chim. Acta* **2023**, *1284*, 341952. [CrossRef] [PubMed]
53. Jesus, F.; Passos, H.; Ferreira, A.M.; Kuroda, K.; Pereira, J.L.; Gonçalves, F.J.M.; Coutinho, J.A.P.; Ventura, S.P.M. Zwitterionic Compounds Are Less Ecotoxic than Their Analogous Ionic Liquids. *Green. Chem.* **2021**, *23*, 3683–3692. [CrossRef]
54. Jordan-Sinisterra, M.; Vargas Medina, D.A.; Lanças, F.M. Microextraction by Packed Sorbent of Polycyclic Aromatic Hydrocarbons in Brewed Coffee Samples with a New Zwitterionic Ionic Liquid-Modified Silica Sorbent. *J. Food Compos. Anal.* **2022**, *114*, 104832. [CrossRef]
55. Faraji, M.; Shirani, M.; Rashidi-Nodeh, H. The Recent Advances in Magnetic Sorbents and Their Applications. *TrAC Trends Anal. Chem.* **2021**, *141*, 116302. [CrossRef]
56. Suliman, M.A.; Sajid, M.; Nazal, M.K.; Islam, M.A. Carbon-Based Materials as Promising Sorbents for Analytical Sample Preparation: Recent Advances and Trends in Extraction of Toxic Metal Pollutants from Various Media. *TrAC Trends Anal. Chem.* **2023**, *167*, 117265. [CrossRef]
57. Vállez-Gomis, V.; Grau, J.; Benedé, J.L.; Chisvert, A. Magnetic Sorbents: Synthetic Pathways and Application in Dispersive (Micro)Extraction Techniques for Bioanalysis. *TrAC Trends Anal. Chem.* **2024**, *171*, 117486. [CrossRef]
58. Li, N.; Jiang, H.L.; Wang, X.; Wang, X.; Xu, G.; Zhang, B.; Wang, L.; Zhao, R.S.; Lin, J.M. Recent Advances in Graphene-Based Magnetic Composites for Magnetic Solid-Phase Extraction. *TrAC Trends Anal. Chem.* **2018**, *102*, 60–74. [CrossRef]
59. Carvalho, A.S.; Oliveira, D.M.; Assis, L.K.C.S.; Rodrigues, A.R.; Guzzo, P.L.; Almeida, L.C.; Padrón-Hernández, E. Synthesis of Nanocomposites Based on Fe_3O_4 Nanoparticles and Nitrogen-Doped Reduced Graphene Oxide Aerogel by Ex-Situ Approach and Their Magnetic Properties. *J. Alloys Compd.* **2023**, *968*, 172038. [CrossRef]

60. Thi Mong Thy, L.; Tan Tai, L.; Duy Hai, N.; Quang Cong, C.; Minh Dat, N.; Ngoc Trinh, D.; Truong Son, N.; Thi Yen Oanh, D.; Thanh Phong, M.; Huu Hieu, N. Comparison of In-Situ and Ex-Situ Methods for Synthesis of Iron Magnetic Nanoparticles-Doped Graphene Oxide: Characterization, Adsorption Capacity, and Fenton Catalytic Efficiency. *FlatChem* **2022**, *33*, 100365. [CrossRef]
61. Kamyab, H.; Chelliapan, S.; Tavakkoli, O.; Mesbah, M.; Bhutto, J.K.; Khademi, T.; Kirpichnikova, I.; Ahmad, A.; ALJohani, A.A. A Review on Carbon-Based Molecularly-Imprinted Polymers (CBMIP) for Detection of Hazardous Pollutants in Aqueous Solutions. *Chemosphere* **2022**, *308*, 136471. [CrossRef] [PubMed]
62. Martins, R.O.; Bernardo, R.A.; Machado, L.S.; Batista Junior, A.C.; Maciel, L.Í.L.; de Aguiar, D.V.A.; Sanches Neto, F.O.; Oliveira, J.V.A.; Simas, R.C.; Chaves, A.R. Greener Molecularly Imprinted Polymers: Strategies and Applications in Separation and Mass Spectrometry Methods. *TrAC Trends Anal. Chem.* **2023**, *168*, 117285. [CrossRef]
63. Jagirani, M.S.; Soylak, M. Green Sorbents for the Solid Phase Extraction of Trace Species. *Curr. Opin. Green Sustain. Chem.* **2024**, *47*, 100899. [CrossRef]
64. Koel, M. Developments in Analytical Chemistry Initiated from Green Chemistry. *Sustain. Chem. Environ.* **2024**, *5*, 100078. [CrossRef]
65. Werner, J.; Zgoła-Grześkowiak, A.; Grześkowiak, T.; Frankowski, R. Biopolymers-Based Sorbents as a Future Green Direction for Solid Phase (Micro)Extraction Techniques. *TrAC Trends Anal. Chem.* **2024**, *173*, 117659. [CrossRef]
66. Da Silva Alves, D.C.; Healy, B.; Yu, T.; Breslin, C.B. Graphene-Based Materials Immobilized within Chitosan: Applications as Adsorbents for the Removal of Aquatic Pollutants. *Materials* **2021**, *14*, 3655. [CrossRef] [PubMed]
67. Guo, Y.; Qiao, D.; Zhao, S.; Liu, P.; Xie, F.; Zhang, B. Biofunctional Chitosan–Biopolymer Composites for Biomedical Applications. *Mater. Sci. Eng. R Rep.* **2024**, *159*, 100775. [CrossRef]
68. Li, C.; Li, F.; Wang, K.; Wang, Q.; Liu, H.; Sun, X.; Xie, D. Synthesis, Characterizations, and Release Mechanisms of Carboxymethyl Chitosan-Graphene Oxide-Gelatin Composite Hydrogel for Controlled Delivery of Drug. *Inorg. Chem. Commun.* **2023**, *155*, 110965. [CrossRef]
69. Chang, Z.; Chen, Y.; Tang, S.; Yang, J.; Chen, Y.; Chen, S.; Li, P.; Yang, Z. Construction of Chitosan/Polyacrylate/Graphene Oxide Composite Physical Hydrogel by Semi-Dissolution/Acidification/Sol-Gel Transition Method and Its Simultaneous Cationic and Anionic Dye Adsorption Properties. *Carbohydr. Polym.* **2020**, *229*, 115431. [CrossRef] [PubMed]
70. Han Lyn, F.; Tan, C.P.; Zawawi, R.M.; Nur Hanani, Z.A. Enhancing the Mechanical and Barrier Properties of Chitosan/Graphene Oxide Composite Films Using Trisodium Citrate and Sodium Tripolyphosphate Crosslinkers. *J. Appl. Polym. Sci.* **2021**, *138*, 50618. [CrossRef]
71. Ayazi, Z.; Farshineh Saei, S.; Pashayi Sarnaghi, S. A Novel Self-Supportive Thin Film Based on Graphene Oxide Reinforced Chitosan Nano-Biocomposite for Thin Film Microextraction of Fluoxetine in Biological and Environmental Samples. *J. Pharm. Biomed. Anal.* **2023**, *236*, 115678. [CrossRef] [PubMed]
72. Marapureddy, S.G.; Thareja, P. Synergistic Effect of Chemical Crosslinking and Addition of Graphene-Oxide in Chitosan— Hydrogels, Films, and Drug Delivery. *Mater. Today Commun.* **2022**, *31*, 103430. [CrossRef]
73. Li, Y.; Liu, F.; Abdiryim, T.; Liu, X. Cyclodextrin-Derived Materials: From Design to Promising Applications in Water Treatment. *Coord. Chem. Rev.* **2024**, *502*, 215613. [CrossRef]
74. Adamkiewicz, L.; Szeleszczuk, Ł. Review of Applications of Cyclodextrins as Taste-Masking Excipients for Pharmaceutical Purposes. *Molecules* **2023**, *28*, 6964. [CrossRef] [PubMed]
75. Maciel, E.V.S.; Pereira dos Santos, N.G.; Medina, D.A.V.; Lanças, F.M. Cyclodextrins-Based Sorbents for Sustainable Sample Preparation Focusing on Food Analysis. *Green Anal. Chem.* **2023**, *7*, 100077. [CrossRef]
76. Gentili, A. Cyclodextrin-Based Sorbents for Solid Phase Extraction. *J. Chromatogr. A* **2020**, *1609*, 460654. [CrossRef] [PubMed]
77. Wang, M.; Wu, H.; Xu, S.; Dong, P.; Long, A.; Xiao, L.; Feng, S.; Chen, C.P. Cellulose Nanocrystal Regulated Ultra-Loose, Lightweight, and Hierarchical Porous Reduced Graphene Oxide Hybrid Aerogel for Capturing and Determining Organic Pollutants from Water. *Carbon* **2023**, *204*, 94–101. [CrossRef]
78. Tabish, M.S.; Hanapi, N.S.M.; Wan Ibrahim, W.N.; Saim, N.; Yahaya, N. Alginate-Graphene Oxide Biocomposite Sorbent for Rapid and Selective Extraction of Non-Steroidal Anti-Inflammatory Drugs Using Micro-Solid Phase Extraction. *Indones. J. Chem.* **2019**, *19*, 684–695. [CrossRef]
79. Makoś-Chełstowska, P.; Gębicki, J. Sorbents Modified by Deep Eutectic Solvents in Microextraction Techniques. *TrAC-Trends Anal. Chem.* **2024**, *172*, 117577. [CrossRef]
80. Werner, J.; Zgoła-Grześkowiak, A.; Płatkiewicz, J.; Płotka-Wasylka, J.; Jatkowska, N.; Kalyniukova, A.; Zaruba, S.; Andruch, V. Deep Eutectic Solvents in Analytical Sample Preconcentration Part B: Solid-Phase (Micro)Extraction. *Microchem. J.* **2023**, *191*, 108898. [CrossRef]
81. González-Campos, J.B.; Pérez-Nava, A.; Valle-Sánchez, M.; Delgado-Rangel, L.H. Deep Eutectic Solvents Applications Aligned to 2030 United Nations Agenda for Sustainable Development. *Chem. Eng. Process.* **2024**, *199*, 109751. [CrossRef]
82. Shen, Y.F.; Zhang, X.; Mo, C.E.; Huang, Y.P.; Liu, Z.S. Preparation of Graphene Oxide Incorporated Monolithic Chip Based on Deep Eutectic Solvents for Solid Phase Extraction. *Anal. Chim. Acta* **2020**, *1096*, 184–192. [CrossRef] [PubMed]
83. Malhotra, N.; Villaflores, O.B.; Audira, G.; Siregar, P.; Lee, J.S.; Ger, T.R.; Hsiao, C.-D. Toxicity Studies on Graphene-Based Nanomaterials in Aquatic Organisms: Current Understanding. *Molecules* **2020**, *25*, 3618. [CrossRef] [PubMed]
84. Ghulam, A.N.; Dos Santos, O.A.L.; Hazeem, L.; Backx, B.P.; Bououdina, M.; Bellucci, S. Graphene Oxide (GO) Materials— Applications and Toxicity on Living Organisms and Environment. *J. Funct. Biomater.* **2022**, *13*, 77. [CrossRef] [PubMed]

85. Kanu, A.B. Recent Developments in Sample Preparation Techniques Combined with High-Performance Liquid Chromatography: A Critical Review. *J. Chromatogr. A* **2021**, *1654*, 4. [CrossRef] [PubMed]
86. He, M.; Wang, Y.; Zhang, Q.; Zang, L.; Chen, B.; Hu, B. Stir Bar Sorptive Extraction and Its Application. *J. Chromatogr. A* **2021**, *1637*, 461810. [CrossRef] [PubMed]
87. Zhang, Q.; You, L.; Chen, B.; He, M.; Hu, B. Reduced Graphene Oxide Coated Nickel Foam for Stir Bar Sorptive Extraction of Benzotriazole Ultraviolet Absorbents from Environmental Water. *Talanta* **2021**, *231*, 122332. [CrossRef] [PubMed]
88. Jafari, M.T.; Rezaei, B.; Bahrami, H. Zirconium Dioxide-Reduced Graphene Oxide Nanocomposite-Coated Stir-Bar Sorptive Extraction Coupled with Ion Mobility Spectrometry for Determining Ethion. *Talanta* **2018**, *182*, 285–291. [CrossRef]
89. Vállez-Gomis, V.; Grau, J.; Benedé, J.L.; Giokas, D.L.; Chisvert, A.; Salvador, A. Fundamentals and Applications of Stir Bar Sorptive Dispersive Microextraction: A Tutorial Review. *Anal. Chim. Acta* **2021**, *1153*, 338271. [CrossRef] [PubMed]
90. Madej, K.; Jonda, A.; Borcuch, A.; Piekoszewski, W.; Chmielarz, L.; Gil, B. A Novel Stir Bar Sorptive-Dispersive Microextraction in Combination with Magnetically Modified Graphene for Isolation of Seven Pesticides from Water Samples. *Microchem. J.* **2019**, *147*, 962–971. [CrossRef]
91. Vállez-Gomis, V.; Grau, J.; Benedé, J.L.; Chisvert, A.; Salvador, A. Reduced Graphene Oxide-Based Magnetic Composite for Trace Determination of Polycyclic Aromatic Hydrocarbons in Cosmetics by Stir Bar Sorptive Dispersive Microextraction. *J. Chromatogr. A* **2020**, *1624*, 461229. [CrossRef] [PubMed]
92. Abdel-Rehim, M. New Trend in Sample Preparation: On-Line Microextraction in Packed Syringe for Liquid and Gas Chromatography Applications: I. Determination of Local Anaesthetics in Human Plasma Samples Using Gas Chromatography–Mass Spectrometry. *J. Chromatogr. B* **2004**, *801*, 317–321. [CrossRef] [PubMed]
93. Granados-Guzmán, G.; Díaz-Hernández, M.; Alvarez-Román, R.; Cavazos-Rocha, N.; Portillo-Castillo, O.J. A Brief Review of the Application of Microextraction by Packed Sorbent for Antibiotics Analysis from Biological, Food, and Environmental Samples. *Rev. Anal. Chem.* **2023**, *42*, 20230057. [CrossRef]
94. Martins, R.O.; de Araújo, G.L.; de Freitas, C.S.; Silva, A.R.; Simas, R.C.; Vaz, B.G.; Chaves, A.R. Miniaturized Sample Preparation Techniques and Ambient Mass Spectrometry as Approaches for Food Residue Analysis. *J. Chromatogr. A* **2021**, *1640*, 461949. [CrossRef] [PubMed]
95. Yang, L.; Said, R.; Abdel-Rehim, M. Sorbent, Device, Matrix and Application in Microextraction by Packed Sorbent (MEPS): A Review. *J. Chromatogr. B* **2017**, *1043*, 33–43. [CrossRef] [PubMed]
96. Casado, N.; Gañán, J.; Morante-Zarcero, S.; Sierra, I. New Advanced Materials and Sorbent-Based Microextraction Techniques as Strategies in Sample Preparation to Improve the Determination of Natural Toxins in Food Samples. *Molecules* **2020**, *25*, 702. [CrossRef] [PubMed]
97. Ahmadi, M.; Moein, M.M.; Madrakian, T.; Afkhami, A.; Bahar, S.; Abdel-Rehim, M. Reduced Graphene Oxide as an Efficient Sorbent in Microextraction by Packed Sorbent: Determination of Local Anesthetics in Human Plasma and Saliva Samples Utilizing Liquid Chromatography-Tandem Mass Spectrometry. *J. Chromatogr. B* **2018**, *1095*, 177–182. [CrossRef] [PubMed]
98. Karimiyan, H.; Uheida, A.; Hadjmohammadi, M.; Moein, M.M.; Abdel-Rehim, M. Polyacrylonitrile/Graphene Oxide Nanofibers for Packed Sorbent Microextraction of Drugs and Their Metabolites from Human Plasma Samples. *Talanta* **2019**, *201*, 474–479. [CrossRef] [PubMed]
99. Vasconcelos Soares Maciel, E.; Henrique Fumes, B.; Lúcia de Toffoli, A.; Mauro Lanças, F. Graphene Particles Supported on Silica as Sorbent for Residue Analysis of Tetracyclines in Milk Employing Microextraction by Packed Sorbent. *Electrophoresis* **2018**, *39*, 2047–2055. [CrossRef] [PubMed]
100. Seidi, S.; Tajik, M.; Baharfar, M.; Rezazadeh, M. Micro Solid-Phase Extraction (Pipette Tip and Spin Column) and Thin Film Solid-Phase Microextraction: Miniaturized Concepts for Chromatographic Analysis. *TrAC-Trends Anal. Chem.* **2019**, *118*, 810–827. [CrossRef]
101. Sun, H.; Feng, J.; Han, S.; Ji, X.; Li, C.; Feng, J.; Sun, M. Recent Advances in Micro- and Nanomaterial-Based Adsorbents for Pipette-Tip Solid-Phase Extraction. *Microchem. Acta* **2021**, *188*. [CrossRef] [PubMed]
102. Shen, Q.; Yang, H.; Li, Y.; Li, S.; Chen, K.; Wang, H.; Wang, H.; Ma, J. Rapid Determination of Antiviral Drugs in Yellow Catfish (*Pelteobagrus fulvidraco*) Using Graphene/Silica Nanospheres (G/KCC-1) Based Pipette Tip Solid-Phase Extraction with Ultra-Performance Liquid Chromatography-Tandem Mass Spectrometry. *J. Chromatogr. B Anal. Technol. Biomed. Life Sci.* **2022**, *1189*, 123097. [CrossRef]
103. Zhang, P.; Wang, W.; Yin, J.; Wang, M.; Han, Y.; Yan, H. Determination of Alectinib and Its Active Metabolite in Plasma by Pipette-Tip Solid-Phase Extraction Using Porous Polydopamine Graphene Oxide Adsorbent Coupled with High-Performance Liquid Chromatography-Ultraviolet Detection. *J. Chromatogr. A* **2024**, *1714*, 464578. [CrossRef] [PubMed]
104. Tsai, P.C.; Pundi, A.; Brindhadevi, K.; Ponnusamy, V.K. Novel Semi-Automated Graphene Nanosheets Based Pipette-Tip Assisted Micro-Solid Phase Extraction as Eco-Friendly Technique for the Rapid Detection of Emerging Environmental Pollutant in Waters. *Chemosphere* **2021**, *276*, 130031. [CrossRef]
105. Carasek, E.; Morés, L.; Huelsmann, R.D. Disposable Pipette Extraction: A Critical Review of Concepts, Applications, and Directions. *Anal. Chim. Acta* **2022**, *1192*, 339383. [CrossRef] [PubMed]
106. Anastassiades, M.; Lehotay, S.J.; Štajnbaher, D.; Schenck, F.J. Fast and Easy Multiresidue Method Employing Acetonitrile Extraction/Partitioning and "Dispersive Solid-Phase Extraction" for the Determination of Pesticide Residues in Produce. *J. AOAC Int.* **2003**, *86*, 412–431. [CrossRef]

107. Syaleyana Md Shukri, D.; Yahaya, N.; Miskam, M.; Yusof, R.; Husaini Mohamed, A.; Kamaruzaman, S.; Nadhirah Mohamad Zain, N.; Semail, N.F. Advances in Dispersive Solid-Phase Extraction Techniques for Analytical Quantification of Fluoroquinolone Antibiotics. *Microchem. J.* **2023**, *193*, 109154. [CrossRef]
108. Płotka-Wasylka, J.; Jatkowska, N.; Paszkiewicz, M.; Caban, M.; Fares, M.Y.; Dogan, A.; Garrigues, S.; Manousi, N.; Kalogiouri, N.; Nowak, P.M.; et al. Miniaturized Solid Phase Extraction Techniques for Different Kind of Pollutants Analysis: State of the Art and Future Perspectives—PART 1. *TrAC Trends Anal. Chem.* **2023**, *162*, 117034. [CrossRef]
109. Ghorbani, M.; Aghamohammadhassan, M.; Ghorbani, H.; Zabihi, A. Trends in Sorbent Development for Dispersive Micro-Solid Phase Extraction. *Microchem. J.* **2020**, *158*, 105250. [CrossRef]
110. Chisvert, A.; Cárdenas, S.; Lucena, R. Dispersive Micro-Solid Phase Extraction. *TrAC Trends Anal. Chem.* **2019**, *112*, 226–233. [CrossRef]
111. Feist, B. Dispersive Micro-Solid Phase Extraction Using a Graphene Oxide Nanosheet with Neocuproine and Batocuproine for the Preconcentration of Traces of Metal Ions in Food Samples. *Molecules* **2023**, *28*, 4140. [CrossRef] [PubMed]
112. Greda, K.; Welna, M.; Szymczycha-Madeja, A.; Pohl, P. Dispersive Micro-Solid Phase Extraction Based on Graphene Oxide for the Ultrasensitive Determination of Cd by Slurry Sampling Microplasma Optical Emission Spectrometry. *Microchem. J.* **2024**, *196*, 109715. [CrossRef]
113. Nakhonchai, N.; Prompila, N.; Ponhong, K.; Siriangkhawut, W.; Vichapong, J.; Supharoek, S. ang Green Hairy Basil Seed Mucilage Biosorbent for Dispersive Solid Phase Extraction Enrichment of Tetracyclines in Bovine Milk Samples Followed by HPLC Analysis. *Talanta* **2024**, *271*, 125645. [CrossRef] [PubMed]
114. Manousi, N.; Rosenberg, E.; Deliyanni, E.; Zachariadis, G.A.; Samanidou, V. Magnetic Solid-Phase Extraction of Organic Compounds Based on Graphene Oxide Nanocomposites. *Molecules* **2020**, *25*, 1148. [CrossRef] [PubMed]
115. Kalaboka, M.; Sakkas, V. Magnetic Solid-Phase Extraction Based on Silica and Graphene Materials for Sensitive Analysis of Emerging Contaminants in Wastewater with the Aid of UHPLC-Orbitrap-MS. *Molecules* **2023**, *28*, 2277. [CrossRef] [PubMed]
116. Mohammadi, Z.; Sabzehmeidani, M.M.; Ghaedi, M.; Dashtian, K.; Abbasi-Asl, H. Dispersive Micro-Solid Phase Extraction Coupled with Spectrophotometric Using (MgFe CLDH)/GO Magnetically Separable Sorbent for Pre-Concentration of Anionic Food Dyes in Water Samples. *Emerg. Contam.* **2024**, *10*, 100347. [CrossRef]
117. Cao, S.; Chen, J.; Lai, G.; Xi, C.; Li, X.; Zhang, L.; Wang, G.; Chen, Z. A High Efficient Adsorbent for Plant Growth Regulators Based on Ionic Liquid and β-Cyclodextrin Functionalized Magnetic Graphene Oxide. *Talanta* **2019**, *194*, 14–25. [CrossRef] [PubMed]
118. Gong, Y.; Liu, P. A Novel Magnetic β-Cyclodextrin-Modified Graphene Oxide and Chitosan Composite as an Adsorbent for Trace Extraction of Four Bisphenol Pollutants from Environmental Water Samples and Food Samples. *Molecules* **2024**, *29*, 867. [CrossRef] [PubMed]

Disclaimer/Publisher's Note: The statements, opinions and data contained in all publications are solely those of the individual author(s) and contributor(s) and not of MDPI and/or the editor(s). MDPI and/or the editor(s) disclaim responsibility for any injury to people or property resulting from any ideas, methods, instructions or products referred to in the content.

Review

Developments and Applications of Molecularly Imprinted Polymer-Based In-Tube Solid Phase Microextraction Technique for Efficient Sample Preparation

Hiroyuki Kataoka *, Atsushi Ishizaki, Keita Saito and Kentaro Ehara

School of Pharmacy, Shujitsu University, Nishigawara, Okayama 703-8516, Japan
* Correspondence: hkataoka@shujitsu.ac.jp

Abstract: Despite advancements in the sensitivity and performance of analytical instruments, sample preparation remains a bottleneck in the analytical process. Currently, solid-phase extraction is more widely used than traditional organic solvent extraction due to its ease of use and lower solvent requirements. Moreover, various microextraction techniques such as micro solid-phase extraction, dispersive micro solid-phase extraction, solid-phase microextraction, stir bar sorptive extraction, liquid-phase microextraction, and magnetic bead extraction have been developed to minimize sample size, reduce solvent usage, and enable automation. Among these, in-tube solid-phase microextraction (IT-SPME) using capillaries as extraction devices has gained attention as an advanced "green extraction technique" that combines miniaturization, on-line automation, and reduced solvent consumption. Capillary tubes in IT-SPME are categorized into configurations: inner-wall-coated, particle-packed, fiber-packed, and rod monolith, operating either in a draw/eject system or a flow-through system. Additionally, the developments of novel adsorbents such as monoliths, ionic liquids, restricted-access materials, molecularly imprinted polymers (MIPs), graphene, carbon nanotubes, inorganic nanoparticles, and organometallic frameworks have improved extraction efficiency and selectivity. MIPs, in particular, are stable, custom-made polymers with molecular recognition capabilities formed during synthesis, making them exceptional "smart adsorbents" for selective sample preparation. The MIP fabrication process involves three main stages: pre-arrangement for recognition capability, polymerization, and template removal. After forming the template-monomer complex, polymerization creates a polymer network where the template molecules are anchored, and the final step involves removing the template to produce an MIP with cavities complementary to the template molecules. This review is the first paper to focus on advanced MIP-based IT-SPME, which integrates the selectivity of MIPs into efficient IT-SPME, and summarizes its recent developments and applications.

Keywords: molecularly imprinted polymer (MIP); in-tube solid-phase microextraction (IT-SPME); sample preparation

Citation: Kataoka, H.; Ishizaki, A.; Saito, K.; Ehara, K. Developments and Applications of Molecularly Imprinted Polymer-Based In-Tube Solid Phase Microextraction Technique for Efficient Sample Preparation. *Molecules* **2024**, *29*, 4472. https://doi.org/10.3390/molecules29184472

Academic Editor: Minjia Meng

Received: 30 August 2024
Revised: 15 September 2024
Accepted: 18 September 2024
Published: 20 September 2024

Copyright: © 2024 by the authors. Licensee MDPI, Basel, Switzerland. This article is an open access article distributed under the terms and conditions of the Creative Commons Attribution (CC BY) license (https://creativecommons.org/licenses/by/4.0/).

1. Introduction

Although state-of-the-art high-performance analytical instruments with improved sensitivity and selectivity have been developed for qualitative and quantitative analysis of analytes in complex matrices, sample preparation remains a time-consuming task for researchers [1–10]. This step is often considered the bottleneck of the entire analytical process, and it is no exaggeration to say that the efficiency of sample preparation greatly affects the quality of the analytical results. For instance, in most chromatographic systems, target analytes present at trace levels in complex matrices may coelute with other structurally similar compounds that coexist at high concentrations [5]. Additionally, coexisting biological macromolecules may adsorb irreversibly to the inner walls of tubing or the pores of the column's stationary phase, causing flow path blockages and decreased column efficiency [3]. In mass spectrometry, ionization suppression by the matrix can significantly impact detection sensitivity and reproducibility. Consequently, sample pretreatment is an essential and critical step not only for pre-separating and enriching the target analyte, but also for removing interfering components, enhancing detection, eliminating matrix interference, improving analytical sensitivity and accuracy, and reducing instrument maintenance and operating costs.

Various sample preparation methods have been used [1,2], but classical liquid-liquid extraction (LLE) [11] requires large volumes of samples and toxic organic solvents, and is time-consuming, labor-intensive, and costly. Solid-phase extraction (SPE) [12–14] was developed to overcome these drawbacks and is widely used due to its relatively simple and efficient operation, low cost, reduced consumption of organic solvents, and high enrichment capability. Furthermore, to minimize sample size and solvent use, and to improve efficiency and automation, various microextraction techniques have been developed, including micro-SPE [15], dispersive micro-SPE [16], solid-phase microextraction (SPME) [17–21], stir bar sorptive extraction (SBSE) [22–24], micro extraction in a packed syringe [25], pipette tip SPE [26], magnetic bead extraction [27] and liquid-phase microextraction [28–30]. Among these, in-tube SPME (IT-SPME) [3–5,31–37], which uses a capillary column as the extraction device, is particularly useful. IT-SPME requires almost no organic solvent, resulting in less liquid waste, higher throughput, smaller size, online connection to analytical instruments, and automation that saves labor and allows continuous overnight operation. As such, IT-SPME is an advanced sample preparation method that aligns well with the principles of green analytical chemistry.

However, the performance of these early microextraction methods is often dependent on the matrix composition, as they are limited in increasing extraction volumes due to sample loading constraints and may be ineffective in selectively extracting analytes from coexisting substances. To improve extraction efficiency and selectivity, various new functional adsorbents have been developed, including monoliths [38], ionic liquids (ILs)/polymer ILs [39], restricted-access materials (RAMs) [40], molecularly imprinted polymers (MIPs) [6–10,41–49], graphene/graphene oxide [50–52], carbon nanotubes (CNTs) [53], inorganic nanoparticles (NPs) [54,55], metal organic frameworks (MOFs) [56,57], and covalent organic frameworks (COFs) [58,59]. Among these, MIPs, chemically stable polymers that acquire template cavities during synthesis, have superior molecular recognition capabilities for specific analytes or structural analogues, making them "smart adsorbents" with high selectivity and loading capacity. Molecular imprinting technology has thus become a powerful tool in the development of advanced sample preparation methods.

Although many excellent review papers have discussed SPE [60–62] and SPME [8–10,63–66] using MIP in terms of adsorbents and extraction methods, there has been no review focused on MIP IT-SPME, which integrates MIP into the IT-SPME method. In this review of recent developments and applications of MIP IT-SPME, we focus on MIP IT-SPME, which combines the characteristics of both methods, such as simplicity, flexibility, robustness, and selectivity.

2. Overview of IT-SPME

IT-SPME, developed by Eisert and Pawliszyn [67] in 1997, is a microextraction technique designed for efficient sample preparation using open tubular fused silica capillary columns as extraction devices. This technique addresses several drawbacks of the initial SPME [68], which uses fused silica fibers with sorbent coatings on the surface as extraction devices. The limitations of the original approach included: (1) fiber fragility, (2) bleeding from the thick film coating, (3) low adsorption capacity, (4) difficulty in applying the technique to non-volatile and thermally unstable compounds unsuitable for gas chromatography (GC) or GC-mass spectrometry (MS), and (5) low stability in the presence of solvents used in high-performance liquid chromatography (HPLC). IT-SPME is a dynamic in-flow microextraction technique that is particularly useful for automated clean-up and rapid online coupling to liquid chromatography (LC) through column switching [1–5,31,32,69,70].

Analytes in complex matrices can be continuously analyzed by HPLC and MS/MS systems linked to IT-SPME with minimal processing such as filtration.

2.1. IT-SPME Operating System

In IT-SPME, operations like extraction, concentration, desorption, and injection can be easily automated using a programmable autosampler and column switching technology [3,32,33]. First, as the sample solution passes through the extraction capillary, the analyte is extracted and concentrated by adsorption or absorption onto the stationary phase in the capillary, based on the distribution equilibrium. For samples with high levels of coexisting substances, water or an appropriate solvent can be passed through the capillary after loading the sample to selectively wash away unwanted matrix components. The extracted analyte is then desorbed either by introducing a solvent into the capillary via valve switching (static desorption) or through a mobile phase (dynamic desorption) and transferred to the analytical instrument. These operating systems include the draw/eject system, where the sample solution is repeatedly aspirated and discharged into the capillary, and the flow-through system, in which sample solution flows in one direction into the capillary (Figure 1). In the repeated draw/eject system, the amount of adsorption depends on the distribution rate to the capillary's stationary phase and the number of repetitions, since the total amount of compounds in the solution is not fully loaded by repeated extractions from a fixed volume. In contrast, in the flow-through system, the amount of adsorption increases with the volume of the sample solution, but adsorption in a narrow capillary is limited by the thickness of the coating and the distribution equilibrium of the compounds, which may lead to overflow. Other innovative approaches to improve extraction and desorption efficiency include magnetic IT-SPME [71,72], which utilizes magnetic nanomaterial-filled tubes to perform extraction and desorption by switching magnetic fields, electrochemically controlled IT-SPME [73–75], and IT-SPME with temperature control devices [76,77].

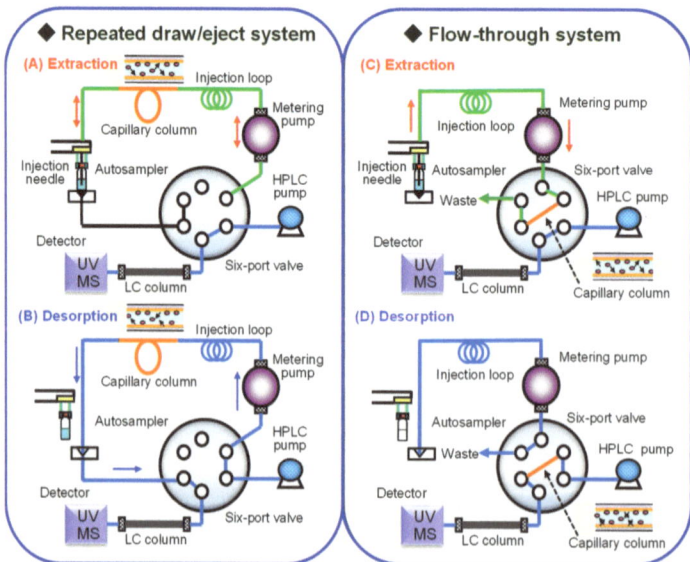

Figure 1. Two operating systems of automated online IT-SPME coupled with HPLC. (**A,C**) are the steps of extracting compounds from the sample solution into the capillary stationary phase, and (**B,D**) are the steps of desorbing the compounds extracted into the capillary. The green and blue lines indicate the flow of the sample solution and the mobile phase, respectively. Reproduced from Figure 5 of Ref. [3] with permission from Elsevier.

IT-SPME offers the advantage of enabling high-throughput analytical systems through process automation and online coupling with analytical instruments. It is primarily used in direct connection with HPLC, but it can also be integrated [78–81] with capillary electrophoresis, capillary electrochromatography, direct MS, or atomic absorption spectrometry. Recently, the power of LC has increased, and coupling with miniaturized chromatography systems, such as ultra-high-pressure LC (UHPLC), capillary LC (capLC), or nanoLC, facilitates integration with MS detectors, enhances column efficiency and sensitivity, reduces solvent consumption, and shortens analysis time. However, online coupling with these miniaturized systems requires specialized interfaces to control flow rates.

2.2. Capillary Tube Configuration for IT-SPME

Capillary tube configurations used in IT-SPME can be classified into four types (Figure 2): (1) inner-surface-coated, (2) particle-packed, (3) fiber-packed, and (4) rod monolith capillaries [3,4,33]. The details are as follows:

1. Inner-surface-coated: includes wall-coated open tubular (WCOT) capillaries and porous layer open tubular (PLOT) capillaries, where the inner wall of the tube is coated with an adsorbent.
2. Particle-packed: capillaries where adsorbent-coated particles are packed inside the tube.
3. Fiber-packed: capillaries in which thin fibers are vertically packed inside the tube.
4. Rod monolith: capillaries where a monolith is formed within the tube.

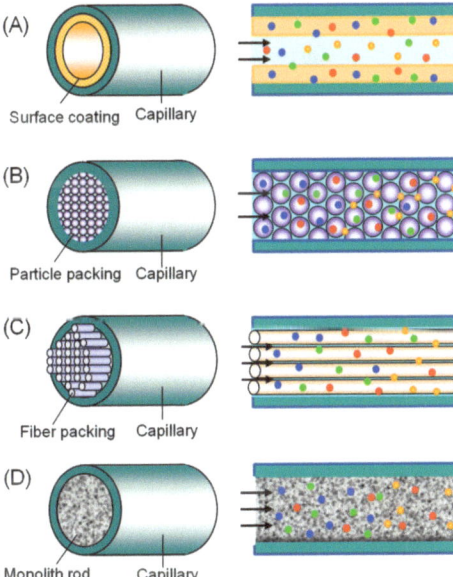

Figure 2. Configurations of capillary tubes for IT-SPME. (**A**) Inner-surface-coated capillary, (**B**) particle-packed capillary, (**C**) fiber-packed capillary, (**D**) monolithic capillary. Reproduced from Figure 4 of Ref. [3] with permission from Elsevier.

Open-tube capillaries offer the advantage of avoiding clogging without increasing the tube's back pressure, even at faster flow rates compared to packed capillaries. They are versatile, interacting with a variety of compounds [3]. However, commercially available GC capillary columns have limitations in selectively extracting and concentrating specific target analytes from complex matrices due to the limited nature and dimensions of the phases, and their infrequent interaction with polar compounds.

For IT-SPME, fused silica tubes with a fixed inner diameter of 1 mm or less and an outer polyimide coating for enhanced flexibility and durability are commonly used. Other materials such as polyetheretherketone (PEEK), polytetrafluoroethylene, polydimethylsiloxane (PDMS), stainless steel, and glass tubes are also utilized [82]. Fused silica capillaries, where silanol groups on the inner wall are ionized with buffer solution, are typically used for adsorbent immobilization. The immobilized phase is formed through polymer coating, polymerization, chemical modification, or electrodeposition. In contrast, due to the difficulty of adsorbent immobilization on the inner walls of plastic or stainless-steel tubes, these materials are often used as packed tubes filled with adsorbent particles or fibers. Fiber-filled capillaries increase the adsorption surface area and improve extraction efficiency by coating the fiber surface. Porous monolithic capillaries enhance extraction efficiency through high permeability, rapid mass transfer, high stability, and high loading capacity.

In IT-SPME, the amount and sensitivity of the extracted analytes depend on the length of the extraction capillary and the adsorption capacity of the adsorbent. Generally, a longer capillary increases the amount extracted, but it also broadens the sample bandwidth, affecting subsequent chromatographic separation. This can delay the target compound's arrival at the detector, leading to excessive peak broadening. Therefore, extraction capillaries typically range from 60 to 80 cm in length. When connecting to chromatography systems with low mobile phase flow rates, such as capLC or nanoLC, these effects should be considered. Additionally, the dimensions of other components necessary for setting up the extraction device (e.g., connectors, injection valves) should be carefully selected to prevent band broadening [32].

2.3. Extraction Phase of Capillary Tube

In IT-SPME, the affinity of the analyte for the extraction phase of the capillary tube affects the extraction efficiency, so selecting an appropriate extractant based on the properties of the target compound is essential. Common sorbents include carbon-based sorbents (e.g., divinylbenzene polymer (DVB), carboxene, carbon molecular sieves, polyethylene glycol, etc.) and silica-based sorbents (e.g., polydimethylsiloxane (PDMS), diphenylpolydimethylsiloxane, cyanopropylphenylmethylpolysiloxane, etc.). These sorbents have been used since the early days of IT-SPME development because they are commercially available capillary columns for GC [3]. These sorbents are used in both hollow coated WCOT and PLOT columns. WCOT columns, such as TRB and CP-Sil 19CB, are liquid-phase types, while PLOT columns, such as Supel Q PLOT and Carboxen 1006 PLOT, are adsorption types. Liquid-phase capillaries, in which the adsorbent is firmly bonded and cross-linked to the capillary inner wall, can be used stably without losing phase when the solvent passes through the capillary. However, their film thickness cannot be made sufficiently thick to achieve a strong extraction effect. On the other hand, adsorption-type capillaries, which are porous with a large surface area, offer high extraction efficiency, but their thick film can deteriorate and exfoliate depending on the mobile phase solvent. A significant advantage of commercially available capillaries is the stable supply of products with various polarities, film thicknesses, and porosities, making them applicable to the extraction of various compounds. Additionally, they offer excellent reproducibility and reusability due to stable extraction performance. However, as mentioned above, their lack of selectivity and relatively low extraction efficiency limit their usefulness.

To improve extraction efficiency and selectivity, various functional extraction phases have been developed [3,35–37]. These include carbon-based nanomaterials such as CNTs and graphene, as well as monoliths, RAM, MIP, ILs, MOF, and deep eutectic solvents (DES). These materials can be coated on the capillary inner wall, or particles or fibers coated and packed in the capillary. For example, CNTs and graphene have high surface areas and provide high adsorption efficiency for target analytes due to interactions such as hydrogen bonding, π-π stacking, electrostatic forces, van der Waals forces, and hydrophobic interactions [83]. IT-SPME using octadecyl silica (C18) monolithic rod capillaries also offers higher extraction efficiency than conventional PDMS-coated capillaries and can be coupled online with capLC systems. Additionally, biocompatible RAM capillaries, MIP capillaries, and immunoaffinity capillaries packed with alkyl diol silica (ADS) particles have been developed to improve selectivity [3,8–10,31,35–37,64,65]. MIPs, in particular, are synthetic polymeric materials that consist of complementary imprinted moieties for specific molecules. They recognize target analytes and compounds with similar molecular structures through a combination of hydrogen bonding, hydrophobicity, and electrostatic interactions, resulting in high extraction selectivity. ILs, MOFs, and DES have also been applied as extractants for IT-SPME. Hybrid materials combining extractants with different functions, such as RAM and MIP, MOF embedded in monoliths, and graphene monoliths embedded in porous polymers, are also promising adsorbents. Highly porous nanoparticle coatings with magnetic hybrid extractant phases have been applied in magnetic IT-SPME as well [70,71]. This review focuses on MIPs, known for their high molecular recognition capacity, which are summarized in detail in the following sections.

3. Fabrication of Molecularly Imprinted Polymers

MIPs were first developed by Wulff [84] and Mosbach [85] in the 1970s to recognize the shape, size, and functional groups of specific target molecules, such as template molecules and their structural analogues. These artificial functional materials mimic biological interactions, such as enzyme-substrate, antigen-antibody and hormone-receptor interactions [6,47]. MIPs are characterized by their unique structural predictability, high specificity and retention, physical and chemical robustness, reusability, and batch-to-batch recognition reproducibility. Their many advantages include relatively simple preparation, low cost, and resistance to high temperature, pressure, acids, bases, organic solvents, and biological degradation [21,62,64,65]. Due to these properties, MIPs are prepared in coated particles, coated fibers, coated stir bars, coated thin films, wall-coated capillaries, monoliths, magnetic beads, and used in various device forms, including SPE [60–62], SPME [8–10,63–66] (fiber, in-tube, monolith, dispersion particles, membranes), SBSE, and others, offering superior functionality.

3.1. Principles of Molecular Imprinting and Synthesis of MIPs

The fabrication process of MIPs generally involves three main steps: prearrangement, polymerization, and template elution (Figure 3) [6–10,21,41–44,47,60,64–66].

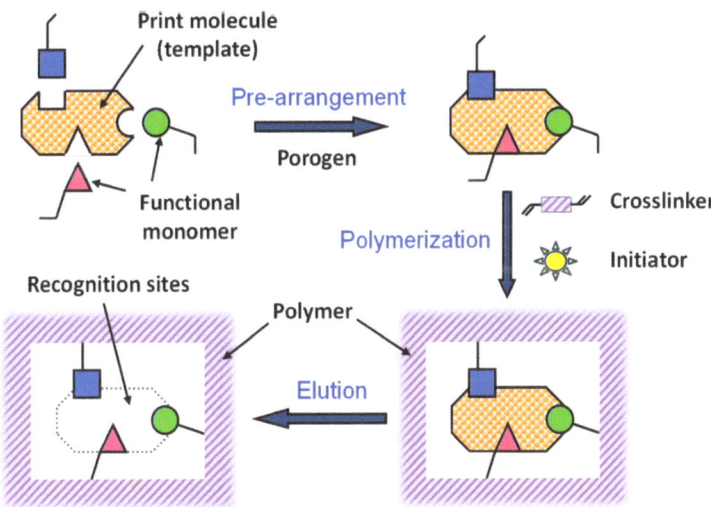

Figure 3. Fabrication process of molecularly imprinted polymer.

1. Prearrangement: In the initial step, template molecules interact with functional monomers through non-covalent, covalent, or semi-covalent bonds, forming a host-guest complex. In non-covalent imprinting, the template and monomer form a monomer-template complex via non-covalent bonds (electrostatic interactions, hydrogen bonds, ion pair formation, van der Waals forces, π-π stacking, metal coordination, etc.). This method facilitates the removal of the template without chemical bond cleavage, preserving the polymer structure. However, the stability of these bonds can be sensitive to changes in the chemical environment, requiring careful optimization of reaction conditions. Excessive use of functional monomers may also introduce non-specific binding sites, reducing selectivity. In contrast, covalent imprinting involves reversible covalent bonding between the template and polymerizable groups. After polymerization, the template is cleaved, leaving functional groups correctly orientated for subsequent re-binding. Although this method requires suitable template-monomer complexes, it ensures accurate uptake of target analytes from aqueous solutions. However, the formation and cleavage of covalent bonds are unlikely to occur under mild conditions. Semi-covalent imprinting combines covalent and non-covalent interactions, offering a balance between template stability and re-binding efficiency.
2. Polymerization: Polymerization is initiated by thermal or UV activation in the presence of cross-linkers and initiators, forming a highly cross-linked polymer network around the template molecules. This step creates the three-dimensional space necessary for molecular recognition.
3. Template elution: In the final step, template molecules are removed from the polymer network through physicochemical methods such as hydrolysis or desorption, leaving MIPs with cavity sites complementary to the template molecules. However, in covalent imprinting, removal of the template by covalent bond cleavage under severe conditions may affect the functionality of the cavity.

Among these methods, non-covalent imprinting is the most popular due to the availability of a wide variety of monomers that interact with different templates [64]. Although it requires time and effort for optimization, non-covalent imprinting allows easy template removal and re-binding under mild conditions. To synthesize non-covalent MIPs with appropriate recognition properties, mechanical and chemical stability, various factors must be optimized, including the chemistry and relative amounts of polymer components (templates, monomers, cross-linkers, initiators, porogens) and polymerization conditions (temperature, initiator activation, etc.) [9,10,18,43,47,62,64]. In trace analysis, when residual templates can cause positive errors, a dummy template—a structurally similar molecule—is used. The ideal dummy template should replicate the shape and functional group orientation of the analyte-template complex. The choice of functional monomers is closely related to the nature of the template, and the monomers must be properly selected to obtain optimal MIP functionality for the target analyte. Acidic monomers are suitable for basic templates and basic monomers for acidic templates, and methacrylic acid (MAA) and 4-vinylpyridine (VP) are commonly used as acidic and basic monomers, respectively. Cross-linkers, such as ethylene glycol dimethacrylate (EGDMA) and trimethylolpropane trimethacrylate (TRIM) play an important role in stabilizing the three-dimensional network of molecular recognition sites and controlling polymer porosity. They affect surface polarity (wettability), area, pore size, and adsorption capacity, but the relationship between adsorption capacity and crosslinker loading is complex affecting the accessibility of the binding site. To obtain a rigid polymer, the minimum required cross-linker must be used, but too much can make the structure too rigid or fill the pore structure, resulting in poor rates of template removal and re-binding. The initiator, typically azobisisobutyronitrile (AIBN), is selected based on the type of polymerization reaction, and the porogen used to solubilize the polymer components influences the physical properties of the MIP, including recognition, enantioselectivity, surface area, pore volume, and swelling. When the main template-monomer interaction is hydrogen bonding, hydrogen bond donors and non-polar, non-protic solvents with low capacity as acceptors are suitable for non-covalent imprinting.

MIPs are synthesized using various imprinting techniques, including surface imprinting, nanoimprinting, living/controlled radical polymerization, multi-template imprinting, multifunctional monomer imprinting, and dummy template imprinting (Table 1) [6]. Common polymerization methods include bulk polymerization, suspension polymerization, emulsion polymerization, precipitation polymerization, multistep swelling polymerization, sol-gel polymerization, multistep swelling polymerization, and in-situ polymerization [6,9,21,41,43–47,62,64,86]. Bulk polymerization, while common, may destroy binding sites during grinding and produce irregularly shaped particles. Precipitation polymerization, based on the growth of polymer chains, and suspension polymerization, which occurs in micelles, yield particles with more regular shapes [43,47]. On the other hand, sol-gel polymerization can be carried out in aqueous media under mild thermal conditions. The resulting MIP has a high degree of cross-linking and excellent thermal and mechanical stability. Therefore, it is useful for forming thin films as an adsorbent with controlled pore size and surface area [41,46].

Table 1. Various imprinting techniques for the fabrication of MIPs. Produced from Ref. [6] with permission from Elsevier.

Surface imprinting	MIPs are typically fabricated in layers on hard particles, forming high affinity recognition sites on the substrate surface. The size of the imprinting cavities on the polymer surface can be effectively controlled, and uniformly distributed sites not only increase the adsorption capacity of the MIP and improve the rebinding rates of the recognition sites to the imprinted molecules, but also enhance the adsorption and separation efficiency of the imprinted material.
Nanoimprinting	Nanoimprinting technology is used to prepare nanostructured MIPs offering advantages such as high resolution, fast processing speed, high throughput, material compatibility, and low cost. These benefits improve the adsorption capacity, recombination rate, and site accessibility of the MIPs.
Living/controlled radical polymerization technology	Common methods include nitroxide-mediated free radical polymerization, atom transfer radical polymerization, and reversible addition cleavage chain transfer polymerization. Advantages include: (1) a wide range of polymerizable monomers, controllable polymer molecular weight, and narrow molecular weight distribution; (2) mild reaction conditions, low polymerization reaction temperatures, and compatibility with various solvents; (3) structural functional control, with the use of "reactive" features and functionalized end groups allowing for the preparation of polymers with complex compositions and structures; (4) a linear increase in polymer molecular weight with the conversion rate.
Multi-template imprinting	This technique uses multiple target molecules as templates to form various recognition sites in a single polymer material. Multiple template MIPs can simultaneously recognize multiple target molecules, allowing for the concurrent extraction, separation, analysis, and detection of different species, thus greatly expanding the practical applications of MIPs.
Multi-functional monomer imprinting	Multifunctional monomer imprinting techniques utilize non-covalent bonds between two or more functional monomers and a template molecule to create different forces with selective adsorption capacity. This improves the selectivity of the MIP for the template molecule, thereby enhancing its enrichment capacity.
Dummy template imprinting	Structural analogues of the target compound are used as template molecules when the target compound is either unsuitable for use as a template molecule or susceptible to degradation.

3.2. Characteristics and Functionalization of MIPs

One of the main advantages of MIPs is that their customizable, allowing specific functional groups to be incorporated and external surface properties to be easily tuned [21,62]. Hybrid MIPs, which combine other extractive phases like RAM and MOF with MIP, have been developed to enhance performance [18]. For example, RAM-MIP adsorbents, made with hydrophilic monomers such as 2-hydroxyethyl methacrylate and glycerol dimethacrylate, selectively exclude macromolecules due to the hydrophilic action of the restricted-access outer layer of RAM, allowing small molecular target analytes to penetrate and be selectively extracted within the MIP phase. This reduces the need for pretreatment, such as protein precipitation, in analytic protocols. MIP-MOF hybrids, where MOF provides a robust polymer structure, increase the surface area and porosity of the MIP, facilitating mass transfer of the target analyte to the binding site [65,66]. MIP adsorbents can also be synthesized with reduced particle size without compromising specificity and selectivity, enhancing reusability and reducing analytical costs. Divinylbenzene cross-linked MIPs exhibit long-term stability and can be reused over 100 times without losing functionality, even under harsh conditions such as high acidity and temperature. However, MIPs derived from acrylates or methacrylates may exhibit a reduced adsorption specificity and capacity due to irreversible hydrolysis or esterification of the polymer regions, particularly under harsh conditions, limiting reusability.

Challenges affecting MIP-based microextraction include template leakage, binding site heterogeneity, delayed analyte diffusion to imprinted sites due to broad pore size distribution, and reduced selective binding capacity in aqueous media [65,66]. Various MIPs have been developed to address these issues. Dummy templates help avoid template leakage, while alternative polymerization methods produce uniform MIP particles, such as spherical particles and monolithic imprinted materials, with improved mass transfer rates [62]. In situ polymerization of MIP monoliths offers lower back pressure and controlled porosity, while precipitation and emulsion polymerization yield microspheres or nanospheres with large specific surface areas, speeding up template removal and improving adsorption and desorption rates. Core-shell materials, such as those combining siloxane cores with MIP shells, have enabled precise control of adsorbate morphology, enhancing the adsorption process. Additionally, CNTs, MWNTs, and magnetic nanoparticles have been used as supports in surface imprinting methods. Core-shell magnetic MIPs (MMIPs) are easy to prepare, chemically stable, and have high binding rates due to their small size and high surface area-to-volume ratio [65]. These materials offer the ability to bind to other components and integrate magnetic properties, not only by use of magnetite, but also cobalt-based magnetic nanoporous carbon, making them useful for selective extraction from complex samples. MIP-based extraction methods have also focused on increasing throughput and efficiency, leading to the development of single, dual, and multi-template MIPs for bioanalytical applications [66].

4. Developments and Applications of MIP IT-SPME

As sample preparation trends towards miniaturization, ease of operation, and automation, making IT-SPME has gained attention for its compatibility with chromatographic instruments. As described in Section 2, there are four main configurations of capillary tubes used in IT-SPME (Figure 2): inner surface-coated capillaries, particle-packed capillaries, fiber-packed capillaries, and rod monolith capillaries [3,4]. These capillaries, containing various adsorbents, have been applied to analyze a wide range of compounds. Among these, the integration of MIPs with IT-SPME has particularly attracted interest, combining the selectivity of MIPs with the practicality of IT-SPME. The first MIP-based IT-SPME device was reported in 2001 [87], where a propranolol-imprinted MIP sorbent was packed into PEEK tubes. Since then, MIP-based IT-SPME methods have expanded, including techniques such as coating the inner wall of capillaries with MIP, packing capillaries with MIP-coated particles or fibers, and using MIP-formed rod-shaped monoliths. These methods have been predominantly applied to the analysis of biological samples, as summarized in Table 2.

Table 2. Various MIP IT-SPME methods developed for sample preparation.

Analyte	Template	Polymerization Composition [1] and Conditions (Monomer/Crosslinker/Initiator/Porogen) [1]	Capillary Tube Configuration	IT-SPME Operation	Enrichment, Sensitivity [2]	IF [3]	Matrix	Detection [4]	Ref.
Estrogen-related compounds	β-Estradiol	VP/EGDMA/AIBN/CH$_2$Cl$_2$, 50 °C for 4.5 h, template/VP/EGDMA (1:6:30). In-situ synthesis of MIP in a fused silica capillary surface by insertion of fluorocarbon yarn (65 cm × 0.20 mm).	Inner-surface-coated fused silica capillary (60 cm × 0.32 mm ID).	Draw/ejection on-line	EF: 1.9–16.4	1.8–3.6	Water	HPLC-UV	-
4-Nitrophenol	4-Nitrophenol	MAA/EGDMA/AIBN/acetonitrile, 60 °C for 4 h. In-situ synthesis of MIP in a glass capillary surface by insertion of a metal rod. MIP: VP/EGDMA/AIBN/acetonitrile in capillary at 50 °C for 4 h, template/monomer (1:4). RAM: hydrophilic monomer-GDMA/PPDS at 70 °C for 20 h.	Inner-surface-coated glass-capillary (100 µL).	Syringe pump off-line	LOD: 0.33 ng mL^{-1}	-	Environmental water	HPLC-DAD	[88]
Parabens	Benzyl-paraben	PY/EGDMA/AIBN/MeOH:H$_2$O (2:1 v/v), cyclic voltammetry in the potential range between −1.0 and −1.0 V during 30 cycles (scan rate: 50 mV/s).	Inner-surface-coated fused silica capillary (50 mm × 0.53 mm ID).	Syringe pump off-line	LLOQ: 3–10 ng mL^{-1}	-	Breast milk	UHPLC-MS/MS	[89]
Indomethacin	Indomethacin		Inner-surface-coated stainless-steel tube (10 cm × 0.75 mm ID).	Electrochemical control flow-through on-line	LOD: 0.6–2.0 ng mL^{-1}	-	Urine, plasma, blood	HPLC-UV	[90]
Carbamazepine	Carbam-azepine	Molecularly imprinted polypyrrole coated on CuO by electrodeposition, cyclic voltammetry in the potential range (0–+3 V during 30 cycles (scan rate: 70 mV/s).	Inner-surface-coated copper tube (10 cm × 0.78 mm ID).	Flow-through on-line	LOQ: 0.1 ng mL^{-1}	-	Urine, plasma	HPLC-UV	[91]
2,4-Dinitroaniline (2,4-DNA)	2,4-DNA	VI/EDMA/AIBN/1-propanol:1,4-butanediol (1:1)/Fe$_3$O$_4$ nanoparticles pre-modified with γ-MAPS at 70 °C for 12 h, template/VI/EDMA (4:1:4).	Inner-surface-coated fused silica capillary (2 cm × 0.53 mm ID).	Magnetic field control flow-through on-line	LOD: 60 pg mL^{-1}	3.1	Environmental water	HPLC-DAD	[92]
Propranolol	Racemic propranolol	MAA/EGDMA/AIBN/toluene at 60 °C for 18 h, template/monomer (1:2).	Particle-packed PEEK tube (80 mm × 0.76 mm ID).	Draw/ejection on-line	LOD: 0.32 µg mL^{-1}	-	Serum	HPLC-UV	[87]
Interferon alpha 2a	Interferon alpha 2a	Two-step sol-gel procedure: APS/TEOS/deionized water:0.1 M HCl: absolute (EtOH) (1:1.4:1.7) + silanes, kept at room temperature for 24 h and dried at 50 °C for 48 h.	Particle-packed PEEK tube (50 mm × 0.02 inch ID).	Draw/ejection on-line	-	-	Plasma	HPLC-FD	[93]
Four fluoroquinolone antibiotics	Ofloxacin, sulfadiazine	MAA/TRIM/AIBN/CH$_3$CN:H$_2$O (α:1, v/v), silica fiber (10 cm × 0.125 mm) coating in glass capillary (10 cm × 1.0 mm ID) at 60 °C for 3 h.	Fiber-packed (6 cm × 6 fibers) PEEK tube (0.5 mm ID).	Flow-through on-line	EF: 69–136, LOD: 16–110 pg mL^{-1}	-	Pork liver	HPLC-UV	[94]
8-Hydroxy-2′-deoxyguanosine (8-OHdG)	Guanosine	Monolith: TEPM/methanol, 40 °C for 12 h; MIP monolith: VP/MBA/AIBN/dodecanol, in capillary at 60 °C for 18 h.	Rod monolith in fused silica capillary (50 mm × 0.53 mm ID).	Syringe pump off-line	EF: 76, LOD: 957 pg mL^{-1}	-	Urine	HPLC-UV	[95]

Table 2. Cont.

Analyte	Template	Polymerization Composition [1] and Conditions (Monomer/Crosslinker/Initiator/Porogen) [1]	Capillary Tube Configuration	IT-SPME Operation	Enrichment, Sensitivity [2]	IF [3]	Matrix	Detection [4]	Ref.
Neurotensin, neuromedin N	Pro-Tyr-Ile-Leu	MIP monolith: MAA/EGDMA/AIBN/MeOH/acetonitrile/isooctane, 60 °C for 16 h template/MAA (1:3).	Inner-surface-coated fused silica capillary (2 cm × 0.53 mm ID).	Syringe pump off-line	LOD: 0.9–1.0 ng mL^{-1}	5.7–13.4	Plasma	HPLC-UV	[96]
Anaesthetics (bupivacaine, mepivacaine, S-ropivacaine)	Bupivacaine, mepivacaine, ropivacaine	Monolith: TRIM/EDMA/BME/2,2,4-trimethylpentane/toluene (80:20, w/w), UV; MIP monolith: MAA/EDMA/AIBN/toluene, Template/MAA/EDMA (0.33:4:20), UV, 1 h Co-precursors: PEG/TMOS/γ-MAPS	Rod monolith in UV transparent capillaries (70 mm × 0.1 mm ID).	Flow-through on-line	-	12–72	Water	HPLC-UV	[97]
Lysozyme	Lysozyme	MIP hybrid monolith: co-precursors + AAm/MBA/AIBN/MeOH:H$_2$O (5:3, v/v), in capillary at 40 and 60 °C for 12 h.	Rod monolith in fused silica capillary (25 cm × 75 μm ID).	Flow-through on-line	-	1.91	Serum, egg white	pCEC-UV (capLC)	[98]
Glycoprotein	Horseradish peroxidase (HRP)	Monolith: VPBA/PETA/AIBN/ethylene glycol:cyclohexanol, in capillary at 75 °C for 12 h; MIP monolith: immobilization of HRP on VPBA-based monolith and poly-dopamine (pDA) coating with DA and APS.	Rod monolith in fused silica capillary (25 cm × 75 μm ID).	Flow-through on-line	-	2.76	Serum	pCEC-UV	[99]
Aflatoxins	5,7-Dimethoxy-coumarin	Monolith: γ-MAPS/TRIM/BME or AIBN/2,2,4-trimethylpentane/toluene (80:20, w/w), UV for 1 h and 60 °C for 24 h MIP monolith: MAA/EGDMA/AIBN/toluene, template/MAA/EGDMA (0.3:4:20), UV, 1 h	Rod monolith in UV transparent fused capillary (70 mm × 0.1 mm ID).	Flow-through on-line	-	-	Water	MicroLC-LIF	[100]
Cocaine and its metabolite	Cocaine	MIP monolith: MAA/EGDMA/AIBN/acetonitrile-isooctane (9:1), 60 °C for 24 h template/MAA/EGDMA (1:4:20).	Inner-surface-coated fused capillary (50 mm × 0.1 mm ID).	Flow-through on-line	LOD: 14.5–6.1 ng mL^{-1}	2.2–3.2	Plasma, saliva	NanoLC-UV	[101]
Cannabinoids	Hydrogenated cannabidiol	MIP monolith: MAA/EGDMA/AIBN/CH$_2$Cl$_2$, 60 °C for 24 h, template/MAA (1:3).	Inner-surface-coated fused capillary (10 cm × 0.53 mm ID).	Flow-through on-line	LLOQ: 10 ng mL^{-1}	-	Plasma	UHPLC-MS/MS	[102]

[1] Functional monomer: MAA, methacrylic acid; VP, 4-vinylpyridine; AAm, acrylamide; APS, 3-aminopropyltriethoxysilane; VPBA: 4-vinylphenylboronic acid; PY, pyrrole; VI, vinylimidazole. Crosslinker: EGDMA, ethylene glycol dimethacrylate; EDMA, ethylene dimethacrylate; MBA, N,N-methylene bisacrylamide; TRIM, trimethylolpropane trimethacrylate; TEOS, tetraethoxysilane; PETA: pentaerythritol triacrylate; TEPM, 3-(triethoxysilyl)propyl methacrylate; GDMA, glycerol dimethacrylate; GMA, glycidyl methacrylate. Initiator: AIBN, 2,2′-azobis-isobutyronitrile; BME, benzoin methyl ether; APS, ammonium persulfate; PPDS, potassium peroxodisulfate; A4-CA, 4,4′-azobis (4-cyanovaleric acid). Co-precursors PEG, poly(ethylene glycol; TMOS, tetramethyloxysilane; γ-MAPS, 3-methacryloxypropyltrimethoxysilane; [2] EF: enrichment factor, LOD: limit of detection; LLOQ: lower limit of quantification; LOQ: limit of quantification; [3] IF: imprinting factor (MIP/NIP, molecularly imprinted capillary/non-imprinted capillary). [4] pCEC: pressurized capillary electrochromatography; DAD, diode-array UV detector; LIF, laser induced fluorescence detector.

4.1. Selectivity and Extraction Efficiency of MIP IT-SPME

MIPs have been employed in IT-SPME systems to selectively recognize and extract compounds based on their template molecules. For instance, MIP capillaries were fabricated using the female hormone β-estradiol as a template, allowing for the selective extraction of estrogens. The MIP was synthesized by dissolving the template, functional monomer VP, and cross-linker EGDMA in a 1:6:30 ratio in dichloromethane and adding the polymerization initiator AIBN. The coating on the inner wall of a fused silica capillary was created by inserting fluorocarbon yarn into the capillary, filling it with the polymer preparation solution, polymerizing it at 50 °C, and then pulling the yarn out. Steroid hormones, environmental estrogens and related chemicals were analyzed using on-line IT-SPME HPLC-UV. The extraction effects were compared using the direct injection method, and IT-SPME methods with uncoated host capillaries, non-imprinted (NMIP) capillaries prepared without templates, and MIP capillaries. As shown in Figure 4, among the steroid hormones, natural and synthetic female hormones were selectively extracted and enriched. For various estrogens and related environmental chemicals, the imprint factor (IF: peak area ratio obtained using MIP and NMIP capillaries, MIP/NMIP) and the enrichment factor (EF: peak area ratio obtained using MIP IT-SPME and direct injection methods, MIP/Direct) were compared. Table 3 shows that molecular recognition ability was high for female hormones but lower for other steroid hormones. Compounds like genistein and bisphenol showed similar IFs, but IFs for nonylphenol and phthalate esters were low. These results suggest that compounds with phenolic hydroxyl groups and polycyclic skeletons are selectively recognized by MIP. Enrichment factors were also high for molecularly recognized compounds, but PCBs and DDT also had high EFs as well, indicating non-selective adsorption due to hydrophobic interactions with the polymer. Therefore, while MIP can extract and concentrate compounds similar in structure to the template in a group-selective manner, non-selective adsorption on the polymer itself must be considered.

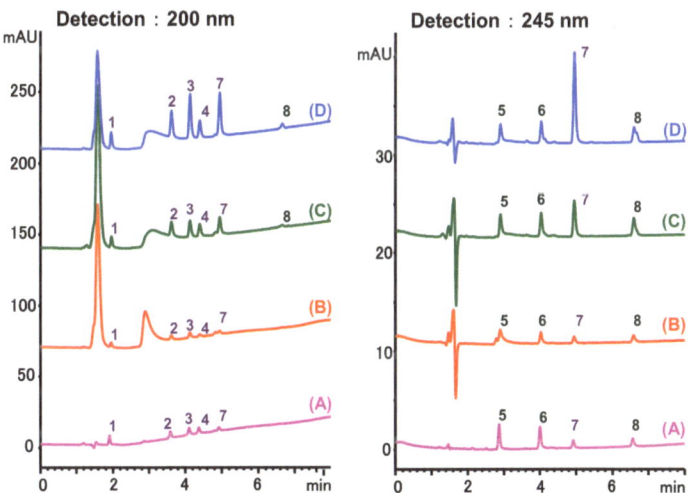

Figure 4. Chromatograms of steroid hormones obtained by HPLC-UV. (A) Direct injection, (B) IT-SPME using host capillary, (C) IT-SPME using NMIP, (D) IT-SPME using MIP. HPLC conditions: column, Eclipse SDB-C8 (150 × 4.6 mm ID, 5 μm particle size, Agilent Technologies, Santa Clara, CA, USA); gradient elution, acetonitrile/H_2O (45/55) 1 mL min^{-1} → acetonitrile/H_2O (65/35) 1.8 mL min^{-1} (8 min); column temperature, 40 °C; detection, UV at 200 and 245 nm. Peaks: 1 = estriol, 2 = β-estradiol, 3 = ethynylestradiol, 4 = diethylsilbestrol, 5 = corticosterone, 6 = testosterone, 7 = estrone, 8 = progesterone.

Table 3. Selectivity and enrichment effects of β-estradiol MIP on various compounds.

Compound	IF [1]	EF [2]	Compound	IF	EF	Compound	IF	EF
Estrone	2.78	16.4	Genistein	3.56	12.2	Nonylphenol	1.39	4.27
β-Estradiol	2.35	3.60	Bisphenol A	2.64	5.26	Di-n-butyl phthalate	1.21	2.54
Estriol	2.43	1.93	Progesterone	1.50	2.63	Di-2-ethylhexyl phthalate	0.86	1.45
Ethinylestradiol	2.29	5.60	Testosterone	0.96	1.03	Polychlorinated biphenyl (PCBs)	0.92	13.7
Diethylstilbestrol (DES)	1.79	2.63	Corticosterone	0.77	0.77	Dichlorodiphenyl-trichloroethane (DDT)	1.16	19.0

[1] IF: Imprinting factor (peak area ratio obtained using MIP and NMIP). [2] EF: Enrichment factor (peak area ratio obtained using MIP IT-SPME and direct injection).

4.2. Fabrications of Various MIP Capillaries and Their Applications to IT-SPME

Inner-surface-coated MIP capillaries have been utilized for both off-line [88,89] and on-line [90–92] IT-SPME. Zarejousheghani et al. [88] developed an open-tubular MIP capillary by inserting a metal rod into a glass capillary, synthesizing the MIP inside the capillary, then removing the metal rod (Figure 5a–c). The metal rod controlled the thickness of the polymer phase during in situ synthesis resulting in a robust and mechanically stable MIP tube (Figure 5d). This technique was applied to a selective off-line IT-SPME method for 4-nitrophenol in water. Souza et al. [89] synthesized a new molecule modified with RAM in an open-tubular fused silica capillary imprinted polymer (RAM-MIP). This extraction capillary, directly connected to a syringe for IT-SPME, was applied to the determination of parabens in breast milk by UHPLC–MS/MS analysis. Asiabia et al. [90] prepared nanostructured copolymers composed of polypyrrole-EGDMA copolymers on the inner surface of stainless-steel tubes through electrochemical synthesis. These were applied to selective analysis of indomethacin in urine and plasma by on-line MIP IT-SPME HPLC-UV. Recently, Song et al. [92] synthesized MIPs mixed with magnetic nanoparticles (Fe_3O_4) in situ in a capillary using 2,4-dinitroaniline (2,4-DNA) as a model template. A magnetic coil wrapped around this MIP-based microextraction tube generated a variable magnetic field during on-line IT-SPME, selectively adsorbing and desorbing 2,4-DNA in environmental water.

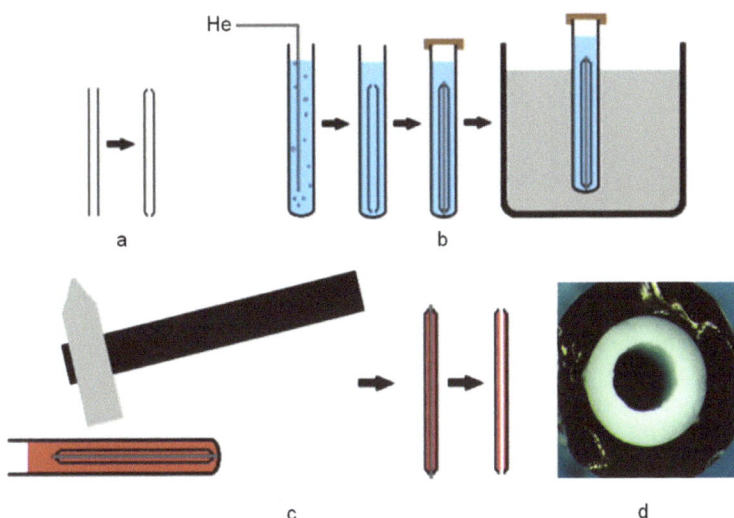

Figure 5. Open-tubular MIP-capillary preparation. (**a**) Both the tips of the glass-capillary were coned with flame to the diameter size of the desired metal rod. (**b**) The prepared assembly was placed in a bigger capillary that contained polymer mixture. (**c**) After the polymerization, the metal rod was removed from the polymer. (**d**) Magnified cross section of the polymer tube inside the 20 μL capillary glass. Reproduced from Figures 1 and 3 of Ref. [88] with permission from Elsevier.

Packed MIP capillaries have been employed in on-line IT-SPME using both particle-packed capillaries [87,93] and fiber-packed capillaries [94]. Chaves et al. [93] synthesized protein-template MIPs using a sol-gel method and developed a gentle template removal method using proteases. MIP particles of interferon alpha 2a were packed in PEEK tubes, leading to the development of an on-line IT-SPME HPLC fluorescence detection method. These MIPs functioned like other selective interferon alfa 2a immobilized phases (e.g., immunosorbents and access-limiting substances), were robust, easy to handle, inexpensive to synthesize, and applicable to the analysis of small plasma samples (50 μL). Hu et al. [94] developed a multi-fiber-packed on-line IT-SPME method, where MIP fibers were packed longitudinally in PEEK tubes (Figure 6). This design reduced back pressure, accelerated reaction rates, and increased extraction capacity by increasing the coating volume. The method was applied to HPLC-UV analysis of fluoroquinolones and sulfonamides in animal food samples, achieving high selectivity and sensitivity.

Monolithic MIP capillaries have been used in both off-line [95,96] and on-line [97–102] IT-SPME. Lei et al. [96] used surface imprinting technology to develop a dummy template (Pro-Tyr-Ile-Leu) to fabricate novel MIP monoliths in capillaries. This off-line method, directly connected to a syringe, exhibited high selectivity for target peptides and low detection limits, and was applied to selective MIP IT-SPME HPLC-UV analysis of neuropeptide neurotensin and neuromedin N in human plasma samples. Lin et al. [98] fabricated a molecularly imprinted inorganic-organic hybrid monolithic capillary using lysozyme as a template in combination with a rigid silica matrix and a flexible organic hydrogel by one-pot process (Figure 7). The resulting highly porous, uniform monolithic matrix is firmly fixed to the inside of the capillary (Figure 8), forming a stable, accessible recognition site, and promoting template re-binding with high imprint coefficients. This method allows selective separation of lysozyme from egg white and human serum through on-line coupling of MIP IT-SPME and pressurized capillary electrochromatography (pCEC)-UV. Szumski et al. [100] synthesized MIPs with high imprinting factors in poly(trimethylolpropane trimethacrylate) (poly-TRIM) core monoliths using 5,7-Dimethoxycoumarin as a dummy template of aflatoxin. This technique was successfully used in the selective separation of aflatoxins B1, B2, G1, and G2 in aqueous solution by microLC-laser induced fluorescence detection. Additionally, Marchioni et al. [102] synthesized durable MIP monoliths in a fused silica capillary using hydrogenated cannabidiol (CBD) as a dummy template through in situ polymerization. An automated analytical method using online coupling of IT-SPME and UHPLC-MS/MS with this MIP adsorbent as the extraction device was successfully applied to determine cannabinoid concentrations in the plasma of patients undergoing CBD treatment.

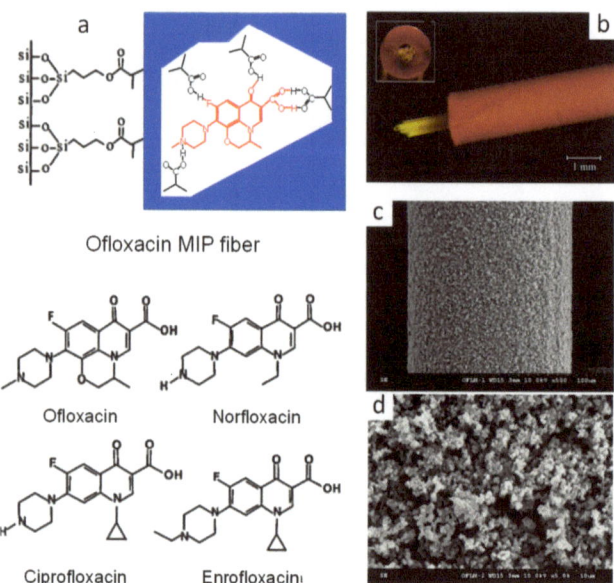

Figure 6. MIP fibers and fiber-packed tubes that recognize fluoroquinolones. (**a**) The chemical structure of fluoroquinolones and the schematic diagram of the resultant MIP structure. (**b**) Micrograph of the multiple-fiber-packed tube. (**c**,**d**) SEM images of the ofloxacin MIP fiber. Reproduced from Figures 1–3 of Ref. [94] with permission from Elsevier.

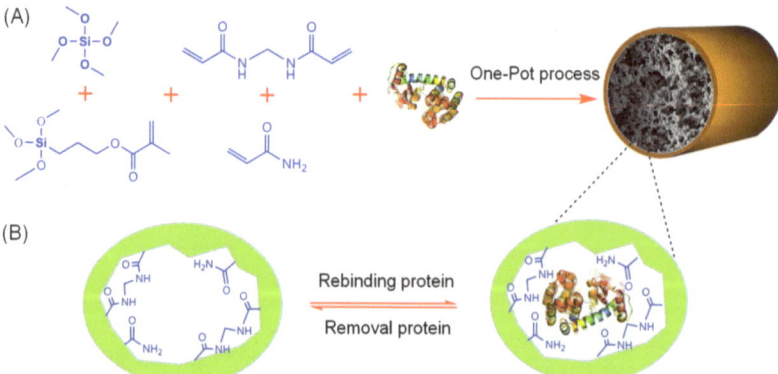

Figure 7. Schematic representation of (**A**) one-pot synthesis of protein-imprinted hybrid monolithic column and (**B**) the recognition mechanism between template protein and functional monomers. Reproduced from Figure S1 of Ref. [98] with permission from Elsevier.

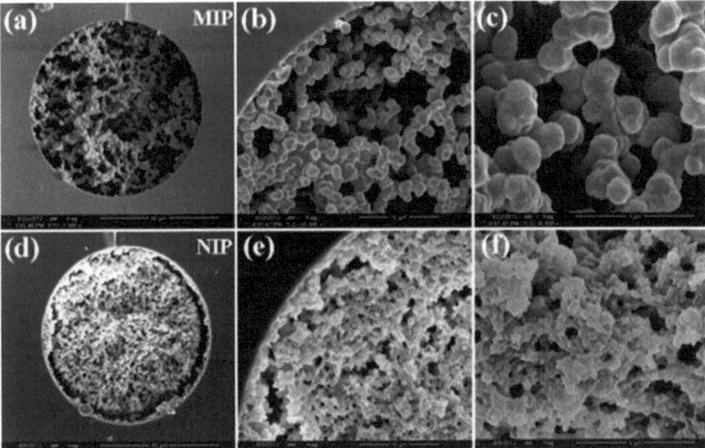

Figure 8. SEM images of the Lyz-MIP (**a**–**c**) and NIP (**d**–**f**) hybrid monolithic columns. (**a**,**d**) 3000×, (**b**,**e**) 10,000×, (**c**,**f**) 30,000×. Reproduced from Figure 1 of Ref. [98] with permission from Elsevier.

Thus, MIPs have been widely employed as a capillary stationary phase in IT-SPME, enabling selective extraction and enrichment of various compounds in complex matrices. As discussed in Section 3, while advancements in MIP design, such as single, dual, multi-template adsorbents, and hybrid materials have improved throughput and efficiency, their application to IT-SPME requires further study. Future, developments are anticipated by integrating novel functional MIP adsorbents with IT-SPME.

5. Conclusions and Perspective

While highly sensitive and high-performance analytical instruments are essential for accurately and precisely measuring trace constituents in complex samples, their effectiveness largely depends on efficient sample pretreatment. The MIP-based IT-SPME method introduced in this review holds great promise as an innovative sample preparation technique due to its various advantageous features:

- High selectivity and extraction efficiency: MIPs have specific binding sites that are complementary to the structure and functional groups of the target analyte, enabling highly selective adsorption. This results in the selective extraction of analytes from complex matrix samples, reducing matrix interference and improving the sensitivity and accuracy of analysis.
- High stability and reusability: MIPs offer superior chemical and physical stability compared to biorecognition materials such as antibodies and enzymes. They can withstand harsh conditions such as exposure to organic solvents and extreme pH environments. They can also be reused after washing, making them suitable for continuous analysis in IT-SPME.
- Ease of multifunctional modification: The outer surface of MIPs can be easily customized by incorporating various monomers to provide specific functional groups. This enables the enhancement of performance by creating hybrid materials, such as the integration of magnetic nanoparticles to for improved recycling or the inclusion of RAM to limit molecular permeability. Combining MIPs with other highly porous materials can also increase surface area and porosity, facilitating mass transfer of target analytes to binding sites.
- Cost reduction due to miniaturization: IT-SPME is cost-effective due to the small extraction phase of the capillary, which allows for rapid and efficient extraction and concentration. The miniaturization of MIP capillaries significantly reduces sample

volume and solvent consumption. Additionally, MIPs are relatively simple and inexpensive to synthesize, facilitating mass production.
- Labor-saving through high throughput: IT-SPME can automate the extraction, desorption, and introduction of compounds into analytical instruments on-line using column-switching techniques. This automation enables high-throughput analysis of a large numbers of samples, saving labor and time.
- Environmental friendliness: The MIP-based IT-SPME method uses minimal organic solvents, reducing health hazards to analysts, minimizing waste, and promoting environmentally friendly sample preparation. However, MIP preparation still requires the use of hazardous organic solvents.

Despite these excellent features, some challenges remain for the effective use of MIP IT-SPME in sample preparation. Issues such as unpredictable component leakage, limited reusability, scale-up constraints, and irreversible adsorption of adsorbates into MIP pores must be addressed. Additionally, IT-SPME faces challenges related to capillary structure and extraction efficiency, including capillary tube clogging, detachment of coating materials, capillary breakage, and the need to improve extraction rates and times.

To further develop and expand the use of MIP IT-SPME, future efforts should focus on overcoming these challenges, improving MIP selectivity and extraction efficiency, enhancing the multifunctionality and speed of the extraction process, facilitating on-line coupling with various analytical instruments, and developing high-throughput, highly sensitive analytical systems. Finally, we hope this review will inspire new ideas and innovations among researchers working in related scientific fields and encourage the application of MIP IT-SPME technology across a broad range of disciplines.

Author Contributions: Conceptualization, H.K.; methodology, H.K.; software, H.K.; validation, H.K.; formal analysis, H.K.; investigation, H.K.; resources, H.K.; data curation, H.K.; writing—original draft preparation, H.K.; writing—review and editing, H.K., A.I., K.S. and K.E.; visualization, H.K.; supervision, H.K.; project administration, H.K.; funding acquisition, H.K. All authors have read and agreed to the published version of the manuscript.

Funding: This research was funded by JSPS KAKENHI (Grant Number JP23K06091, Japan) and the Smoking Research Foundation 2024 (Tokyo, Japan).

Institutional Review Board Statement: Not applicable.

Informed Consent Statement: Not applicable.

Data Availability Statement: No new data were created or analyzed in this study.

Conflicts of Interest: The authors declare no conflicts of interest.

References

1. Kataoka, H.; Saito, K.; Yokoyama, A. Sampling and Sample Preparation for Clinical and Pharmaceutical Analysis. In *Handbook of Sample Preparation*; Pawliszyn, J., Lord, H.L., Eds.; John Wiley and Sons: Hoboken, NJ, USA, 2011; pp. 285–311, ISBN 978-0470099346. [CrossRef]
2. Kataoka, H. Sample Preparation for Liquid Chromatography. In *Handbooks in Separation Science: Liquid Chromatography*, 3rd ed.; Fanali, S., Chankvetadze, B., Haddad, P.R., Poole, C., Riekkola, M.-L., Eds.; Elsevier: Amsterdam, The Netherlands, 2023; Chapter 1; Volume 2, pp. 1–49, ISBN 978-0323999694. [CrossRef]
3. Kataoka, H. In-tube solid-phase microextraction: Current trends and future perspectives. *J. Chromatogr. A* **2021**, *1636*, 461787. [CrossRef] [PubMed]
4. Xu, L.; Hu, Z.S.; Duan, R.; Wang, X.; Yang, Y.S.; Dong, L.Y.; Wang, X.H. Advances and applications of in-tube solid-phase microextraction for analysis of proteins. *J. Chromatogr. A* **2021**, *1640*, 461962. [CrossRef] [PubMed]
5. Manousi, N.; Tzanavaras, P.D.; Zacharis, C.K. Bioanalytical HPLC Applications of In-Tube Solid Phase Microextraction: A Two-Decade Overview. *Molecules* **2020**, *25*, 2096. [CrossRef] [PubMed]
6. Zhao, G.; Zhang, Y.; Sun, D.; Yan, S.; Wen, Y.; Li, G.; Liu, H.; Li, J.; Song, Z. Recent Advances in Molecularly Imprinted Polymers for Antibiotic Analysis. *Molecules* **2023**, *28*, 335. [CrossRef] [PubMed]
7. Turiel, E.; Martín-Esteban, A. Molecularly imprinted polymers for sample preparation: A review. *Anal. Chim. Acta* **2010**, *668*, 87–99. [CrossRef]

8. Sarafraz-Yazdi, A.; Razavi, N. Application of molecularly-imprinted polymers in solid-phase microextraction techniques. *TrAC Trends Anal. Chem.* **2015**, *73*, 81–90. [CrossRef]
9. Ansari, S.; Karim, M. Recent progress, challenges and trends in trace determination of drug analysis using molecularly imprinted solid-phase microextraction technology. *Talanta* **2017**, *164*, 612–625. [CrossRef]
10. Turiel, E.; Martín-Esteban, A. Molecularly imprinted polymers-based microextraction techniques. *TrAC Trends Anal. Chem.* **2019**, *118*, 574–586. [CrossRef]
11. Hammad, S.F.; Abdallah, I.A.; Bedair, A.; Mansour, F.R. Homogeneous liquid-liquid extraction as an alternative sample preparation technique for biomedical analysis. *J. Sep. Sci.* **2022**, *45*, 185–209. [CrossRef]
12. Poole, C.F. Core concepts and milestones in the development of solid-phase extraction. In *Solid-Phase Extraction*; Poole, C.F., Ed.; Elsevier: Amsterdam, The Netherlands, 2020; Chapter 1; pp. 1–36, ISBN 978-0-12-816906-3. [CrossRef]
13. Hamidi, S. Recent Advances in Solid-Phase Extraction as a Platform for Sample Preparation in Biomarker Assay. *Crit. Rev. Anal. Chem.* **2023**, *53*, 199–210. [CrossRef]
14. Mahdavijalal, M.; Petio, C.; Staffilano, G.; Mandrioli, R.; Protti, M. Innovative Solid-Phase Extraction Strategies for Improving the Advanced Chromatographic Determination of Drugs in Challenging Biological Samples. *Molecules* **2024**, *29*, 2278. [CrossRef] [PubMed]
15. Hamidi, S.; Taghvimi, A.; Mazouchi, N. Micro Solid Phase Extraction Using Novel Adsorbents. *Crit. Rev. Anal. Chem.* **2021**, *51*, 103–114. [CrossRef] [PubMed]
16. Bedair, A.; Abdelhameed, R.M.; Hammad, S.F.; Abdallah, I.A.; Mansour, F.R. Applications of metal organic frameworks in dispersive micro solid phase extraction (D-µ-SPE). *J. Chromatogr. A* **2024**, *1732*, 465192. [CrossRef] [PubMed]
17. Souza-Silva, É.A.; Jiang, R.; Rodríguez-Lafuente, A.; Gionfriddo, E.; Pawliszyn, J. A critical review of the state of the art of solid-phase microextraction of complex matrices I. Environmental analysis. *TrAC Trends Anal. Chem.* **2015**, *71*, 224–235. [CrossRef]
18. Li, N.; Zhang, Z.; Li, G. Recent advance on microextraction sampling technologies for bioanalysis. *J. Chromatogr. A* **2024**, *1720*, 464775. [CrossRef]
19. Boyacı, E.; Rodríguez-Lafuente, Á.; Gorynski, K.; Mirnaghi, F.; Souza-Silva, É.A.; Hein, D.; Pawliszyn, J. Sample preparation with solid phase microextraction and exhaustive extraction approaches: Comparison for challenging cases. *Anal. Chim. Acta* **2015**, *873*, 14–30. [CrossRef]
20. Owczarzy, A.; Kulig, K.; Piordas, K.; Piśla, P.; Sarkowicz, P.; Rogóż, W.; Maciążek-Jurczyk, M. Solid-phase microextraction—A future technique in pharmacology and coating trends. *Anal Methods* **2024**, *16*, 3164–3178. [CrossRef]
21. Jin, H.-F.; Shi, Y.; Cao, J. Recent advances and applications of novel advanced materials in solid-phase microextraction for natural products. *TrAC Trends Anal. Chem.* **2024**, *178*, 117858. [CrossRef]
22. Camino-Sánchez, F.J.; Rodríguez-Gómez, R.; Zafra-Gómez, A.; Santos-Fandila, A.; Vílchez, J.L. Stir bar sorptive extraction: Recent applications, limitations and future trends. *Talanta* **2014**, *130*, 388–399. [CrossRef]
23. Hasan, C.K.; Ghiasvand, A.; Lewis, T.W.; Nesterenko, P.N.; Paull, B. Recent advances in stir-bar sorptive extraction: Coatings, technical improvements, and applications. *Anal. Chim. Acta* **2020**, *1139*, 222–240. [CrossRef]
24. He, M.; Wang, Y.; Zhang, Q.; Zang, L.; Chen, B.; Hu, B. Stir bar sorptive extraction and its application. *J. Chromatogr. A* **2021**, *1637*, 461810. [CrossRef] [PubMed]
25. Yang, L.; Said, R.; Abdel-Rehim, M. Sorbent, device, matrix and application in microextraction by packed sorbent (MEPS): A review. *J. Chromatogr. B Anal. Technol. Biomed. Life Sci.* **2017**, *1043*, 33–43. [CrossRef] [PubMed]
26. Sun, H.; Feng, J.; Han, S.; Ji, X.; Li, C.; Feng, J.; Sun, M. Recent advances in micro- and nanomaterial-based adsorbents for pipette-tip solid-phase extraction. *Mikrochim. Acta* **2021**, *188*, 189. [CrossRef] [PubMed]
27. Chen, L.X.; Yang, F.Q. Applications of magnetic solid-phase extraction in the sample preparation of natural product analysis (2020–2023). *J. Sep. Sci.* **2024**, *47*, e2400082. [CrossRef] [PubMed]
28. Spietelun, A.; Marcinkowski, Ł.; de la Guardia, M.; Namieśnik, J. Green aspects, developments and perspectives of liquid phase microextraction techniques. *Talanta* **2014**, *119*, 34–45. [CrossRef]
29. Hansen, F.; Øiestad, E.L.; Pedersen-Bjergaard, S. Bioanalysis of pharmaceuticals using liquid-phase microextraction combined with liquid chromatography-mass spectrometry. *J. Pharm. Biomed. Anal.* **2020**, *189*, 113446. [CrossRef]
30. Bayatloo, M.R.; Tabani, H.; Nojavan, S.; Alexovič, M.; Ozkan, S.A. Liquid-Phase Microextraction Approaches for Preconcentration and Analysis of Chiral Compounds: A Review on Current Advances. *Crit. Rev. Anal. Chem.* **2023**, *53*, 1623–1637. [CrossRef]
31. Moliner-Martinez, Y.; Herráez-Hernández, R.; Verdú-Andrés, J.; Molins-Legua, C.; Campíns-Falcó, P. Recent advances of in-tube solid-phase microextraction. *TrAC Trends Anal. Chem.* **2015**, *71*, 205–213. [CrossRef]
32. Fernández-Amado, M.; Prieto-Blanco, M.C.; López-Mahía, P.; Muniategui-Lorenzo, S.; Prada-Rodríguez, D. Strengths and weaknesses of in-tube solid-phase microextraction: A scoping review. *Anal. Chim. Acta* **2016**, *906*, 41–57. [CrossRef]
33. Moliner-Martinez, Y.; Ballester-Caudet, A.; Verdú-Andrés, J.; Herráez-Hernández, R.; Molins-Legua, C.; Campíns-Falcó, P. In-tube solid-phase microextraction. In *Solid-Phase Extraction*; Poole, C.F., Ed.; Elsevier: Amsterdam, The Netherlands, 2019; pp. 387–427, ISBN 978-0-12-816906-3. [CrossRef]
34. Queiroz, M.E.; Melo, L.P. Selective capillary coating materials for in-tube solid-phase microextraction coupled to liquid chromatography to determine drugs and biomarkers in biological samples: A review. *Anal. Chim. Acta* **2014**, *826*, 1–11. [CrossRef]

35. Ponce-Rodríguez, H.D.; Verdú-Andrés, J.; Herráez-Hernández, R.; Campíns-Falcó, P. Innovations in Extractive Phases for In-Tube Solid-Phase Microextraction Coupled to Miniaturized Liquid Chromatography: A Critical Review. *Molecules* **2020**, *25*, 2460. [CrossRef] [PubMed]
36. Grecco, C.F.; de Souza, I.D.; Queiroz, M.E.C. Novel materials as capillary coatings for in-tube solid-phase microextraction for bioanalysis. *J. Sep. Sci.* **2021**, *44*, 1662–1693. [CrossRef] [PubMed]
37. Souza, I.D.; Oliveira, I.G.C.; Queiroz, M.E.C. Innovative extraction materials for fiber-in-tube solid phase microextraction: A review. *Anal. Chim. Acta* **2021**, *1165*, 238110. [CrossRef] [PubMed]
38. Souza, I.D.; Queiroz, M.E.C. Organic-silica hybrid monolithic sorbents for sample preparation techniques: A review on advances in synthesis, characterization, and applications. *J. Chromatogr. A* **2024**, *1713*, 464518. [CrossRef] [PubMed]
39. Clark, K.D.; Emaus, M.N.; Varona, M.; Bowers, A.N.; Anderson, J.L. Ionic liquids: Solvents and sorbents in sample preparation. *J. Sep. Sci.* **2018**, *41*, 209–235. [CrossRef]
40. de Faria, H.D.; Abrão, L.C.; Santos, M.G.; Barbosa, A.F.; Figueiredo, E.C. New advances in restricted access materials for sample preparation: A review. *Anal. Chim. Acta* **2017**, *959*, 43–65. [CrossRef]
41. Hu, Y.; Pan, J.; Zhang, K.; Lian, H.; Li, G. Novel applications of molecularly-imprinted polymers in sample preparation. *TrAC Trends Anal. Chem.* **2013**, *43*, 37–52. [CrossRef]
42. Moein, M.M.; Abdel-Rehim, M. Molecularly imprinted polymers for on-line extraction techniques. *Bioanalysis* **2015**, *7*, 2145–2153. [CrossRef]
43. Gama, M.R.; Bottoli, C.B. Molecularly imprinted polymers for bioanalytical sample preparation. *J. Chromatogr. B Anal. Technol. Biomed. Life Sci.* **2017**, *1043*, 107–121. [CrossRef]
44. Speltini, A.; Scalabrini, A.; Maraschi, F.; Sturini, M.; Profumo, A. Newest applications of molecularly imprinted polymers for extraction of contaminants from environmental and food matrices: A review. *Anal. Chim. Acta* **2017**, *974*, 1–26. [CrossRef]
45. Huang, S.; Xu, J.; Zheng, J.; Zhu, F.; Xie, L.; Ouyang, G. Synthesis and application of magnetic molecularly imprinted polymers in sample preparation. *Anal. Bioanal. Chem.* **2018**, *410*, 3991–4014. [CrossRef] [PubMed]
46. Moein, M.M.; Abdel-Rehim, A.; Abdel-Rehim, M. Recent Applications of Molecularly Imprinted Sol-Gel Methodology in Sample Preparation. *Molecules* **2019**, *24*, 2889. [CrossRef] [PubMed]
47. Azizi, A.; Bottaro, C.S. A critical review of molecularly imprinted polymers for the analysis of organic pollutants in environmental water samples. *J. Chromatogr. A* **2020**, *1614*, 460603. [CrossRef] [PubMed]
48. Martín-Esteban, A. Green molecularly imprinted polymers for sustainable sample preparation. *J. Sep. Sci.* **2022**, *45*, 233–245. [CrossRef] [PubMed]
49. Suzaei, F.M.; Daryanavard, S.M.; Abdel-Rehim, A.; Bassyouni, F.; Abdel-Rehim, M. Recent molecularly imprinted polymers applications in bioanalysis. *Chem. Zvesti.* **2023**, *77*, 619–655. [CrossRef]
50. de Toffoli, A.L.; Maciel, E.V.S.; Fumes, B.H.; Lanças, F.M. The role of graphene-based sorbents in modern sample preparation techniques. *J. Sep. Sci.* **2018**, *41*, 288–302. [CrossRef]
51. Maciel, E.V.S.; Mejía-Carmona, K.; Jordan-Sinisterra, M.; da Silva, L.F.; Vargas Medina, D.A.; Lanças, F.M. The Current Role of Graphene-Based Nanomaterials in the Sample Preparation Arena. *Front. Chem.* **2020**, *8*, 664. [CrossRef]
52. Borsatto, J.V.B.; Lanças, F.M. Recent Trends in Graphene-Based Sorbents for LC Analysis of Food and Environmental Water Samples. *Molecules* **2023**, *28*, 5134. [CrossRef]
53. Herrero-Latorre, C.; Barciela-García, J.; García-Martín, S.; Peña-Crecente, R.M.; Otárola-Jiménez, J. Magnetic solid-phase extraction using carbon nanotubes as sorbents: A review. *Anal. Chim. Acta* **2015**, *892*, 10–26. [CrossRef]
54. Ng, N.T.; Kamaruddin, A.F.; Wan Ibrahim, W.A.; Sanagi, M.M.; Abdul Keyon, A.S. Advances in organic-inorganic hybrid sorbents for the extraction of organic and inorganic pollutants in different types of food and environmental samples. *J. Sep. Sci.* **2018**, *41*, 195–208. [CrossRef]
55. Öztürk, E.E.; Dalgıç Bozyiğit, G.; Büyükpınar, Ç.; Bakırdere, S. Magnetic Nanoparticles Based Solid Phase Extraction Methods for the Determination of Trace Elements. *Crit. Rev. Anal. Chem.* **2022**, *52*, 231–249. [CrossRef] [PubMed]
56. Pérez-Cejuela, H.M.; Herrero-Martínez, J.M.; Simó-Alfonso, E.F. Recent Advances in Affinity MOF-Based Sorbents with Sample Preparation Purposes. *Molecules* **2020**, *25*, 4216. [CrossRef] [PubMed]
57. Vardali, S.C.; Manousi, N.; Barczak, M.; Giannakoudakis, D.A. Novel Approaches Utilizing Metal-Organic Framework Composites for the Extraction of Organic Compounds and Metal Traces from Fish and Seafood. *Molecules* **2020**, *25*, 513. [CrossRef] [PubMed]
58. González-Sálamo, J.; Jiménez-Skrzypek, G.; Ortega-Zamora, C.; González-Curbelo, M.Á.; Hernández-Borges, J. Covalent Organic Frameworks in Sample Preparation. *Molecules* **2020**, *25*, 3288. [CrossRef] [PubMed]
59. Torabi, E.; Mirzaei, M.; Bazargan, M.; Amiri, A. A critical review of covalent organic frameworks-based sorbents in extraction methods. *Anal. Chim. Acta* **2022**, *1224*, 340207. [CrossRef]
60. Hu, T.; Chen, R.; Wang, Q.; He, C.; Liu, S. Recent advances and applications of molecularly imprinted polymers in solid-phase extraction for real sample analysis. *J. Sep. Sci.* **2021**, *44*, 274–309. [CrossRef]
61. Silva, M.S.; Tavares, A.P.M.; de Faria, H.D.; Sales, M.G.F.; Figueiredo, E.C. Molecularly Imprinted Solid Phase Extraction Aiding the Analysis of Disease Biomarkers. *Crit. Rev. Anal. Chem.* **2022**, *52*, 933–948. [CrossRef]
62. Sobiech, M.; Luliński, P. Molecularly imprinted solid phase extraction—Recent strategies, future prospects and forthcoming challenges in complex sample pretreatment process. *TrAC Trends Anal. Chem.* **2024**, *174*, 117695. [CrossRef]

63. Zhang, M.; Zeng, J.; Wang, Y.; Chen, X. Developments and trends of molecularly imprinted solid-phase microextraction. *J. Chromatogr. Sci.* **2013**, *51*, 577–586. [CrossRef]
64. Shahhoseini, F.; Azizi, A.; Bottaro, C.S. A critical evaluation of molecularly imprinted polymer (MIP) coatings in solid phase microextraction devices. *TrAC Trends Anal. Chem.* **2022**, *156*, 116695. [CrossRef]
65. Díaz-Álvarez, M.; Turiel, E.; Martín-Esteban, A. Recent advances and future trends in molecularly imprinted polymers-based sample preparation. *J. Sep. Sci.* **2023**, *46*, 2300157. [CrossRef] [PubMed]
66. Martins, R.O.; Batista Junior, A.C.; Brito, C.C.S.M.; Rocha, Y.A.; Chaves, A.R. Molecularly imprinted polymers in solid-phase microextraction: Enhancing the analysis of pharmaceutical compounds across diverse sample matrices. *J. Pharm. Biomed. Anal. Open* **2024**, *4*, 100037. [CrossRef]
67. Eisert, R.; Pawliszyn, J. Automated In-Tube Solid-Phase Microextraction Coupled to High-Performance Liquid Chromatography. *Anal. Chem.* **1997**, *69*, 3140–3147. [CrossRef]
68. Pawliszyn, J.; Arthur, C.L. Solid Phase Microextraction with Thermal Desorption Using Fused Silica Optical Fibers. *Anal. Chem.* **1990**, *62*, 2145–2148. [CrossRef]
69. Kataoka, H. Automated sample preparation using in-tube solid-phase microextraction and its application—A review. *Anal. Bioanal. Chem.* **2002**, *373*, 31–45. [CrossRef]
70. Kataoka, H.; Ishizaki, A.; Saito, K. Online column-switching sample preparation for liquid chromatography. In *Reference Module in Chemistry, Molecular Sciences and Chemical Engineering, Comprehensive Sampling and Sample Preparation*, 2nd ed.; Soylak, M., Ed.; Elsevier: Amsterdam, The Netherlands, 2024; CH 00061 Chapter 14547, ISBN 978-0-12-409547-2. [CrossRef]
71. Moliner-Martínez, Y.; Prima-Garcia, H.; Ribera, A.; Coronado, E.; Campíns-Falcó, P. Magnetic in-tube solid phase microextraction. *Anal. Chem.* **2012**, *84*, 7233–7240. [CrossRef]
72. Safari, M.; Yamini, Y. Application of magnetic nanomaterials in magnetic in-tube solid-phase microextraction. *Talanta* **2021**, *221*, 121648. [CrossRef]
73. Ahmadi, S.H.; Manbohi, A.; Heydar, K.T. Electrochemically controlled in-tube solid phase microextraction. *Anal. Chim. Acta* **2015**, *853*, 335–341. [CrossRef]
74. Asiabi, H.; Yamini, Y.; Shamsayei, M. Development of electrochemically controlled packed-in-tube solid phase microextraction method for sensitive analysis of acidic drugs in biological samples. *Talanta* **2018**, *185*, 80–88. [CrossRef]
75. Shamsayei, M.; Yamini, Y.; Asiabi, H. Electrochemically controlled fiber-in-tube solid-phase microextraction method for the determination of trace amounts of antipsychotic drugs in biological samples. *J. Sep. Sci.* **2018**, *41*, 3598–3606. [CrossRef]
76. Yu, Q.-W.; Ma, Q.; Feng, Y.-Q. Temperature-response polymer coating for in-tube solid-phase microextraction coupled to high-performance liquid chromatography. *Talanta* **2011**, *84*, 1019–1025. [CrossRef] [PubMed]
77. Yang, Y.; Rodriguez-Lafuente, A.; Pawliszyn, J. Thermoelectric-based temperature-controlling system for in-tube solid-phase microextraction. *J. Sep. Sci.* **2014**, *37*, 1617–1621. [CrossRef] [PubMed]
78. Zhang, S.W.; Zou, C.J.; Luo, N.; Weng, Q.F.; Cai, L.S.; Wu, Y.C.; Xing, J. Determination of urinary 8-hydroxy-2′-deoxyguanosine by capillary electrophoresis with molecularly imprinted monolith in-tube solid phase microextraction. *Chin. Chem. Lett.* **2010**, *21*, 85–88. [CrossRef]
79. Shuo, Z.H.A.O.; Hao-Tian, W.A.N.G.; Ke, L.I.; Jing, Z.; Xia-Yan, W.A.N.G.; Guang-Sheng, G.U.O. Fast Determination of Residual Sulfonamides in Milk by In-Tube Solid-Phase Microextraction Coupled with Capillary Electrophoresis-Laser Induced Fluorescence. *Chin. J. Anal. Chem.* **2018**, *46*, e1810–e1816. [CrossRef]
80. Ren, Y.; Zhang, W.; Lin, Z.; Bushman, L.R.; Anderson, P.L.; Ouyang, Z. In-capillary microextraction for direct mass spectrometry analysis of biological samples. *Talanta* **2018**, *189*, 451–457. [CrossRef]
81. Asiabi, H.; Yamini, Y.; Seidi, S.; Shamsayei, M.; Safari, M.; Rezaei, F. On-line electrochemically controlled in-tube solid phase microextraction of inorganic selenium followed by hydride generation atomic absorption spectrometry. *Anal. Chim. Acta* **2016**, *922*, 37–47. [CrossRef]
82. Maciel, E.V.S.; de Toffoli, A.L.; Lanças, F.M. Current status and future trends on automated multidimensional separation techniques employing sorbent-based extraction columns. *J. Sep. Sci.* **2019**, *42*, 258–272. [CrossRef]
83. Lashgari, M.; Yamini, Y. An Overview of the Most Common Lab-Made Coating Materials in Solid Phase Microextraction. *Talanta* **2019**, *191*, 283–306. [CrossRef]
84. Wulff, G.; Sarhan, A. Über die Anwendung von enzymanalog gebauten Polymeren zur Racemattrennung. *Angew. Chem.* **1972**, *84*, 364. [CrossRef]
85. Arshady, R.; Mosbach, K. Synthesis of substrate-selective polymers by host-guest polymerization. *Die Makromol. Chem.* **1981**, *182*, 687–692. [CrossRef]
86. Hashemi, B.; Zohrabi, P.; Shamsipur, M. Recent developments and applications of different sorbents for SPE and SPME from biological samples. *Talanta* **2018**, *187*, 337–347. [CrossRef] [PubMed]
87. Mullett, W.M.; Martin, P.; Pawliszyn, J. In-tube molecularly imprinted polymer solid-phase microextraction for the selective determination of propranolol. *Anal. Chem.* **2001**, *73*, 2383–2389. [CrossRef] [PubMed]
88. Zarejousheghani, M.; Möder, M.; Borsdorf, H. A new strategy for synthesis of an in-tube molecularly imprinted polymer-solid phase microextraction device: Selective off-line extraction of 4-nitrophenol as an example of priority pollutants from environmental water samples. *Anal. Chim. Acta* **2013**, *798*, 48–55. [CrossRef] [PubMed]

89. Souza, I.D.; Melo, L.P.; Jardim, I.C.S.F.; Monteiro, J.C.S.; Nakano, A.M.S.; Queiroz, M.E.C. Selective molecularly imprinted polymer combined with restricted access material for in-tube SPME/UHPLC-MS/MS of parabens in breast milk samples. *Anal. Chim. Acta* **2016**, *932*, 49–59. [CrossRef] [PubMed]
90. Asiabi, H.; Yamini, Y.; Seidi, S.; Ghahramanifard, F. Preparation and evaluation of a novel molecularly imprinted polymer coating for selective extraction of indomethacin from biological samples by electrochemically controlled in-tube solid phase microextraction. *Anal. Chim. Acta* **2016**, *913*, 76–85. [CrossRef]
91. Kefayati, H.; Yamini, Y.; Shamsayei, M.; Abdi, S. Molecularly imprinted polypyrrole@CuO nanocomposite as an in-tube solid-phase microextraction coating for selective extraction of carbamazepine from biological samples. *J. Pharm. Biomed. Anal.* **2021**, *204*, 114256. [CrossRef]
92. Song, X.; Li, X.; Wang, J.; Huang, X. Adoption of new strategy for molecularly imprinted polymer based in-tube solid phase microextraction to improve specific recognition performance and extraction efficiency. *Microchem. J.* **2023**, *194*, 109224. [CrossRef]
93. Chaves, A.R.; Queiroz, M.E.C. In-tube solid-phase microextraction with molecularly imprinted polymer to determine interferon alpha 2a in plasma sample by high performance liquid chromatography. *J. Chromatogr. A* **2013**, *1318*, 43–48. [CrossRef]
94. Hu, Y.; Song, C.; Li, G. Fiber-in-tube solid-phase microextraction with molecularly imprinted coating for sensitive analysis of antibiotic drugs by high performance liquid chromatography. *J. Chromatogr. A* **2012**, *1263*, 21–27. [CrossRef]
95. Zhang, S.-W.; Xing, J.; Cai, L.-S.; Wu, C.-Y. Molecularly imprinted monolith in-tube solid-phase microextraction coupled with HPLC/UV detection for determination of 8-hydroxy-2′-deoxyguanosine in urine. *Anal. Bioanal. Chem.* **2009**, *395*, 479–487. [CrossRef]
96. Lei, X.; Huang, T.; Wu, X.; Mangelings, D.; Eeckhaut, A.V.; Bongaerts, J.; Terryn, H.; Heyden, Y.V. Fabrication of a molecularly imprinted monolithic column via the epitope approach for the selective capillary microextraction of neuropeptides in human plasma. *Talanta* **2022**, *243*, 123397. [CrossRef] [PubMed]
97. Courtois, J.; Fischer, G.; Sellergren, B.; Irgum, K. Molecularly imprinted polymers grafted to flow through poly(trimethylolpropane trimethacrylate) monoliths for capillary-based solid-phase extraction. *J. Chromatogr. A* **2006**, *1109*, 92–99. [CrossRef] [PubMed]
98. Lin, Z.; Lin, Y.; Sun, X.; Yang, H.; Zhang, L.; Chen, G. One-pot preparation of a molecularly imprinted hybrid monolithic capillary column for selective recognition and capture of lysozyme. *J. Chromatogr. A* **2013**, *1284*, 8–16. [CrossRef] [PubMed]
99. Lin, Z.; Wang, J.; Tan, X.; Sun, L.; Yu, R.; Yang, H.; Chen, G. Preparation of boronate-functionalized molecularly imprinted monolithic column with polydopamine coating for glycoprotein recognition and enrichment. *J. Chromatogr. A* **2013**, *1319*, 141–147. [CrossRef] [PubMed]
100. Szumski, M.; Grzywiński, D.; Prus, W.; Buszewski, B. Monolithic molecularly imprinted polymeric capillary columns for isolation of aflatoxins. *J. Chromatogr. A* **2014**, *1364*, 163–170. [CrossRef] [PubMed]
101. Bouvarel, T.; Delaunay, N.; Pichon, V. Selective extraction of cocaine from biological samples with a miniaturized monolithic molecularly imprinted polymer and on-line analysis in nano-liquid chromatography. *Anal. Chim. Acta* **2020**, *1096*, 89–99. [CrossRef]
102. Marchioni, C.; Vieira, T.M.; Crotti, A.E.M.; Crippa, J.A.; Queiroz, M.E.C. In-tube solid-phase microextraction with a dummy molecularly imprinted monolithic capillary coupled to ultra-performance liquid chromatography-tandem mass spectrometry to determine cannabinoids in plasma samples. *Anal. Chim. Acta* **2020**, *1099*, 145–154. [CrossRef]

Disclaimer/Publisher's Note: The statements, opinions and data contained in all publications are solely those of the individual author(s) and contributor(s) and not of MDPI and/or the editor(s). MDPI and/or the editor(s) disclaim responsibility for any injury to people or property resulting from any ideas, methods, instructions or products referred to in the content.

Article

Synthesis and Characterization of a Multi-Walled Carbon Nanotube–Ionic Liquid/Polyaniline Adsorbent for a Solvent-Free In-Needle Microextraction Method

Soyoung Ahn and Sunyoung Bae *

Department of Chemistry, Seoul Women's University, 621 Hwarang-ro, Nowon-gu, Seoul 01797, Republic of Korea; ahnso7179@naver.com
* Correspondence: sbae@swu.ac.kr; Tel.: +82-2-970-5652

Abstract: Sample preparation is an essential process when handling complex matrices. Extraction without using a solvent requires the direct transfer of analytes from the sample to the adsorbent either in the gas or liquid phase. In this study, a wire coated with a new adsorbent was fabricated for in-needle microextraction (INME) as a solvent-free sample extraction method. The wire inserted into the needle was placed in the headspace (HS), which was saturated with volatile organic compounds from the sample in a vial. A new adsorbent was synthesized via electrochemical polymerization by mixing aniline with multi-walled carbon nanotubes (MWCNTs) in the presence of an ionic liquid (IL). The newly synthesized adsorbent using IL is expected to achieve high thermal stability, good solvation properties, and high extraction efficiency. The characteristics of the electrochemically synthesized surfaces coated with MWCNT–IL/polyaniline (PANI) adsorbents were characterized using Fourier transform infrared (FTIR) spectroscopy, scanning electron microscopy (SEM), thermogravimetric analysis (TGA), and atomic force microscopy (AFM). Then, the proposed HS–INME–MWCNT–IL/PANI method was optimized and validated. Accuracy and precision were evaluated by analyzing replicates of a real sample containing phthalates, showing spike recovery between 61.13% and 108.21% and relative standard deviations lower than 15%. The limit of detection and limit of quantification of the proposed method were computed using the IUPAC definition as 15.84~50.56 µg and 52.79~168.5 µg, respectively. We concluded that HS–INME using a wire coated with the MWCNT–IL/PANI adsorbent could be repeatedly used up to 150 times without degrading its extraction performance in an aqueous solution; it constitutes an eco-friendly and cost-effective extraction method.

Keywords: polyaniline; in-needle microextraction; multi-walled carbon nanotube; ionic liquid; electrochemical polymerization

1. Introduction

Chemical analysis involves several steps, including sampling, extraction purification, and sample introduction [1–3]. Major issues are caused by interferences and low concentrations of analytes in the sample matrix. An extraction procedure may be required to separate and enrich the target analyte from the complicated matrixes [4]. The extraction methods used for sample analysis in the aqueous phase are liquid–liquid extraction, solid-phase extraction, solid phase microextraction, and in-needle microextraction [5–11].

In-needle microextraction (INME) was developed as a simple sample preparation method [11–16]. To fabricate the INME needle, a layer of adsorbent is coated inside the needle or outside the wire to be inserted into the needle [11–16]. The fabricated needle is attached to a syringe and directly placed in an aqueous solution or exposed in the headspace of the vial containing the analytes during the extraction process. The adsorbent can be synthesized using sol–gel polymerization [12–14] or electrochemical deposition [11,15–17].

Among the various adsorbents containing activated carbon, neutral zeolites, alumina, biochar, and clay minerals, polymeric substances have attracted a great deal of attention due to their simple synthesis reaction, adjustable morphology, and different functional groups [18]. For INME adsorbents, a polyaniline-based adsorbent has been investigated because it shows conducting properties, good mechanical stability, and high efficiency of adsorption, and it is abundant in nature. The chemical oxidative polymerization of aniline is initiated by adding an oxidant in a strong acidic solution. The mixture of multi-wall carbon nanotubes (MWCNTs) with PANI during synthesis exhibits a large surface area, high electrical conductivity, rich stacking π electrons, high mechanical strength, and outstanding chemical and thermal stability [11,19]. Due to the unique properties of these mixtures, MWCNTs show promise in a variety of applications such as sensors [20], sampling [21], solid-phase extraction [22], solid-phase microextraction (SPME) [23], chromatography [24], and the INME method [11]. However, widespread use of CNTs has been limited due to their agglomeration and weak coating stability [25].

In this study, a new adsorbent, synthesized with MWCNT–IL/PANI, was coated on a stainless-steel wire and inserted into the INME needle. IL was used as an additive to achieve a well-dispersed MWCNT due to its distinct characteristics, which include low toxicity, low melting points, high thermal stability, and good solvation properties [26–28]. CNT–IL hybrids are ideal for sampling applications [29–31]. The physicochemical properties of the synthesized adsorbent were characterized using Fourier transform infrared (FTIR) spectroscopy, thermogravimetric analysis (TGA), and atomic force microscopy (AFM). For the feasibility study, the HS–INME method followed by GC/MS was applied to hot water covered by low-density polyethylene (LLDPE) wrap, as an example of the real sample. The proposed HS–INME–MWCNT–IL/PANI method was optimized and validated according to metrics including the limit of detection (LOD), limit of quantification (LOQ), dynamic range, recovery, and reproducibility.

2. Results and Discussion

2.1. Optimization of the HS–INME–MWCNT–IL/PANI Coating Layer

Extraction efficiency can be affected by various parameters, including MWCNT–IL/PANI synthesis and the extraction conditions when using HS–INME followed by GC/MS. Synthesis conditions, such as the mixing ratio of MWCNTs and IL, the polymerization potential, the electrochemical deposition time, and the coating layer's length, were varied to compare the peak areas as well as standard deviations of the standard compounds. The HS–INME–GC/MS analysis conditions were investigated by changing the saturation time, extraction temperature, and adsorption time.

It can be inferred that the higher the peak area obtained, the higher the adsorption efficiency of the analyte on the adsorbent. The working solution used in the optimization experiment was run in an aqueous solution containing 20 µg mL^{-1} phthalates and all measurements were repeated three times.

2.1.1. Effect of Synthesis Conditions on Extraction

The peak area of phthalates for the HS–INME–MWCNT–IL/PANI method was determined according to the percentage (% w/v) of MWCNTs and IL within the range of 5% to 15%. As shown in Figure S1, the adsorption of phthalates was largely similar at all rates investigated in this study. However, the standard deviation of the peak areas of 5% and 15% was relatively higher than that of 10%. In addition, impurities were observed at 15%. As a result, 10% MWCNT–IL (% w/v) was determined to be the optimal mixing ratio.

The polymerization potential was investigated within the range of 1.0 V and 3.0 V. As shown in Figure S2, all phthalates showed higher peak areas at 2.0 V than at 1.0 V and 3.0 V. Even though good reproducibility was shown at 3.0 V, this was not considered to be an optimal condition because the stainless-steel wire might itself react [11]. Therefore, the potential of 2.0 V was determined to be the optimal polymerization potential for coulometry.

The effect of the electrochemical deposition time was investigated from 150 s to 700 s in consideration of the time as well as the extraction efficiency. As shown in Figure S3, the extraction efficiency of the phthalates increased up to 350 s and then decreased at 500 s.

The high standard deviation of the peak areas at 150 s and 350 s suggest that this may not be sufficient to solidify the hybrid composite on a wire. Therefore, the optimum deposition time was set to 500 s, which showed the best reproducibility based on the peak area value among the subsequent deposition times.

The effect of adsorbent length on the phthalate extraction efficiency was investigated in the range of 0.5 cm to 1.5 cm. As shown in Figure S4, the longer the coating length, the more phthalates were adsorbed. However, it was noted that the standard deviation of the peak area value was much larger, with a length of more than 1 cm. Therefore, the coating length of 1 cm was set as the optimum condition; similar values have been reported in other studies [11,15,16].

2.1.2. Effect of the Extraction Conditions of the HS–INME Method on Extraction

The extraction efficiency of the phthalates was determined according to the saturation time from 15 min to 60 min. As shown in Figure S5, the adsorption efficiency appeared to be reasonable at the initial saturation time of 15 min, but the standard deviations were very large. After 15 min, the peak areas of all phthalates were shown to be similar, but with different standard deviations. The highest reproducibility was obtained at 60 min saturation, so this was determined to be the optimal saturation time.

For the extraction efficiency of phthalates, the extraction temperature was investigated at various temperatures from 25 °C to 80 °C. Figure S6 shows that the peak area of the phthalates, except for benzyl butyl phthalate and di(2-ethylhexyl) phthalate, which have high molecular weights, is large at 25 °C, with very high standard deviation values. Therefore, the peak area for compounds with a higher molecular weight at 50 °C with a small standard deviation was determined to be the optimal extraction temperature. It is believed that increasing the extraction temperature increases the diffusion coefficient and decreases the distribution constant, resulting in faster equilibrium times than those achieved at room temperature [32].

The adsorption time is a critical parameter for reaching equilibrium in the distribution of analytes between the MWCNT–IL/PANI coating layers and samples in HS. The adsorption time was investigated in the range of 10 min to 60 min. Figure S7 shows that the peak area for all phthalate compounds increases as the adsorption time increases. Even though the adsorption amount was the highest at 60 min, this was not determined to be the optimal condition due to the high standard deviation. As a result, the optimal adsorption time was selected as 30 min, with high peak areas for phthalates.

The effect of the desorption time on the extraction efficiency of the phthalates was evaluated. The desorption percentages for the desorption times of 30 s, 1 min, 3 min, and 5 min at 230 °C were compared, as shown in Figure S8. The phthalate compounds were desorbed by almost 100% as the desorption time increased, except for dibutyl phthalate and di(2-ethylhexyl) phthalate. Finally, the optimized desorption time was set at 3 min.

2.2. Synthesis and Characteristics of the MWCNT–IL/PANI Layer on a Wire

Synthesis of the MWCNT–IL/PANI via a number of steps was confirmed by FT–IR and TGA, and coating of the hybrid composite was confirmed by SEM and EDS mapping.

The FT–IR analysis was conducted to confirm that the target functional groups were those generated in the process of combining the MWCNT hybrid composite described in Section 3.2. Figure 1 shows the FT–IR spectra of (a) pristine MWCNTs, (b) oxidized MWCNTs, (c) MWCNT–IL, and (d) the MWCNT–IL/PANI composite.

From the oxidation of the MWCNT composite, a C=C stretching band appeared at 1627.20 cm^{-1}; C=O and C–O stretching bands were observed at 1714.69 cm^{-1} and 1215.97 cm^{-1}, respectively, due to the carboxyl group, while the pristine MWCNTs showed a C=C stretching peak at 1624.46 cm^{-1}. Finally, the MWCNT composite combined with the IL confirmed the presence of C=O and C=C stretching band peaks at 1713.90 cm^{-1} and 1630.56 cm^{-1}, respectively; moreover, C–N stretching band peaks from the imidazolium cation were identified through the 1425.58 cm^{-1}, 1213.70 cm^{-1}, 1045.32 cm^{-1}, and

927.63 cm^{-1} vibration bands. As a result, it was confirmed that the MWCNT–IL hybrid composite was successfully synthesized and identified by several major peaks generated in each preparation step.

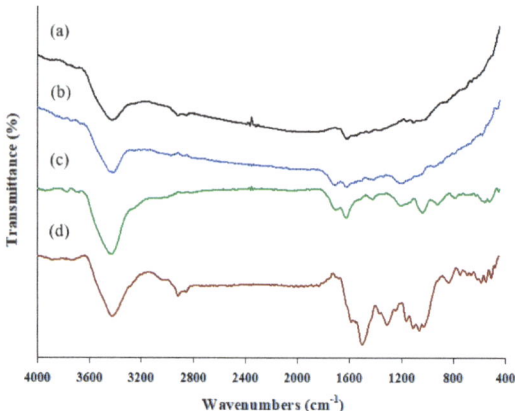

Figure 1. The FT-IR spectra of (a) pristine MWCNTs, (b) oxidized MWCNT, (c) MWCNT–IL, and (d) the MWCNT–IL/PANI composite.

For the MWCNT–IL/PANI composite at optimum conditions, the benzenoid ring stretching peaks of 1504.68 cm^{-1} and the quinonoid ring stretching peaks from 1591 cm^{-1} were observed in the polyaniline structure, which was oxidized by electrochemical polymerization [33,34]. The in-plane and out-of-plane deformation of C–H bands at 1066.77 cm^{-1}, 836.64 cm^{-1}, and 752.32 cm^{-1} confirmed the presence of a quinonoid ring [35]. Additionally, we identified a C–N stretching band at 1314.91 cm^{-1} from the secondary aromatic amine [36] and a N–H stretching peak at 590.16 cm^{-1} from the primary amino group [37,38]. The peaks identified in the MWCNT–IL composite and the representative peaks of the PANI were simultaneously identified, indicating that the electrochemically synthesized MWCNT–IL and PANI were successfully cured under optimum conditions.

The TGA was used to determine the thermal stability of MWCNT–IL/PANI, a wire-coated adsorbent at optimum conditions. From Figure S9, it was found that the MWCNT–IL/PANI adsorbent is thermally stable up to 368 °C. Considering that the decomposition temperature of PANI is usually about 200 °C [28,30], thermal stability is improved with the interaction of PANI with MWCNT–IL. This enhancement might be attributed to the long conjugate π–π bond between MWCNT and PANI, similar to that reported in previous studies [11,39,40].

The morphological image of the surface was obtained using AFM (Figure S10). The R_q value of the surface was 0.084 nm, while the R_a value was 0.038 nm. The root mean square (RMS) roughness (R_q) refers to the square root of the surface height distribution, which is considered more sensitive than the average roughness (R_a) for large deviations from the mean line or plane [41]. It can be inferred that the coating of the adsorbent consisting of MWCNT–IL/PANI was evenly spread on the surface of the stainless-steel wire.

The cross-section image of the MWCNT–IL/PANI coated on a wire in Figure 2 shows that the adsorbent was evenly distributed on the wire with a thickness of about 2.25 μm. In the elemental composition of the uncoated part (Figure 2f), large quantities of Fe (54.05%) and Cr (27.64%) were observed from the stainless-steel part, while the coated part (Figure 2g) consisted of C (75.9%), N (20.33%), and O (3.76%) from MWCNT–IL/PANI. Elemental mapping analysis at the coating interface shows that C, N, and O are evenly distributed on the INME wire.

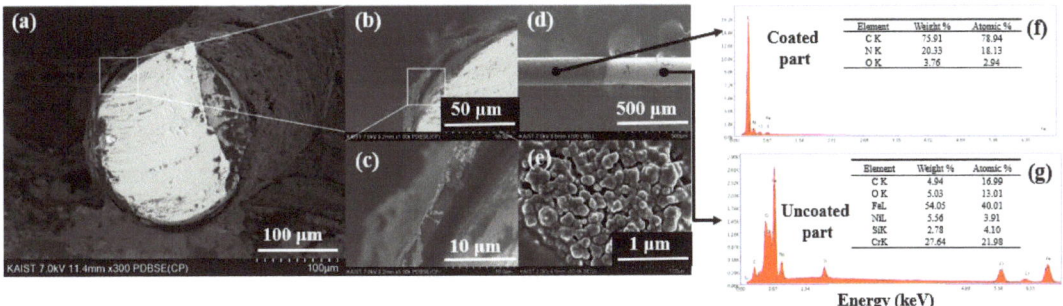

Figure 2. Cross-section SEM images of MWCNT–IL/PANI–coated wire (**a**–**e**) and EDS spectrum of the coated wire (**f**) and uncoated wire with MWCNT–IL/PANI (**g**).

2.3. Validation of the Analytical Method

The adsorption efficiency was investigated using phthalate compounds. The HS–INME–MWCNT–IL/PANI method used in this study was verified by the regression equation of each calibration curve, as well as the LOD, LOQ, and dynamic ranges, and precision and accuracy tests (Table 1). Using the IUPAC definition, the LOD and LOQ were obtained from the mean of the blank measurement and the standard deviation of the blank measurement with a ratio. The values of the ratio were 3 for LOD and 10 for LOQ. Most of phthalates investigated in this study showed very good linearity, with $r^2 = 0.99$. Each regression equation of 5 points was calculated by taking measurements three times at each concentration under the optimum conditions. LOD and LOQ were calculated according to the IUPAC definition [42,43], resulting in values of 15.84~50.56 µg for LOD and 52.79~168.5 µg for LOQ, while the dynamic range was between 52.79 µg and 1.00×10^3 µg.

Table 1. Validation data of HS–INME using MWCNT–IL/PANI and GC/MS ($n = 3$).

Compound	Regression Equation	Coefficient of Determination (r^2)	LOD (µg)	LOQ (µg)	Dynamic Range (µg)
Dimethyl phthalate	y = 2.5326x + 0.0233	0.9907	19.57	65.23	65.23~2.50 × 10²
Diethyl phthalate	y = 6.5383x + 0.0390	0.9967	15.84	52.79	52.79~1.00 × 10³
Diallyl phthalate	y = 2.1476x + 0.0653	0.9959	16.17	53.91	53.91~5.00 × 10²
Dibutyl phthalate	y = 5.1304x + 0.1313	0.9734	18.9	63.01	63.01~5.00 × 10²
Benzyl butyl phthalate	y = 2.6991x + 0.0557	0.9931	42.03	140.1	140.1~1.00 × 10³
Di(2-ethylhexyl) phthalate	y = 2.6567x + 0.0994	0.9918	50.56	168.5	168.5~1.00 × 10³

The accuracy of the method was assessed through spike recovery. The sample was investigated by adding 200 µg of phthalate standards. The accuracy results ranged from 61.13% to 108.21%. The identified absolute values and the recovery values obtained by the spikes of the standards are listed in Table S3.

The precision in reproducibility was calculated using the relative standard deviation value. The intra assay (run-to-run) was repeated five times using the same needle, while the inter assay (needle-to-needle) was repeated five times using five different needles. Relative standard deviations for intra assays between the same needles showed an average of less than 10%, and inter-assay experiments comparing five different needles showed an average of less than 15% (Table S3).

The LOD and LOQ of HS–INME–MWCNT–IL/PANI were found to be higher than in the previous study [11,15]. However, the addition of IL might contribute to the homogenous dispersion of MWCNTs in mixture, which could lead to improved recovery and reproducibility compared to HS–INME–MWCNT–PANI [11].

2.4. Comparison of Extraction Efficiency

Extraction efficiency was compared according to the dynamic and static INME methods. The difference between static and dynamic HS lies in the presence of a pumping application that can circulate the upper part of the sample during the adsorption process. Phthalates of 200 µg were analyzed under optimized conditions for each method, and the enrichment factor (EF) was calculated using the peak areas of all phthalates investigated in this study (Equation (1)). EF refers to the enriched concentration of the target compound in the adsorbent during the extraction process. Higher EF values indicate better extraction efficiencies than lower EF values.

$$EF = A_1/A_0 \quad (1)$$

In the equation, A_1 is the area of the GC/MS peak obtained using the dynamic HS method, and A_0 is the area of the GC/MS peak obtained using the static HS method. EF values were obtained as 1.60 (±0.11) for dimethyl phthalate, 1.63 (±0.13) for diethyl phthalate, 3.10 (±0.41) for diallyl phthalate, 3.76 (±0.40) for dibutyl phthalate, 4.05 (±0.28) for benzyl butyl phthalate, and 2.61 (±0.33) for di(2-ethylhexyl) phthalate. For all phthalate components, the dynamic HS method has an average EF value of at least 1.5 times that of the static HS method. In particular, benzyl butyl phthalate had the highest value of EF, implying that dynamic extraction with the INME adsorbent synthesized in this study was effective for heavy molecules with benzene substitutes. A higher molecular weight tends to produce higher EF values. However, it was confirmed that the EF value of di(2-ethylhexyl) phthalate does not increase significantly, even though its molecular weight is higher than that of benzyl butyl phthalate. This might contribute to the fact that benzyl butyl phthalate has more benzene ring structures than di(2-ethylhexyl) phthalate, which has a significant effect on producing a better π–π interaction with the MWCNT–IL/PANI used as adsorbent. If the molecular weight is very high, it can be inferred that it may affect the difference between the dynamic and static HS methods. The value of EFs was found to be between 1.60 (for dimethyl phthalate) and 4.05 (for benzyl butyl phthalate).

2.5. Application of HS–INME–MWCNT–IL/PANI to an Aqueous Sample

The HS–INME–MWCNT–IL/PANI was applied to phthalates in aqueous samples for the feasibility study. To simulate hot water covered by plastic wrap, it was immersed into the water at 80 °C. Phthalates might be leached out into hot water. Figure 3a shows a chromatogram of the phthalates eluted from an industrial LLDPE wrap (sample 1). A similar result for household LLDPE wrap (sample 2) is shown in Figure S11. The measurements of the phthalates are summarized in Table S3. Phthalates from the industrial wrap (sample 1) and a household LLDPE wrap were detected but not quantified due to having lower values than the LOD, except for that of dibutyl phthalate. The values of the dibutyl phthalate concentration were measured as 26.05 µg ± 3.12 µg for the industrial wrap and 24.87 µg ± 1.65 µg for the household LLDPE wrap. Similar data were obtained regarding the migration of plasticizers from LDPE films as food packaging [44]. The chromatogram of the spike sample (Figure 3b) indicated the increase in the peak area by the known amount added. The results showed reasonable recovery and reproducibility, implying that the proposed method could be used as an HS–INME for aqueous samples.

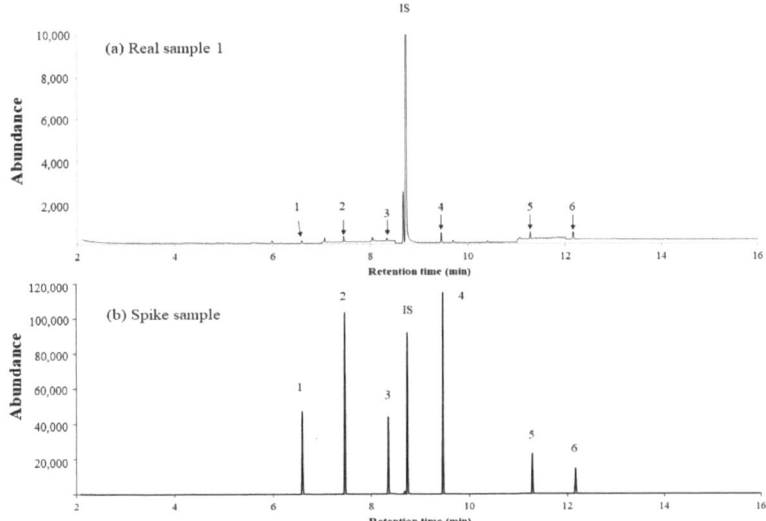

Figure 3. Chromatogram obtained from (**a**) sample 1, and (**b**) spiked sample 1 using HS–INME–MWCNT–IL/PANI. Peak 1, dimethyl phthalate; 2, diethyl phthalate; 3, diallyl phthalate; 4, dibutyl phthalate; 5, benzyl butyl phthalate; 6, di(2-ethylhexyl) phthalate; and IS, anthracene.

3. Materials and Methods

3.1. Reagents and Materials

Aniline (>99.5%) as a monomer and nitric acid (64.0~66.0%) were purchased from Duksan Pure Chemical Co. (Ansan, Republic of Korea). The MWNCTs were about 1.0 μm~2.0 μm in length with 10.0 nm in diameter, and the specific surface area was 350 ± 5 m^2 g^{-1}; they were obtained from NTP (Shenzhen, China). A supply of 1-(2-Hydroxyethyl)-3-methylimidazolium tetrafluoroborate ([HOEMIm]BF4) (>98.0%) was purchased from TCI (Tokyo, Japan) and N, N-dimethylformamide (DMF, >99.0%) was obtained from Samchun Pure Chemical Co. (extra pure grade, Pyeongtaek, Republic of Korea). Anthracene of chemical pure grade, to be used as an internal standard, was obtained from Junsei Chemical Co. (Tokyo, Japan).

Six different phthalates were used for the feasibility study. Dimethyl phthalate (>99.0%), diethyl phthalate (>98.0%), diallyl phthalate (>98.0%), dibutyl phthalate (>97.0%), benzyl butyl phthalate (>97.0%), and di(2-ethylhexyl) phthalate (>98.0%) were obtained from TCI (Tokyo, Japan). The physical properties and chemical structure of all the standard compounds are listed in Table S1. The phthalate stock solutions were prepared with n-hexane (HPLC grade, Samchun Pure Chemical Co., Pyeongtaek, Republic of Korea) at a concentration of 10,000 mg L^{-1}, and they were subjected to step-by-step dilution with n-hexane for further use. These solutions were stored in a refrigerator (5 °C) until use. The working solution was used for the experiment by making 20 mg L^{-1} phthalate mixture by diluting each phthalate stock solution. The water used was ultrapure water (18.1 MΩ cm^{-1}, Pure Water Co., Namyangju, Republic of Korea).

The homemade INME device consisted of a 22-gauge stainless-steel needle (Hamilton 90,022, metal hub syringe needle, 718 μm O.D., 413 μm I.D., 51 mm length, bevel tip, Hamilton, Reno, NV, USA), a 1 mL Luer lock gas-tight syringe barrel (Hamilton 1001N), and a polytetrafluoroethylene (PTFE Teflon) plunger. The device was used as a headspace device in the needle microextraction (HS–INME) for the adsorption and desorption tests. Stainless-steel wire (0.22 mm O.D.) was obtained from a local company (09one Science, Gyeonggi-do, Republic of Korea) and used as a working electrode for the coulometric coating. The working solution for the coulometric coating was filtrated using a 0.45 μm syringe filter (polyvinylidene fluoride membrane, 25 mm I.D. × 0.45 μm pore size, SV25P045NL, Hyundai Micro, Seoul, Republic of Korea).

3.2. Preparation of the MWCNT–IL/PANI Layer Coated on a Wire

MWCNT of 300 mg was refluxed in 21 mL of concentrated nitric acid at 115 °C for 3 h according to a previous report [45]. The resulting oxidized MWCNT solid was filtered through filter paper (No 20, 5 μm~8 μm pore size, Hyundai Micro, Seoul, Republic of Korea), and washed with distilled water until the acidic pH was neutralized; it was then dried at 80 °C overnight. The MWCNT–IL composite was prepared by grinding the mixture of 1.6 mL [HOEMIm] BF4 and 160 mg oxidized MWCNTs with agate mortar [29,46], followed by sonication for 30 min with 18 mL DMF. Then, it was sonicated for an additional 1 h after adding 32 mL distilled water.

The MWCNT–IL/PANI composites were deposited on the surface of a stainless-steel wire using electrochemical polymerization in aqueous solutions containing aniline as a monomer and MWCNT–IL as the carbon nanomaterials, based on a potential difference during the electrochemical polymerization [47]. The combined MWCNT–IL (80 mg) composite was dispersed in 40 mL distilled water for 1 h at room temperature with a sonicator, followed by the addition of 1 mL aniline and further sonication for 15 min. The partially ionizable carboxylic acid groups generated during the oxidation of the MWCNTs with nitric acid can both serve as charge carriers in the solution and charge balance dopants in the polymers. Therefore, no additional supporting electrolyte was used, to avoid penetration of dopants other than the ionized MWCNTs [48].

Coulometric polymerization was performed using a three-electrode arrangement (WonAtech, ZIVE-SP2 electrochemical workstation, Seoul, Republic of Korea) with the filtered MWCNT–IL hybrid composite and an aniline solution at room temperature. Before the electrochemical polymerization step, the stainless-steel wire used as the working electrode was cleaned with acetone, and one end of this wire was coiled up, as mentioned in a previous study [13]. Then, the stainless-steel wire was immersed in the solution; a platinum wire was used as the counter electrode and we proceeded using a RE–5B Ag/AgCl reference electrode (Bioanalytical Systems Inc., W. Lafayette, IN, USA).

A constant potential at 2.0 V was applied to the electricity measurement system for 500 s. The potential range tested was determined so as to avoid the oxidation of unwanted materials. After the coulometric process, the coating layer was rinsed several times with distilled water to remove unreacted chemicals such as residual MWCNT–IL and aniline. The coated wire was dried in an oven at 80 °C for 30 min and then thermally purified at 230 °C for 1 h. As shown in Figure 4, a wire coated with the MWCNT–IL/PANI adsorbent was inserted into the INME needle. Prior to the extraction test, a needle containing the coated wire was inserted into the GC injector at 230 °C for 30 min for thermal purification, then kept in a desiccator at room temperature. For the extraction process, the INME needle was simply connected to the Luer lock gas-tight syringe barrel and plunger.

3.3. Characterization of the MWCNT–IL/PANI Coating Layer

Using the Perkin Elmer Spectrometer 100 (Waltham, MA, USA), FT-IR (Fourier transform infrared) spectra of pristine MWCNT, oxidized MWCNT, and MWCNT–IL complexes were obtained using the KBr pellet method ranging from 450 cm^{-1} to 4000 cm^{-1}, and each functional group was identified.

The thermal stability of the MWCNT–IL/PANI layer was determined by thermogravimetric analysis (TGA, SDT Q600, TA Instruments, New Castle, DE, USA). The temperature was raised from room temperature to 800 °C at a heating rate of 10 °C min^{-1} in a nitrogen atmosphere.

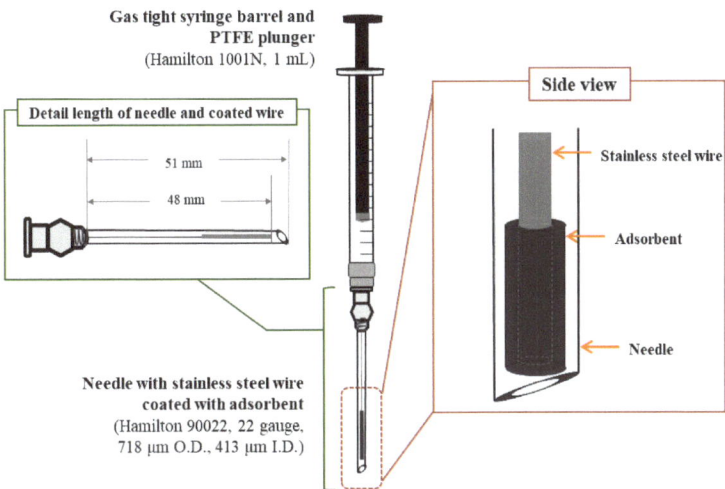

Figure 4. A schematic illustration of the side view and details of each length of a needle containing a stainless-steel wire coated with MWCNT–IL/PANI.

The roughness of the adsorbent layer was analyzed using an atomic force microscope (AFM; Park NX10, Suwon, Republic of Korea). The scan area was 20 μm × 20 μm. Then, 3D images of the adsorbent's upper surface were taken, and the related parameters of surface roughness were obtained using the supporting software. Morphology and energy dispersive spectroscopy (EDS) mapping images of the MWCNT–IL/PANI coating layer surface were obtained using scanning electron microscopy (SEM, Hitachi, SU8230, Tokyo, Japan).

3.4. Headspace In–Needle Microextraction Procedure

The test solution prepared for the headspace (HS) extraction test was a mixture of 9.00 mL of water and 1.00 mL of the phthalate working solution in a 50 mL vial. The experiment was performed by exposing the INME needle to the HS of the vial containing the standard solutions. The extraction and adsorption processes were performed by sucking out the analytes in the upper part of the sample vial by automatic compression and aspiration using a homemade high-efficiency extraction reciprocating pump. The speed of the reciprocating pump was 6 cycles/min (10 s/cycle) [49]. After the adsorption was completed, the needle with the analyte adsorbed was immediately connected to another gas-tight syringe and inserted into the GC injection port at 230 °C for 3 min. Then, the analytes were injected into the GC column to be separated at the same time. The INME needle used in the adsorption process was washed with acetone, and the remaining impurities were desorbed after 30 min of conditioning at 230 °C before the next adsorption experiment.

3.5. Optimization of HS–INME–MWCNT–IL/PANI

To evaluate the extraction capacity of the MWCNT–IL/PANI coating on a wire for the HS–INME method and to validate the proposed INME method, phthalate standards were saturated in the vial for HS–INME extraction. Various parameters affecting the extraction efficiency were investigated, including the weight/volume percentage, polymerization potential, electrochemical deposition time, and the length of the MWCNT adsorbent for the synthesis of the adsorbent, and the saturation time, extraction time, adsorption time, and desorption time for the HS–INME method. The peak areas of the standard compounds at various conditions were obtained three times and compared for the selection of the optimal conditions based on the amount of adsorption as well as reproducibility. The parameters and conditions for the adsorbent consisting of the MWCNT–IL/PANI layer and the HS–INME method are shown in Table 2.

Table 2. The parameters of HS–INME using an MWCNT–IL/PANI-coated stainless-steel wire investigated in this study.

MWCNT–IL/PANI Layer Parameter	Conditions
Percentage of MWCNT-ionic liquid (%, w/v)	5, 10, 15
Polymerization potential (V)	1, 2, 3
Electrochemical deposition time (s)	150, 350, 500, 700
Adsorbent surface length (cm)	0.5, 1.0, 1.5
HS–INME parameter	**Conditions**
Saturation time (min)	15, 30, 45, 60
Extraction temperature (°C)	25, 50, 80
Adsorption time (min)	10, 20, 30, 60
Desorption time (min)	0.5, 1, 3, 5

3.6. Validation of HS–INME–MWCNT–IL/PANI

To verify the developed analytical method, a calibration curve measured in triplicate was established to determine the LOD, LOQ, precision, accuracy, and recovery using the phthalate standard solution. Recovery and reproducibility tests were conducted to confirm the accuracy and precision of the analytical method. Quantitative analysis was performed with the internal standard method using anthracene as an internal standard.

The proposed HS–INME–MWCNT–IL/PANI method with GC/MS was used for the feasibility study in the aqueous samples. The aqueous samples were prepared for two commercially available wraps, considering the risk of phthalate influx into food [50]. The wraps investigated in this study were an industrial linear low-density polyethylene (LLDPE) wrap (Okong stretch film, Okong, Republic of Korea) and a household LLDPE wrap (Bio wrap, Comex latex, Republic of Korea). Each wrap (1%, w/v) was put in water (80 °C, 200 mL) for 3 h and then removed from the water, and an aliquot of the remaining water (9 mL) was taken for phthalate analysis.

3.7. Gas Chromatography/Mass Spectrometry (GC/MS)

Optimization and validation of the analytical methods, a comparison of extraction efficiencies, and a quantitative analysis of the real samples were performed using GC/MS. The analyses were carried out using the GC system (Agilent 7820A) and an MSD 5977E Mass Spectrometer in selected ion monitoring (SIM) mode. The separation of phthalates was performed using an HP-5MS analytical column (30 m × 0.25 mm × 0.25 μm, (5%-Phenyl)-methylpolysiloxane, Agilent Technologies, Santa Clara, CA, USA) in splitless mode. Detailed operating conditions of the GC/MS are summarized in Table S2.

4. Conclusions

In this study, we successfully synthesized an MWCNT–IL/PANI adsorbent to be coated on stainless steel using the electrochemical polymerization of aniline after combining MWCNTs with IL for INME needle fabrication. The MWCNT–IL hybrid composites showed that the MWCNTs and IL were well-formed from carboxyl groups and an imidazolium cation. As a result, the MWCNT–IL/PANI composite used to coat the stainless-steel wire was evenly coated on the wire's surface; its high thermal stability was confirmed by the TGA, AFM, and SEM results.

The optimization of the adsorbent design process was performed, and the HS–INME analysis conditions were determined. The optimal weight per volume percent of MWCNTs and IL was 10%, and the MWCNT–IL/PANI composite of 1.0 cm length was effectively electrochemically synthesized on the surface of the stainless-steel wire when a constant potential of 2.0 V was applied for 500 s. The optimum conditions for HS–INME analysis were 60 min of saturation time, a 50 °C extraction temperature, 30 min of adsorption time, and 3 min of desorption time. To validate the HS–INME–MWCNT–IL/PANI, calibration curves, the LOD, the LOQ, the recovery, the reproducibility, and the enrichment factor were determined for optimum conditions using a GC/MS system.

In conclusion, a newly synthesized and characterized MWCNT–IP/PANI was used as an adsorbent for the HS–INME method. The INME needle with an MWCNT–IL/PANI-coated wire layer was used repeatedly up to 150 times without losing performance. This proposed adsorbent and the INME method could be implemented as an environmentally friendly, solvent-free extraction method.

Supplementary Materials: The following supporting information can be downloaded at: https://www.mdpi.com/article/10.3390/molecules28083517/s1, Figure S1: Influence of percentage of MWCNTs and ionic liquid (% w/v) on the peak area of target analytes; Figure S2: Influence of applied polymerization potential on the peak area of target analytes; Figure S3: Influence of electrochemical deposition time on the peak area of target analytes; Figure S4: Influence of adsorbent surface length on the peak area of target analytes; Figure S5: Influence of saturation time on the peak area of target analytes; Figure S6: Influence of applied extraction temperature on the peak area of target analytes; Figure S7: Influence of adsorption time on the peak area of target analytes; Figure S8. Influence of applied desorption time on the peak area of target analytes; Figure S9. Thermogravimetric analysis curve of MWCNT–IL/PANI adsorbent; Figure S10: AFM images of MWCNTs IL/PANI deposited on the surface of stainless steel wire; Figure S11: Chromatogram obtained from (a) sample 2, and (b) spiked sample 2 by HS INME MWCNT–IL/PANI. Peak 1, dimethyl phthalate; 2, diethyl phthalate; 3, diallyl phthalate; 4, dibutyl phthalate; 5, benzyl butyl phthalate; 6, di(2 ethylhexyl) phthalate; and IS, anthracene; Table S1. Physical properties and chemical structures for each of the phthalates used in the target compounds in this study; Table S2. The operating conditions of gas chromatograph/mass spectrometer (GC/MS); Table S3. Recovery of HS–INME using MWCNT–IL/PANI coating layer followed GC/MS.

Author Contributions: Conceptualization, S.B.; methodology, S.A.; validation, S.A.; writing—original draft preparation, S.A.; writing—review and editing, S.B.; supervision, S.B. All authors have read and agreed to the published version of the manuscript.

Funding: This research was funded by Seoul Women's University (2023-0032).

Institutional Review Board Statement: Not applicable.

Informed Consent Statement: Informed consent was obtained from all subjects involved in the study.

Data Availability Statement: Not applicable.

Conflicts of Interest: The authors declare no conflict of interest.

Sample Availability: Not available.

References

1. Aly, A.A.; Górecki, T. Green Approaches to Sample Preparation Based on Extraction Techniques. *Molecules* **2020**, *25*, 1719. [CrossRef] [PubMed]
2. Câmara, J.S.; Perestrelo, R.; Berenguer, C.V.; Andrade, C.F.P.; Gomes, T.M.; Olayanju, B.; Kabir, A.; MR Rocha, C.; Teixeira, J.A.; Pereira, J.A.M. Green Extraction Techniques as Advanced Sample Preparation Approaches in Biological, Food, and Environmental Matrices: A Review. *Molecules* **2022**, *27*, 2953. [CrossRef] [PubMed]
3. Picó, Y. *Comprehensive Sampling and Sample Preparation*; Pawliszyn, J., Ed.; Academic Press: Cambridge, MA, USA, 2012; Volume 3, Chapter 3; p. 569.
4. Jon, C.-S.; Meng, L.-Y.; Li, D. Recent review on carbon nanomaterials functionalized with ionic liquids in sample pretreatment application. *TrAC Trends Anal. Chem.* **2019**, *120*, 115641. [CrossRef]
5. Prokůpková, G.; Holadová, K.; Poustka, J.; Hajšlová, J. Development of a solid-phase microextraction method for the determination of phthalic acid esters in water. *Anal. Chim. Acta* **2002**, *457*, 211–223. [CrossRef]
6. Serôdio, P.; Nogueira, J. Considerations on ultra-trace analysis of phthalates in drinking water. *Water Res.* **2006**, *40*, 2572–2582. [CrossRef]
7. Jobling, S.; Reynolds, T.; White, R.; Parker, M.G.; Sumpter, J.P. A variety of environmentally persistent chemicals, including some phthalate plasticizers, are weakly estrogenic. *Environ. Health Perspect.* **1995**, *103*, 582–587. [CrossRef]

8. Castillo, M.; Oubina, A.; Barceló, D. Evaluation of ELISA kits followed by liquid chromatography-atmospheric pressure chemical ionization-mass spectrometry for the determination of organic pollutants in industrial effluents. *Environ. Sci. Technol.* **1998**, *32*, 2180–2184. [CrossRef]
9. Potter, D.W.; Pawliszyn, J. Rapid determination of polyaromatic hydrocarbons and polychlorinated biphenyls in water using solid-phase microextraction and GC/MS. *Environ. Sci. Technol.* **1994**, *28*, 298–305. [CrossRef]
10. Colon, I.; Dimandja, J.D. High-throughput analysis of phthalate esters in human serum by direct immersion SPME followed by isotope dilution–fast GC/MS. *Anal. Bioanal. Chem.* **2004**, *380*, 275–283. [CrossRef]
11. Lee, S.Y.; Yoon, J.H.; Bae, S.; Lee, D.S. In-needle microextraction coupled with gas chromatography/mass spectrometry for the analysis of phthalates generating from food containers. *Food Anal. Methods* **2018**, *11*, 2767–2777. [CrossRef]
12. Jeon, H.L.; Son, H.H.; Bae, S.; Lee, D.S. Use of polyacrylic acid and polydimethylsiloxane mixture for in-needle microextraction of volatile aroma compounds in essential oils. *Bull. Korean Chem. Soc.* **2015**, *36*, 2730–2739. [CrossRef]
13. Lee, E.J.; Lee, D.S. Fabrication of in-needle microextraction device using nichrome wire coated with poly(ethylene glycol) and poly(dimethylsiloxane) for determination of volatile compounds in lavender oils. *Bull. Korean Chem. Soc.* **2014**, *35*, 211–217. [CrossRef]
14. Bang, Y.J.; Hwang, Y.R.; Lee, S.Y.; Park, S.M.; Bae, S. Sol–gel-adsorbent-coated extraction needles to detect volatile compounds in spoiled fish. *J. Sep. Sci.* **2017**, *40*, 3839–3847. [CrossRef]
15. Hwang, Y.; Lee, Y.; Ahn, S.; Bae, S. Electrochemically polyaniline-coated microextraction needle for phthalates in water. *Anal. Sci. Technol.* **2020**, *33*, 76–85.
16. Kim, S.; Bae, S.; Lee, D.-S. Characterization of scents from Juniperus chinensis by headspace in-needle microextraction using graphene oxide-polyaniline nanocomposite coated wire followed by gas chromatography-mass spectrometry. *Talanta* **2022**, *245*, 123463. [CrossRef]
17. Kim, S.; Bae, S. In vitro and in vivo human body odor analysis method using GO:PANI/ZNRs/ZIF-8 adsorbent followed by GC/MS. *Molecules* **2022**, *27*, 4795. [CrossRef]
18. Motitswe, M.G.; Badmus, K.O.; Khotseng, L. Development of Adsorptive Materials for Selective Removal of Toxic Metals in Wastewater: A Review. *Catalysts* **2022**, *12*, 1057. [CrossRef]
19. Qu, L.; Dai, L. Substrate-enhanced electroless deposition of metal nanoparticles on carbon nanotubes. *J. Am. Chem. Soc.* **2005**, *127*, 10806–10807. [CrossRef]
20. Randriamahazaka, H.; Ghilane, J. Electrografting and controlled surface functionalization of carbon based surfaces for electro-analysis. *Electroanalysis* **2016**, *28*, 13–26. [CrossRef]
21. Dobrowolski, R.; Mróz, A.; Dąbrowska, M.; Olszański, P. Solid sampling high-resolution continuum source graphite furnace atomic absorption spectrometry for gold determination in geological samples after preconcentration onto carbon nanotubes. *Spectrochim. Acta Part B At. Spectrosc.* **2017**, *132*, 13–18. [CrossRef]
22. Guan, Z.; Huang, Y.; Wang, W. Carboxyl modified multi-walled carbon nanotubes as solid-phase extraction adsorbents combined with high-performance liquid chromatography for analysis of linear alkylbenzene sulfonates. *Anal. Chim. Acta* **2008**, *627*, 225–231. [CrossRef] [PubMed]
23. Du, W.; Zhao, F.; Zeng, B. Novel multiwalled carbon nanotubes-polyaniline composite film coated platinum wire for headspace solid-phase microextraction and gas chromatographic determination of phenolic compounds. *J. Chromatogr. A* **2009**, *1216*, 3751–3757. [CrossRef] [PubMed]
24. Stege, P.W.; Lapierre, A.V.; Martinez, L.D.; Messina, G.A.; Sombra, L.L. A combination of single-drop microextraction and open tubular capillary electrochromatography with carbon nanotubes as stationary phase for the determination of low concentration of illicit drugs in horse urine. *Talanta* **2011**, *30*, 278–283. [CrossRef] [PubMed]
25. Jun, L.Y.; Mubarak, N.M.; Yee, M.J.; Yon, L.S.; Bing, C.H.; Khalid, M.; Abdullah, E.C. An Overview of functionalised carbon nanomaterial for organic pollutant removal. *J. Ind. Eng. Chem.* **2018**, *67*, 175–186. [CrossRef]
26. Tunckol, M.; Durand, J.; Serp, P. Carbon nanomaterial-ionic liquid hybrids. *Carbon* **2012**, *50*, 4303–4334. [CrossRef]
27. Bellayer, S.; Gilman, J.W.; Eidelman, N.; Bourbigot, S.; Flambard, X.; Fox, D.M.; De Long, H.C.; Trulove, P.C. Preparation of homogeneously dispersed multiwalled carbon nanotube/polystyrene nanocomposites via melt extrusion using trialkyl imidazolium compatibilizer. *Adv. Funct. Mater.* **2005**, *5*, 910–916. [CrossRef]
28. Shim, Y.; Kim, H.J. Solvation of carbon nanotubes in a room-temperature ionic liquid. *ACS Nano* **2009**, *3*, 1693–1702. [CrossRef]
29. Li, L.; Wu, M.; Feng, Y.; Zhao, F.; Zeng, B. Doping of three-dimensional porous carbon nanotube-graphene-ionic liquid composite into polyaniline for the headspace solid-phase microextraction and gas chromatography determination of alcohols. *Anal. Chim. Acta* **2016**, *948*, 48–54. [CrossRef]
30. Wu, M.; Zhang, H.; Zhao, F.; Zeng, B. A novel poly(3,4-ethylenedioxythiophene)-ionic liquid composite coating for the headspace solid-phase microextraction and gas chromatography determination of several alcohols in soft drinks. *Anal. Chim. Acta* **2014**, *850*, 41–48. [CrossRef]
31. Fukushima, T.; Kosaka, A.; Ishimura, Y.; Yamamoto, T.; Takigawa, T.; Ishii, N.; Aida, T. Molecular ordering of organic molten salts triggered by single-walled carbon nanotubes. *Science* **2003**, *3000*, 2072–2074. [CrossRef]
32. Pawliszyn, J. Theory of solid-phase microextraction. *J. Chromatogr. Sci.* **2000**, *38*, 270–278. [CrossRef]
33. Ping, Z. In situ FT-IR-attenuated total reflection spectroscopic investigations on the base-acid transitions of polyaniline. base-acid transition in the emeraldine form of polyaniline. *J. Chem. Soc. Faraday Trans.* **1996**, *92*, 3063–3067. [CrossRef]

34. Furukawa, Y.; Ueda, F.; Hyodo, Y.; Harada, I.; Nakajima, T.; Kawagoe, T. Vibrational spectra and structure of polyaniline. *Macromolecules* **1988**, *21*, 1297–1305. [CrossRef]
35. Trchová, M.; Stejskal, J. Polyaniline: The infrared spectroscopy of conducting polymer nanotubes (IUPAC technical report). *Pure Appl. Chem.* **2011**, *83*, 1803–1817. [CrossRef]
36. Boyer, M.I.; Quillard, S.; Rebourt, E.; Louarn, G.; Buisson, J.P.; Monkman, A.; Lefrant, S. Vibrational Analysis of Polyaniline: A model compound approach. *J. Phys. Chem. B* **1998**, *102*, 7382–7392. [CrossRef]
37. Zheng, W.; Angelopoulos, M.; Epstein, A.J.; MacDiarmid, A.G. Experimental evidence for hydrogen bonding in polyaniline: Mechanism of aggregate formation and dependency on oxidation state. *Macromolecules* **1997**, *30*, 2953–2955. [CrossRef]
38. Tammer, M.G. Sokrates: Infrared and Raman characteristic group frequencies: Tables and charts. *Colloid Polym. Sci.* **2004**, *283*, 235. [CrossRef]
39. Dhand, C.; Arya, S.K.; Singh, S.P.; Singh, B.P.; Datta, M.; Malhotra, B. Preparation of polyaniline/multiwalled carbon nanotube composite by novel electrophoretic route. *Carbon* **2008**, *46*, 1727–1735. [CrossRef]
40. Ngo, C.L.; Le, Q.T.; Ngo, T.T.; Nguyen, D.N.; Vu, M.T. Surface modification and functionalization of carbon nanotube with some organic compounds. *Adv. Nat. Sci. Nanosci. Nanotechnol.* **2013**, *4*, 035017.
41. Kumar, B.R.; Rao, T.S. AFM Studies on Surface Morphology, Topography and Texture of Nanostructured Zinc Aluminum Oxide Thin Films. *Dig. J. Nanomater. Biostruct.* **2012**, *7*, 1881–1889.
42. Thompson, M.; Ellison, S.L.R.; Wood, R. Harmonized guidelines for single-laboratory (IUPAC technical report). *Pure Appl. Chem.* **2002**, *74*, 835–855. [CrossRef]
43. Currie, L.A. Nomenclature in evaluation of analytical methods including detection and quantification capabilities. *Pure Appl. Chem.* **1995**, *67*, 1699–1723. [CrossRef]
44. Fasano, E.; Bono-Blay, F.; Cirillo, T.; Montuori, P.; Lacorte, S. Migration of phthalates, alkylphenols, bisphenol A and di(2-ethylhexyl)adipate from food packaging. *Food Control* **2012**, *27*, 132–138. [CrossRef]
45. Asadollahzadeh, H.; Noroozian, E.; Maghsoudi, S. Solid-phase microextraction of phthalate esters from aqueous media by electrochemically deposited carbon nanotube/polypyrrole composite on a stainless steel fiber. *Anal. Chim. Acta* **2010**, *669*, 32–38. [CrossRef]
46. Sun, Y.; Fang, Z.; Wang, C.; Zhou, A.; Duan, H. Incorporating nanoporous polyaniline into layer-by-layer ionic liquid-carbon nanotube-graphene paper: Towards freestanding flexible electrodes with improved supercapacitive performance. *Nanotechnology* **2015**, *26*, 374002. [CrossRef]
47. El Rhazi, M.; Majid, S.; Elbasri, M.; Salih, F.E.; Oularbi, L.; Lafdi, K. Recent progress in nanocomposites based on conducting polymer: Application as electrochemical sensors. *Int. Nano Lett.* **2018**, *8*, 79–99. [CrossRef]
48. Chen, G.Z.; Shaffer, M.S.P.; Coleby, D.; Dixon, G.; Zhou, W.; Fray, D.J.; Windle, A.H. Carbon nanotube and polypyrrole composites: Coating and doping. *Adv. Mater.* **2000**, *12*, 522–526. [CrossRef]
49. Son, H.H.; Bae, S.; Lee, D.S. New needle packed with polydimethylsiloxane having a micro-bore tunnel for headspace in-needle microextraction of aroma components of citrus oils. *Anal. Chim. Acta* **2012**, *751*, 86–93. [CrossRef]
50. Qian, S.; Ji, H.; Wu, X.X.; Li, N.; Yang, Y.; Bu, J.; Zhang, X.; Qiao, L.; Yu, H.; Xu, N.; et al. Detection and quantification analysis of chemical migrants in plastic food contact products. *PLoS ONE* **2018**, *13*, e0208467. [CrossRef]

Disclaimer/Publisher's Note: The statements, opinions and data contained in all publications are solely those of the individual author(s) and contributor(s) and not of MDPI and/or the editor(s). MDPI and/or the editor(s) disclaim responsibility for any injury to people or property resulting from any ideas, methods, instructions or products referred to in the content.

Article

Vacuum-Assisted MonoTrap™ Extraction for Volatile Organic Compounds (VOCs) Profiling from Hot Mix Asphalt

Stefano Dugheri [1,*], Giovanni Cappelli [1], Niccolò Fanfani [1], Donato Squillaci [1], Ilaria Rapi [1], Lorenzo Venturini [1], Chiara Vita [2], Riccardo Gori [3], Piero Sirini [3], Domenico Cipriano [4], Mieczyslaw Sajewicz [5] and Nicola Mucci [1]

[1] Department of Experimental and Clinical Medicine, University of Florence, 50134 Florence, Italy; giovanni.cappelli@unifi.it (G.C.); niccolo.fanfani@unifi.it (N.F.); donato.squillaci@unifi.it (D.S.); ilaria.rapi@edu.unifi.it (I.R.); lorenzo.venturini@unifi.it (L.V.); nicola.mucci@unifi.it (N.M.)
[2] PIN—University Center "Città di Prato" Educational and Scientific Service, University of Florence, 59100 Prato, Italy; chiara.vita@pin.unifi.it
[3] Department of Civil and Environmental Engineering, University of Florence, 50139 Florence, Italy; riccardo.gori@unifi.it (R.G.); piero.sirini@unifi.it (P.S.)
[4] Ricerca sul Sistema Energetico (RSE), 20134 Milan, Italy; domenico.cipriano@rse-web.it
[5] Institute of Chemistry, University of Silesia, 40006 Katowice, Poland; mieczyslaw.sajewicz@us.edu.pl
* Correspondence: stefano.dugheri@unifi.it

Citation: Dugheri, S.; Cappelli, G.; Fanfani, N.; Squillaci, D.; Rapi, I.; Venturini, L.; Vita, C.; Gori, R.; Sirini, P.; Cipriano, D.; et al. Vacuum-Assisted MonoTrap™ Extraction for Volatile Organic Compounds (VOCs) Profiling from Hot Mix Asphalt. *Molecules* **2024**, *29*, 4943. https://doi.org/10.3390/molecules29204943

Academic Editors: Constantinos K. Zacharis and Hiroyuki Kataoka

Received: 12 August 2024
Revised: 2 October 2024
Accepted: 17 October 2024
Published: 18 October 2024

Copyright: © 2024 by the authors. Licensee MDPI, Basel, Switzerland. This article is an open access article distributed under the terms and conditions of the Creative Commons Attribution (CC BY) license (https://creativecommons.org/licenses/by/4.0/).

Abstract: MonoTrap™ was introduced in 2009 as a novel miniaturized configuration for sorptive sampling. The method for the characterization of volatile organic compound (VOC) emission profiles from hot mix asphalt (HMA) consisted of a two-step procedure: the analytes, initially adsorbed into the coating in no vacuum- or vacuum-assistance mode, were then analyzed following an automated thermal desorption (TD) step. We took advantage of the theoretical formulation to reach some conclusions on the relationship between the physical characteristics of the monolithic material and uptake rates. A total of 35 odor-active volatile compounds, determined by gas chromatography-mass spectrometry/olfactometry analysis, contributed as key odor compounds for HMA, consisting mainly of aldehydes, alcohols, and ketones. Chemometric analysis revealed that MonoTrap™ RGC18-TD was the better coating in terms of peak area and equilibrium time. A comparison of performance showed that Vac/no-Vac ratios increased, about an order of magnitude, as the boiling point of target analytes increased. The innovative hybrid adsorbent of silica and graphite carbon monolith technology, having a large surface area bonded with octadecylsilane, showed effective adsorption capability, especially to polar compounds.

Keywords: hot mix asphalt; monolithic material sorptive extraction; under vacuum extraction; volatile organic compounds; odor emission; gas chromatography-mass spectrometry

1. Introduction

Anthropic emissions of odorous compounds can strongly limit the use of the territory [1]. Therefore, the attempt to link the emissions of pollutants in the atmosphere, not only with concentration limits but also with limits for the odor impact, arises from the need to ensure that activities with significant osmogenic flows do not hinder the usability of the territory. Volatile chemicals are emitted from a variety of sources and, in recent years, the attention toward volatile organic compounds (VOCs) has increased, due to the consistent number of anthropogenic processes emitting them, for their environmental and health impact. In addition, VOC emissions can produce odor annoyance, which can reduce life quality [2].

Asphalt mixture plants represent common sources of odor-active VOCs, due to the high temperature (ranging from 150 to 180 °C) of the final hot mix asphalt (HMA) [3]. The characterization of odor impact from these plants is challenging; some phases of their production processes involve large and open areas, and transient emissions are common. Moreover, odors can be released from different plant areas, resulting in a complex source localization and a difficult estimation and measure of odors.

Generally, VOCs are defined based on their physical–chemical properties [4]. For Wang et al. [5], they are compounds with a boiling point (BP) below 100 °C at 101 kPa and vapor pressure (VP) higher than 13.3 Pa at 25 °C. In another definition provided by the European Union solvents directive (1999/13/EC) [6], VOCs are compounds with a VP of at least 10 Pa at 20 °C. Other example volatility criteria proposed by the United States (US) Environmental Protection Agency [7] and New Jersey Department of Environmental Protection [8] define volatile chemicals based on a VP greater than 133 Pa at 25 °C. Health Canada [9] classifies VOCs as compounds that have a BP roughly in the range of 50 to 250 °C.

The presence of odor-related VOCs in ambient air can result in the discomfort of the plant's employees and the residents of neighboring areas. Despite the odor threshold (OT) of VOCs, the perception of an unpleasant odor is related to their concentration and relative hedonic tone; however, their influence on the perception is specific for each compound [10]. In addition, the understanding of these factors may also explain why, despite the enormous effort invested in creating odor parameters, governing bodies have had difficulty establishing fair and effective regulations that address community needs. European countries, like Portugal, Greece, and Austria, currently have no specific odor legislation, while other countries, like Germany, only present regulations for waste management activities.

The Italian Government gave its regions the power to regulate an odor impact [11], and on 28 June 2023, it published a Directorial Decree n. 309-MinAmbiente [12], integrating article 272-bis, in which the HMA is identified among anthropic activities as having a potential odor impact and whose application for authorization must therefore involve the description and evaluation of odor emissions. Like other European regulations, this guideline is based on dynamic olfactometry and dispersion modeling; however, even though it defines the requirements of the odor impact studies by simulation, it does not set any acceptability criteria.

Currently, environmental monitoring solutions are characterized by high costs and the need for extensive resources. Odor impact is generally determined from concentration data, expressed in odor units per cubic meter, i.e., the number of dilutions necessary so that 50% of human sensory panels no longer perceive the smell of the sample analyzed [13]. This type of analysis presents some limitations, such as high costs of air sampling-laboratory management and, mostly, the impossibility of continuous measurements. Electronic noses could represent the best solution for meeting the expectations for environmental issues regarding odor annoyance. Nevertheless, their use is still limited due to technological problems (e.g., sensor drift, variability due to atmospheric conditions, etc.).

Thus, for the qualitative and/or quantitative determination of VOCs in complex matrices, the technique of choice is represented by hyphenated methods such as gas chromatography (GC) with mass spectrometry (MS). To identify odor characteristics and the intensity of the detected VOCs, an accompanying investigation by GC–Olfactometry (GC–O) is binding [14]. In principle, a GC–O is a sniffing device with a split at the end of the GC column. Trained sensory panelists sniff the eluate—combined with a heated and additionally humidified inert gas, in parallel with MS detection—describing each perceived odor and its relative intensity in parallel to the detection of the substances by GC–MS.

However, it is still difficult to directly inject the sample into a GC system to identify small molecules. So, the sample preparation step, consisting of the extraction, purification, and enrichment of analytes, is primary to obtain accurate results. Recently, thanks to automation developments, mainly due to an increasing demand for Green Analytical Chemistry (GAC) [15], monolithic silica and polymers were modified to suit devices for the extraction and enrichment of analytes in various matrices (environmental, food, and biological). This approach contributed to miniaturization and automation by on-line preconcentration, which can reduce the time and the cost of sample preparation [16–18].

Monolithic material was introduced in 1989 by Hjerten et al. [19]. Its preparation is performed through polymerization of a monomer mixture with a porogen solvent. The Monolithic Material Sorptive Extraction (MMSE) technology offers chemical stability over a wide pH range, with pores in a monolithic structure having a large surface area of at least 150 m^2/g to allow for a simple method of sample preparation. In 2009 [20,21], a commercial miniaturized monolithic hybrid adsorptive device, called MonoTrap^{TM} (GL Science Inc., Tokyo, Japan) [22]—made of high-purity silica and/or silica with activated carbon, graphite, or chemically modified octadecyl silane (ODS)—was introduced as a sampling device, particularly recommended for polar compounds. More recently, polydimethylsiloxane (PDMS) with graphitic carbon has been used as an additional sorbent phase.

Using vacuum as a pre-equilibration step, the degree of headspace (HS) partitioning for VOCs reducing the pressure increases. In 2001, Brunton et al. [23] reported the positive effects of low-pressure HS–solid phase microextraction (SPME) sampling of food aroma volatiles from raw turkey, while in 2012 Psillakis et al. [24] evacuated air from a sampling container by using a tailor-made closure before introducing a liquid sample, designating the method as vacuum-assisted HS-SPME (Vac-HS-SPME). Currently, custom-designed closures offer gastight seals to commercial 20 mL HS vials, allowing for microextraction sampling under vacuum conditions. If, to date, a few studies have focused on Vac-HS sampling for liquid samples using the SPME technique [25,26], even fewer are proposed for solid samples [27,28].

The challenge faced by this work include the chemical characterization of the odorous emissions from HMA via new, innovative sampling by MonoTrap^{TM}. To contribute to the growing use of this technology, this study explores the vial preparation method involved in Vac-HS-MonoTrap^{TM} sampling from solid matrices and the related analysis performed by GC–MS/O. Likewise, the optimization of analytical parameters was performed throughout the application of design of experiments (DoE) that allowed us to carry out fewer experiments for method development.

2. Results and Discussion

To date, there are few analytical methods [29–32] for the determination of the odor-active compounds of HMA; the gap especially occurs for polar VOCs. Vac-HS-MMSE-MonoTrap^{TM} sampling and following GC–MS/O analysis was investigated as a possible alternative to conventional methods for odorous compound determination, to provide a simple, fast, sensitive, and solvent-free innovative procedure for HMA fingerprint. Ultimately, this technique allowed us to distinguish 35 odor-active compounds in HMA. These compounds are represented by aldehydes, alcohols, and ketones, having an OT that decreases markedly to sub-ppb as the number of carbon atoms increases.

Considering what is indicated above, Table 1 shows the results as odorous compounds identified by mass spectrum (compared from the mass spectra library), GC–O analysis, the retention times, and LTPRI (see Section 3.7). The VOC cut-off was based on the retention time of the tridecane (RT 25.8 min), a C_{13} n-alkane. All the odor-active compounds considered have a BP and VP lower than those of tridecane—232 °C and 10 Pa, respectively—except for 1-decanol (1 Pa), due to its low (0.7 ppb) odor threshold (OT).

Table 1. Olfactometrically detected VOCs in HMA by Vac-HS-MonoTrapTM sampling and GC-MS/O analysis.

Num.	Compound ** (Name/Formula)	CAS n.	MW Da	BP [a,d,e] °C	VP [a,d,e] Pa	RVD *,[a] Air = 1	LTPRI [b] Estimated	Retention Times (RTs)	Peak Area Score Units [f]	Odor Smell	OT [c] ppb
1	Acetaldehyde/C_2H_4O	75-07-0	44	20	101,000	1.5	412	8.342	+	Pungent, fruity	1.5
2	Ethanol/C_2H_6O	64-17-5	46	78	5800	1.6	458	8.501	+	Weak	520
3	Propanal/C_3H_6O	123-38-6	58	49	31,000	2.0	471	9.117	+++	Pungent, choking	1.0
4	tert-Butyl alcohol/$C_4H_{10}O$	75-65-0	74	83	4100	2.6	476	9.304	+	Camphorous	4500
5	Acetone/C_3H_6O	67-64-1	58	56	24,000	2.0	478	9.398	+	Fruity	42,000
6	1-Propanol/C_3H_8O	71-23-8	60	97	2000	2.1	533	10.087	+	Weak	94
7	Acetic acid/$C_2H_4O_2$	64-19-7	60	118	1500	2.1	543	10.228	+	Strong, vinegar-like	6
8	2-Butanone/C_4H_8O	78-93-3	72	79	10,500	2.41	548	10.431	++	Mint	440
9	Butanal/C_4H_8O	123-72-8	72	75	12,200	2.5	556	10.535	+++	Pungent	0.6
10	Butanedione/$C_4H_6O_2$	431-03-8	86	88	7600	3.0	567	11.079	++	Chlorine-like	0.05
11	Butanol/$C_4H_{10}O$	71-36-3	74	117	580	2.6	607	11.187	+	Harsh	38
12	2-Pentanone/$C_5H_{10}O$	107-87-9	86	101	1600	3.0	647	11.328	+++	Aceton-like	28
13	n-Pentanal/$C_5H_{10}O$	110-62-3	86	133	3400	3.0	664	11.488	++++	Acrid, pungent	0.41
14	1-Pentanol/$C_5H_{12}O$	71-41-0	88	138	600	3.0	742	12.136	+	Fusel-like	100
15	2-Ethylfuran/C_6H_8O	3208-16-0	96	92	6666	-	756	12.375	++	Smoky burn	-
16	2-Ethylbutanal/$C_6H_{12}O$	97-96-1	100	116	2000	-	762	13.108	++	Pungent	-
17	2-Hexanone/$C_6H_{12}O$	591-78-6	100	126	360	3.5	770	13.247	+++	Sharp	24
18	1-Hexanal/$C_6H_{12}O$	66-25-1	100	129	1100	-	780	13.378	++++	Strong, green grass	0.28
19	2-Hexanol/$C_6H_{14}O$	626-93-7	102	136	2300	3.5	795	14.087	+	Sweet	6
20	2-Heptanone/$C_7H_{14}O$	110-43-0	114	151	200	3.9	859	14.252	++	Penetrating-spicy	6.8
21	Heptanal/$C_7H_{14}O$	111-71-7	114	153	3500	-	878	15.143	++++	Pungent, fatty	0.18

aldehydes can exhibit sweet and fruity aromas [36–39]. Conversely, alcohols above C_9 have fattier and more unpleasant oily odor notes [37], differently from most straight-chain C_4–C_9 alcohols with a fruit-like aroma.

2.1. Heat Transfer Theory

The MonoTrap^{TM} technology was selected based on its capacity to load the highest mass of analytes. The adsorption processes of solid, porous phases occur on the sorbent surface; the substantial thickness of the coating allows the analyte to be retained exclusively within the pores of the solid phase. This theoretical framework can be effectively employed to minimize the number of experiments required to predict trends in MonoTrap^{TM} analysis; however, the assumption of ideal conditions necessary for mathematical modeling should be verified. To calculate n, the mass (ng) of the adsorbed analyte in a sampling time t (s), using a porous coating, the theory of heat transfer can be applied [40–42]:

$$n = \frac{D_g \times A}{\delta} \times C_g \times t \quad (1)$$

where D_g represents the diffusion coefficient in air (cm^2 s^{-1}), A the surface of the sorbent phase (i.e., 2.67 cm^2), C_g the concentration of the analyte in the 20 mL vial (0.5 ng mL^{-1}), and δ the thickness of the boundary layer surrounding the MonoTrap^{TM} (cm), defined as follows:

$$\delta = 9.52 \times \frac{b}{R_e^{0.62} \times S_c^{0.38}} \quad (2)$$

with the Reynolds number (Re) expressed as $2ubv^{-1}$, where u is the linear air speed (cm s^{-1}), v is the air viscosity (0.014607 cm^2 s^{-1}), b is the radius of the MonoTrap^{TM} (0.19 cm), and Schmidt's number (Sc) is defined by vD_g^{-1}. In theory, linear air speeds exceeding 10 cm s^{-1} result in a δ value approaching zero, thereby rendering Equation (1) invalid. By means of Equation (1), we calculated the theoretical uptake (ng s^{-1}) and the theoretical SR (mL min^{-1}) for each of the 35 substances surveyed.

Table 2 presents the D_g, the theoretical uptakes and theoretical SRs for each substance, both at atmospheric pressure and in vacuum condition. Conversely, the SRs at 11.6 mbar are significantly enhanced, by a factor of approximately more than two orders of magnitude. This result indicates that the performance of MonoTrap^{TM} improves, suggesting that the signal enhancement observed when working at reduced pressure is likely due to an enhanced sorbent capacity, in addition to a more efficient stripping from the matrix.

Table 2. Diffusion coefficient [a,b], theoretical uptake and theoretical sampling rate (SR) for each substance surveyed, calculated at atmospheric pressure and in vacuum conditions (i.e., 11.6 mbar).

Num.	Compound Name	Atmospheric Pressure			Vacuum		
		D_g cm^2/s	Uptake ng/s	SR mL/min	D_g cm^2/s	Uptake ng/s	SR mL/min
1	Acetaldehyde	0.13	0.08	9.10	12.10	1.26	151
2	Ethanol	0.12	0.07	8.66	11.40	1.21	146
3	Propanal	0.11	0.07	8.20	10.00	1.12	134
4	tert-Butyl alcohol	0.091	0.06	7.29	8.40	1.01	121
5	Acetone	0.11	0.07	8.20	10.00	1.12	134
6	1-Propanol	0.1	0.06	7.73	9.50	1.08	130
7	Acetic acid	0.11	0.07	8.20	10.40	1.15	138
8	2-Butanone	0.094	0.06	7.44	8.70	1.03	123
9	Butanal	0.094	0.06	7.44	8.70	1.03	123

Table 2. Cont.

Num.	Compound Name	Atmospheric Pressure			Vacuum		
		D_g cm²/s	Uptake ng/s	SR mL/min	D_g cm²/s	Uptake ng/s	SR mL/min
10	Butanedione	0.092	0.06	7.34	8.40	1.01	121
11	Butanol	0.091	0.06	7.29	8.30	1.00	120
12	2-Pentanone	0.084	0.06	6.94	7.70	0.95	114
13	n-Pentanal	0.084	0.06	6.94	7.70	0.95	114
14	1-Pentanol	0.082	0.06	6.84	7.50	0.94	112
15	2-Ethylfuran	0.08	0.06	6.73	7.40	0.93	112
16	2-Ethylbutanal	0.077	0.05	6.58	7.10	0.91	109
17	2-Hexanone	0.077	0.05	6.58	7.00	0.90	108
18	1-Hexanal	0.077	0.05	6.58	7.00	0.90	108
19	2-Hexanol	0.075	0.05	6.47	6.90	0.89	107
20	2-Heptanone	0.071	0.05	6.25	6.50	0.86	103
21	Heptanal	0.071	0.05	6.25	6.50	0.86	103
22	Cyclohexanone	0.078	0.06	6.63	7.20	0.91	110
23	1-Heptanol	0.069	0.05	6.14	6.30	0.84	101
24	6-Methyl-2-heptanone	0.066	0.05	5.98	6.10	0.82	98.9
25	3-Octanone	0.066	0.05	5.98	6.10	0.82	98.9
26	Octanal	0.066	0.05	5.98	6.10	0.82	98.9
27	1-Octanol	0.065	0.05	5.92	5.90	0.81	96.9
28	2-Nonanone	0.062	0.05	5.75	5.70	0.79	94.9
29	Nonanal	0.062	0.05	5.75	5.70	0.79	94.9
30	1-Nonanol	0.061	0.05	5.69	5.60	0.78	93.8
31	2-Decanone	0.058	0.05	5.52	5.40	0.76	91.7
32	Decanal	0.058	0.05	5.52	5.40	0.76	91.7
33	1-Decanol	0.058	0.05	5.52	5.30	0.76	90.7
34	2-Undecanone	0.056	0.04	5.40	5.10	0.74	88.5
35	Undecanal	0.056	0.04	5.40	5.10	0.74	88.5

[a] Advamacs—TriMen Chemicals (Łodz, Poland). [b] U.S. Environmental Protection Agency (EPA)—EPA On-line Tools for Site Assessment Calculation.

2.2. Choice of the MonoTrap™ Adsorbent Phase Through a Chemometric Approach

The automation with a three-axis autosampler on-line with the GC allowed for a high-throughput analytical procedure, and it also delivered the ability to define the best conditions for the development of the method regarding the choice of the adsorbent phase to adopt.

The optimization of the extraction procedure on HMA, conducted by applying the 2^3 full factorial design [43–45], was performed by investigating three main variables: the RGPSTD and RGC18-TD MonoTrap™ (x_1), the extraction step conducted under vacuum (x_2), and the equilibration step (x_3). From preliminary studies, it was decided to exclude the use of RSC18-TD from the experimental matrix because it was able to extract less analytes from the matrix compared to the other adsorbent phases, and it could not study equilibration times like 12 min due to the decrease in sensitivity. These parameters were optimized to maximize the peak area intensities (y_1, y_2, y_3, and y_4) of four compounds (butanal, 1-hexanal, heptanal, decanal, respectively), performing only eight experiments. The results highlighted that the models computed for y_2, y_3, and y_4 were optimized when the three variables (x_1, x_2, and x_3) were in the high level (equal to 1), as shown in Figure 3. This implies that the MonoTrap^{TM} used to perform the extraction must be the RGC18-TD ($x_1 = 1$), the Vac extraction must be employed ($x_2 = 1$), and lastly the equilibration step must last 4 min ($x_3 = 1$).

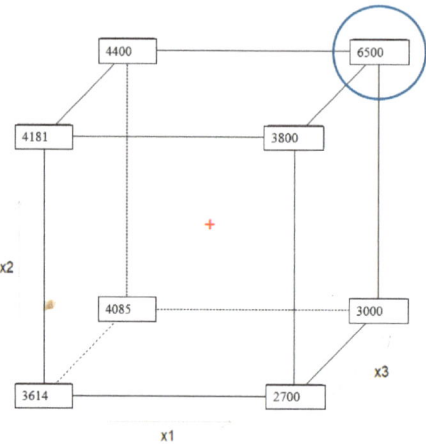

Figure 3. Experimental domain reporting the areas under the peak area intensities obtained for the eight experiments for the 1-hexanal. In blue, experiment number eight is circled, which allows one to obtain the highest sensitivity.

Instead, the model for y_1, which describes the sensitivity for butanal, has a different behavior compared to the other molecules, showing higher responses with the extraction performed at atmospheric pressure ($x_2 = -1$) and the equilibration time set at 8 min ($x_3 = -1$); see Figure 4 for further details.

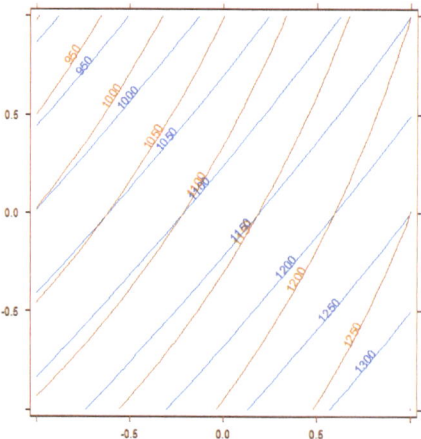

Figure 4. Overlapping of the contour plots obtained for the model describing y_1, peak areas of butanal. Blue lines describe the variables x_1 vs. x_2, and red lines describe the variable x_1 vs. x_3.

Thus, the experimental conditions selected to simultaneously determine the presence of VOCs at cut-off in HMA are those reported in Table 3 (experiment number eight) and highlighted in Figure 4, which optimize the signals for most of the analytes assessed. Finally, comparing Figures 4 and 5, it is understandable that the authors chose to select the optimal experimental conditions as $x_1 = 1$, $x_2 = 1$, and $x_3 = 1$ (top right corner of Figures 3 and 5) instead of $x_1 = 1$, $x_2 = -1$, $x_3 = -1$ (bottom right corner of same figures): the decrease in sensitivity for y_1 is lower compared to the one that other analytes would have.

Table 3. Description of the experiments performed to optimize the analytical method by DoE.

	Experimental Matrix			Experimental Plan		
Exp	x_1	x_2	x_3	MonoTrap^{TM}	Vac	Equilibration min
1	−1	−1	−1	RGPS TD	No	8
2	1	−1	−1	RGC18 TD	No	8
3	−1	1	−1	RGPS TD	Yes	8
4	1	1	−1	RGC18 TD	Yes	8
5	−1	−1	1	RGPS TD	No	4
6	1	−1	1	RGC18 TD	No	4
7	−1	1	1	RGPS TD	Yes	4
8	1	1	1	RGC18 TD	Yes	4

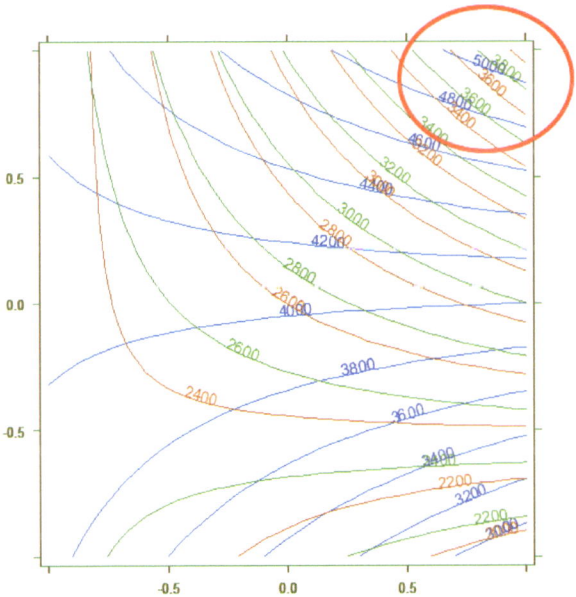

Figure 5. Overlapping of the contour plots obtained, computing variables x_1 vs. x_2 for the models of y_2 (blue lines), y_3 (red lines), and y_4 (green lines).

2.3. Vacuum Effect on HS-MMSE-MonoTrap^{TM} Sampling

The research conducted on HMA by Vac-HS-MMSE-MonoTrap^{TM} demonstrated that the utilization of a higher phase volume and layer thickness enables the successful coupling of sample heating at elevated temperatures (160 °C) and vacuum sampling. This approach reduces the equilibrium time, thereby maximizing VOC extraction, particularly for those with a higher molecular weight. This outcome aligns with the observations made at lower temperatures in other matrices [46]. In vacuum conditions, elevated sampling temperatures were demonstrated to result in a reduction in extraction efficiency compared to atmospheric pressure. This phenomenon was found to be associated with increased humidity levels occurring during the heating of the sample, with a more pronounced effect observed when an absorbent type (such as PDMS) is used [47]. Indeed, extraction efficiencies obtained under vacuum and at a mild sampling temperature are comparable with those at regular pressure and a much higher sample temperature; this is coherent with the general decrease in boiling point, observable in vacuum conditions, for organic compounds [40,48–50]. Lastly, elevated temperatures might reduce the effect of vacuum; the VP of analytes and volatile matrix components increases exponentially, therefore enhancing the total pressure inside the sample container. Following the theory of ideal gas, in an empty 20 mL HS crimp-top vial, evacuated at an absolute pressure of 8 mbar and then heated to 160 °C, the pressure increases up to 11.6 mbar. This condition can be extended even when the 20 mL HS crimp-top vial is loaded with a small amount of dry solid sample which also does not release large amounts of VOCs, since in the presence of water or other volatile compounds, the maximum vacuum that can be achieved depends on the liquid–vapor equilibrium in the phase diagram.

The analytes present within solid samples (adsorbed, dissolved, and/or in gaseous phases) result in greater resistance to volatilization compared with liquid samples [51]. In a modified form of Fick's law of diffusion, Yiantzi et al. [52] stated that a reduction in total pressure would result in a vapor flux increase at the solid surface, thereby accelerating the volatilization rate and shifting the equilibrium towards a higher analyte concentration in the HS. We found that the operational fundamentals of the sample for an ideal workflow using 20 mL HS vials included evacuation at 8 mbar for 10 s.

The greater response produced by volatiles was revealed in HMA 1-hexanal, heptanal, octanal, nonanal, and decanal using Vac assistance, while propanal and butanal with no-Vac. A comparison of performance, denoted as Vac/no-Vac ratios, can be calculated by dividing the compound response from Vac-HS-MMSE-MonoTrap^{TM} sampling by the compound response from HS-MMSE-MonoTrap^{TM} without vacuum assistance (Vac/no-Vac ratio), which resulting in the following values: hexanal 4.1, 3-octanone 7.7, octanal 8.1, 1-octanol 8.8, 2-nonanone 9.1, nonanal 9.2, 1-nonanol 9.8, and decanal 10.9. As the BP of target analytes increased, the Vac/no-Vac response ratios also generally increased in agreement with Solomou et al. [53]. When Vac is removed, there is poor response for 2-decanone, 2-undecanone, and undecanal.

3. Materials and Methods

3.1. Reagents

The *n*-alkanes butane (99%, CAS n. 106-97-8), pentane (99%, CAS n. 109-66-0), hexane (>95%, CAS n. 110-54-3), heptane (99%, CAS n. 142-82-5), octane (>99%, CAS n. 111-65-9), nonane (>99%, CAS n. 111-84-2), decane (99%, CAS n. 124-18-5), undecane (>99%, CAS n. 1120-21-4), dodecane (99%, CAS n. 112-40-3), tridecane (>99%, CAS n. 629-50-5), and the internal standard 2,4,6-trimethylpyridine (99%, CAS n. 108-75-8) were purchased from Supelco (Merck KGaA, Darmstadt, Germany) and used for evaluation of the (*i*) LTPRI and (*ii*) VOC cut-off based on the RT of the tridecane.

3.2. Vac-HS-MMSE-Monotrap^{TM} TD Procedure

A Kit Vac-SPME-Fiber (part no. 20-102) by ExtraTECH Analytical Solutions equipped with a conditioned/ready-to-use cylindrical Thermogreen^{TM} LB-2 septa (part no. 20608, Supelco) was purchased from Markes International Ltd. (Bridgend, UK) and used for all experiments for closure of the 20 mL HS crimp-top vial (Markes International Ltd.). A LABOPORT® N820G (KNF Service GmbH, Freiburg im Breisgau, Germany) pumping unit (8 mbar ultimate vacuum without gas ballast) connected to an 18-gauge (1.219 mm external diameter) needle was used to evacuate the air inside the 20 mL HS crimp-top vial. The needle was also used to support the MonoTrap^{TM}.

Disposable, ready-to-use, and preconditioned MonoTrap™ TD rods (external diameter 2.9 mm, internal diameter 1 mm, length 10 mm) were purchased from GL Sciences (Shinjuku, Tokyo, Japan) in ampoules for single-use: (*i*) RSC18-TD (part no. 1050-73201) as silica gel with ODS as the functional group for hydrophobic analytes with medium (250 °C) to high (300 °C) BP, (*ii*) RGC18-TD (part no. 1050-74201) as graphite carbon with ODS for polar or hydrophobic analytes with low (200 °C) to medium BP, and (*iii*) RGPS-TD (part no. 1050-74202) as graphite carbon with PDMS for polar or hydrophobic analytes with a low to medium BP.

3.3. Sample and Sampling

The HMA analyzed was the conventional Surface Layer (SL) made (w/w) of 5% bitumen (50/70 type), 65% aggregate, 8% filler, and 22% Reclaimed Asphalt Pavement (RAP). The relative humidity (RH) of the HMA-SL declared by the manufacturer was less than 1%. Two grams of HMA-SL samples and MonoTrap™ rod were loaded into a 20 mL HS crimp-top vial before pulling a vacuum. Then, the vial was heated at the bottom at 160 °C and cooled at the head with compressed air at 25 °C. After adsorption, the MonoTrap™ rod was placed in a liner and desorbed.

3.4. Three-Axis Autosampler and Multi-Mode GC Inlet Systems

A three-axis Shimadzu AOC-6000 Plus Multifunctional Autosampler (Shimadzu, Kyoto, Japan) was used, on-line with GC for a fully automated analysis. After the sampling, the MonoTrap™ containing the collected analytes was placed in a GC liner for thermal desorption and sealed with the Capping-De-Capping (CDC) station (GL Sciences) moved by the Automatic LINer EXchanger (LINEX, GL Sciences); both are accessories of the OPTIC-4 (GL Science) multi-mode GC inlet system. The analytes adsorbed on MonoTrap™ were desorbed into the GC–MS directly into the CryoFocus-4 (GL Sciences), a cryo-trap at the head of the GC column cooled by liquid nitrogen to sub-ambient temperature (−150 °C for 300 s). After trapping, the analytes were released from the cryo-trap using a fast heating (60 °C s^{-1} up to 290 °C), ensuring that they were introduced onto the capillary column in a very sharp band.

3.5. GC–MS/O

The GC instrument used was a Shimadzu GC-2030 with QP2020 NX (Shimadzu) MS detector. To select olfactometrically detected analytes from volatile analytes, a sniffing port PHASER Pro (GL Sciences) with Olfactory Voicegram and a GC-O Aroma Palette was configured in-line to GC-MS by a transfer-line. The sample was split 1:20 v/v; the effluent from the capillary column was equally split between the detection systems A J&W GC column VF-5ms (part no. CP8949, length 60 m × internal diameter 0.25 mm × film thickness 1 µm) provided by Agilent Inc. (Santa Clara, CA, USA) was used. Helium was used as carrier gas at 0.9 mL min^{-1}. The oven temperature program was 45 °C (5 min hold) to 320 °C (10 °C min^{-1}), with a final hold of 8 min. MS conditions were as follows: detector interface temperature 250 °C, ion source temperature 230 °C, ionization energy 70 eV, and mass range 28.5–300 amu. Based on the chromatogram data obtained from GC–MS, we used the National Institute of Standards and Technology 11 Mass Spectral Library and Flavor & Fragrance Natural & Synthetic Compounds GC–MS library (both from Shimadzu) for searching for mass spectra with a similarity score of 90% or higher to identify the analytes.

A panel of four sensory panelists (2 men and 2 women, aged from 25 to 34) from the laboratory staff with previous GC–O sniffing experience was assembled in the GC–MS/O study. When an aroma was being detected, the sensory panelists were asked to press a specific button on the instrument keypad to record the time, the description, and odor intensity (noted by a voice recording system). The scale used for intensity was 1–4 (1 = weak, 2 = moderate, 3 = strong, 4 = very strong). The analysis followed the guidelines of Pollien et al. [54]. Each sample was evaluated consecutively by each of the sensory panelists.

Figure 6 reports the scheme of the GC-MS instrument, equipped with the olfactory port, as well as a brief flow chart of the sampling process.

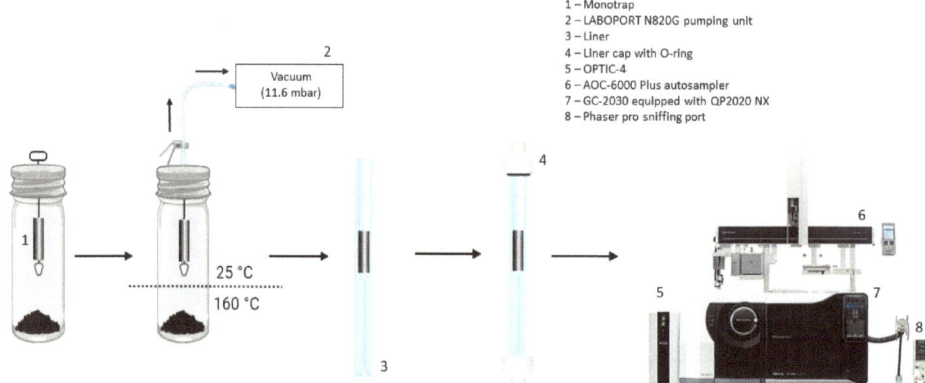

Figure 6. Flow chart and image of xyz-autosampler for the automated Vac-HS-MonoTrap^{TM} sampling on-line with GC-MS/O instrumentation.

3.6. Chemometric Tool

Microsoft Excel (version 18.0) was used to collect data, while an open source and R-based software (version 4.3.3), Chemometric Agile Tool (CAT) [55], was used to process it.

First, a 2^3 full factorial design was applied to optimize the analytical method developed on four compounds considered (butanal, 1-hexanal, heptanal, decanal), which means that three factors were studied at two levels each. The first factor (x_1) considered the use of RGPS-TD (low level, -1) and RGC18-TD (high level, 1) MonoTrap^{TM} in the extraction step. The second one (x_2) explained the use of Vac during the extraction step, low level for the extraction at atmospheric pressure and high level for the extraction under Vac. And last, the third factor (x_3) referred to the duration of the equilibration step, which was low level for 8 min and high level for 4 min.

3.7. Identification of VOCs by LTPRI

An important tool for the identification of compounds is the use of Retention Indexes that were developed originally by Kovats [56] for isothermal analysis and modified by van den Dool and Kratz [57] for linear temperature-programmed analysis. The most used is the latter, named the Linear Temperature-Programmed Retention Index (LTPRI) [58–62]. The LTPRI was defined under identical gas chromatographic conditions of the sample as follows:

$$LTPRI = 100 \times \frac{t_{R(A)} - t_{R(C)}}{t_{R(C+1)} - t_{R(C)}} + 100 \times C \qquad (3)$$

where $t_{R(A)}$ is the analyte retention time, $t_{R(C)}$ is the retention time of the n-alkane eluting immediately before the analyte, $t_{R(C+1)}$ is the retention time of the n-alkane eluting immediately after the analyte, and C is the number of carbon atoms for $t_{R(C)}$.

Volatile compounds were identified by matching the mass spectra to the database and the LTPRI of each compound with its reference values.

4. Conclusions

The introduction of miniaturized analytical solutions in recent years is noteworthy and consistent with the needs of Green Analytical Chemistry, a virtuous trend of continuous improvement in the framework of an increasing preservation of the environment. Several miniaturized techniques are currently available on the market and integrated with new analytical solutions, such as MonoTrap™ coupled with vacuum-assisted extraction.

As for the monitoring of VOC emissions from anthropic sources, i.e., HMA from industrial plans, the Vac-HS-MonoTrap™ with the GC–MS/O analytical approach has proved to be successful. The main advantages reside in a larger surface area, a high sensitivity, a high uptake, especially for the polar VOCs, and no need for derivatization steps. Moreover, reducing the pressure in the sample vial as a pre-equilibration step increases the degree of HS partitioning for dry solid samples. A chemometric approach was used to optimize the method with as few experiments as possible. The developed method, tested on real HMA samples, allowed for the generation of an emission fingerprint, represented by an MS chromatogram and a matching odorgram.

The main compounds associated with the HMA odor fingerprint result in aldehydes, ketones, and alcohols. Alcohols do not show evidence of toxic activity on reproductive systems or developing organisms, but their inhalation could provoke irritation. Although the potential risks associated with aldehyde and ketone exposure are well documented, the toxic mechanisms remain poorly understood.

The toxicological implications will require careful quantification of the compounds present in the emission, broadening the panorama regarding the odor component investigated here.

The combination of analytical chemistry, engineering, and biomedical science has enabled significant advances in the understanding of odorous emissions. Future progress has the potential to safeguard public health and environmental well-being while simultaneously supporting the achievement of sustainable development goals.

Author Contributions: Conceptualization, S.D., G.C. and R.G.; methodology, L.V., M.S. and D.C.; software, C.V. and I.R.; validation, R.G. and I.R.; formal analysis, N.F.; investigation, R.G. and P.S.; resources, N.M.; data curation, D.S.; writing—original draft preparation, S.D.; writing—review and editing, D.S.; visualization, N.F.; supervision, N.M. and P.S.; project administration, S.D.; funding acquisition, N.M. All authors have read and agreed to the published version of the manuscript.

Funding: This manuscript has been developed in the concept of the project FIN.E.ODOR. (FINgerprint ed Emissioni ODORigene di conglomerati bituminosi) by SITEB—associazione Strade ITaliane E Bitumi (Bologna, Italy) and the PIN—Polo Universitario Città di Prato (Prato, Italy) of the University of Florence.

Institutional Review Board Statement: Not applicable.

Informed Consent Statement: Not applicable.

Data Availability Statement: Data will be made available on request.

Acknowledgments: The authors of this work would like to thank Stefano Ravaioli and Alessandro Pesaresi, respectively, for the precious support and the collaboration in harvesting the HMA samples analyzed.

Conflicts of Interest: Author Domenico Cipriano is employed by the company Ricerca sul Sistema Energetico. The remaining authors declare that the research was conducted in the absence of any commercial or financial relationships that could be construed as a potential conflict of interest.

References

1. Polvara, E.; Roveda, L.; Invernizzi, M.; Capelli, L.; Sironi, S. Estimation of Emission Factors for Hazardous Air Pollutants from Petroleum Refineries. *J. Atmos.* **2021**, *12*, 1531. [CrossRef]
2. Wojnarowska, M.; Plichta, G.; Sagan, A.; Plichta, J.; Stobiecka, J.; Sołtysik, M. Odour nuisance and urban residents quality of life: A case study in Kraków's Plaszow district. *Urban Clim.* **2020**, *34*, 100704. [CrossRef]
3. Milad, A.; Babalghaith, A.M.; Al-Sabaeei, A.M.; Dulaimi, A.; Ali, A.; Reddy, S.S.; Bilema, M.; Yusoff, N.I.M. A Comparative Review of Hot and Warm Mix Asphalt Technologies from Environmental and Economic Perspectives: Towards a Sustainable Asphalt Pavement. *Int. J. Environ. Res. Public Health* **2022**, *19*, 14863. [CrossRef]
4. Heidari, M.; Bahrami, A.; Ghiasvand, A.R.; Shahna, F.G.; Soltanian, A.R. A novel needle trap device with single wall carbon nanotubes sol–gel sorbent packed for sampling and analysis of volatile organohalogen compounds in air. *Talanta* **2012**, *101*, 314–321. [CrossRef] [PubMed]

5. Wang, L.K.; Pereira, N.C.; Hung, Y.-T. *Air Pollution Control Engineering*; Humana Press Inc.: Totowa, NJ, USA, 2004.
6. Demeestere, K.; Dewulf, J.; De Witte, B.; Van Langenhove, H. Sample preparation for the analysis of volatile organic compounds in air and water matrices. *J. Chromatogr. A* **2007**, *1153*, 130–144. [CrossRef]
7. Technical Overview of Volatile Organic Compounds. EPA. 3 May 2024. Available online: https://www.epa.gov/indoor-air-quality-iaq/technical-overview-volatile-organic-compounds#definition (accessed on 3 July 2024).
8. Site Remediation and Waste Management Program and Department of Environmental Protection. Capping of Volatile Contaminants for the Impact to Ground Water Pathway. 2019. Available online: https://www.nj.gov/dep/srp/guidance/rs/igw_vo_capping.pdf (accessed on 3 July 2024).
9. Health Canada. Government of Canada. Available online: https://www.canada.ca/en/health-canada.html (accessed on 3 July 2024).
10. Bokowa, A.; Diaz, C.; Koziel, J.A.; McGinley, M.; Barclay, J.; Schauberger, G.; Guillot, J.-M.; Sneath, R.; Capelli, L.; Zorich, V.; et al. Summary and Overview of the Odour Regulations Worldwide. *Atmosphere* **2021**, *12*, 206. [CrossRef]
11. Li, J.; Zou, K.; Li, W.; Wang, G.; Yang, W. Olfactory Characterization of Typical Odorous Pollutants Part I: Relationship Between the Hedonic Tone and Odor Concentration. *Atmosphere* **2019**, *10*, 524. [CrossRef]
12. Ministero dell'Ambiente e della Sicurezza Energetica. Decreto Direttoriale MinAmbiente 28 Giugno 2023, n. 309. Rete Ambiente: Osservatorio Normativa Ambientale. Available online: https://www.reteambiente.it/normativa/51893/decreto-direttoriale-minambiente-28-giugno-2023-n-309/ (accessed on 3 July 2024).
13. Emissioni da Sorgente Fissa—Determinazione della Concentrazione di Odore Mediante Olfattometria Dinamica e della Portata di Odore. UNI EN 13725: 2022. Available online: https://store.uni.com/uni-en-13725-2022 (accessed on 7 August 2024).
14. Brattoli, M.; Cisternino, E.; Dambruoso, P.R.; De Gennaro, G.; Giungato, P.; Mazzone, A.; Palmisani, J.; Tutino, M. Gas Chromatography Analysis with Olfactometric Detection (GC-O) as a Useful Methodology for Chemical Characterization of Odorous Compounds. *Sensors* **2013**, *13*, 16759–16800. [CrossRef]
15. Kurowska-Susdorf, A.; Zwierżdżyński, M.; Bevanda, A.M.; Talić, S.; Ivanković, A.; Płotka-Wasylka, J. Green analytical chemistry: Social dimension and teaching. *TrAC Trends Anal. Chem.* **2019**, *111*, 185–196. [CrossRef]
16. Dugheri, S.; Bonari, A.; Gentili, M.; Cappelli, G.; Pompilio, I.; Bossi, C.; Arcangeli, G.; Campagna, M.; Mucci, N. High-Throughput Analysis of Selected Urinary Hydroxy Polycyclic Aromatic Hydrocarbons by an Innovative Automated Solid-Phase Microextraction. *Molecules* **2018**, *23*, 1869. [CrossRef]
17. Dugheri, S.; Mucci, N.; Bonari, A.; Marrubini, G.; Cappelli, G.; Ubiali, D.; Campagna, M.; Montalti, M.; Arcangeli, G. Solid Phase Microextraction Techniques Used for Gas Chromatography: A Review. *Acta Chromatogr.* **2020**, *32*, 1–9. [CrossRef]
18. Dugheri, S.; Mucci, N.; Cappelli, G.; Trevisani, L.; Bonari, A.; Bucaletti, E.; Squillaci, D.; Arcangeli, G. Advanced Solid-Phase Microextraction Techniques and Related Automation: A Review of Commercially Available Technologies. *J. Anal. Methods Chem.* **2022**, *2022*, 8690569. [CrossRef] [PubMed]
19. Hjertén, S.; Liao, J.-L.; Zhang, R. High-Performance Liquid Chromatography on Continuous Polymer Beds. *J. Chromatogr. A* **1989**, *473*, 273–275. [CrossRef]
20. China-Japan-Korea Symposium on Analytical Chemistry: Novel Approach for Aroma Components Analysis Using Monolithic Material Sorptive Extraction Method. In Proceedings of the JAIMA Conference at Makuhari, Chiba, Japan, 31 August–2 September 2009. Available online: https://www.jsac.or.jp/~gc/pdf/2009CJK/2009CJKsymposiumabstract.pdf (accessed on 3 June 2024).
21. Hashi, Y.; Son, H.H.; Cha, E.J.; Lee, D.S. China-Japan-Korea Symposium on Analytical Chemistry. In Proceedings of the JAIMA Conference: New Approach of Simple Pretreatment for GCMS Analysis by Using Monolithic Material Sorption Extraction for Determination of 16 PAHs, Chiba, Japan, 31 August–2 September 2009. Available online: http://www.jsac.or.jp/~gc/pdf/2009CJK/2009CJKProgramtable_oral.pdf (accessed on 3 June 2024).
22. Sato, A.; Terashima, H.; Takei, Y. Monolith Adsorbent and Method and Apparatus for Adsorbing Samples with the Same—EP2161573A1. European Patent Office. 2008. Available online: https://patents.google.com/patent/EP2161573A1/en (accessed on 3 June 2024).
23. Brunton, N.P.; Cronin, D.A.; Monahan, F.J. The Effects of Temperature and Pressure on the Performance of Carboxen/PDMS Fibres During Solid Phase Microextraction (SPME) of Headspace Volatiles from Cooked and Raw Turkey Breast. *Flavour Fragr. J.* **2001**, *16*, 294–302. [CrossRef]
24. Psillakis, E.; Yiantzi, E.; Sanchez-Prado, L.; Kalogerakis, N. Vacuum-Assisted Headspace Solid Phase Microextraction: Improved Extraction of Semivolatiles by Non-Equilibrium Headspace Sampling Under Reduced Pressure Conditions. *Anal. Chim. Acta* **2012**, *742*, 30–36. [CrossRef]
25. Psillakis, E. The Effect of Vacuum: An Emerging Experimental Parameter to Consider During Headspace Microextraction Sampling. *Anal. Bioanal. Chem.* **2020**, *412*, 5989–5997. [CrossRef]
26. Trujillo-Rodríguez, M.J.; Pino, V.; Psillakis, E.; Anderson, J.L.; Ayala, J.H.; Yiantzi, E.; Afonso, A.M. Vacuum-Assisted Headspace-Solid Phase Microextraction for Determining Volatile Free Fatty Acids and Phenols. Investigations on the Effect of Pressure on Competitive Adsorption Phenomena in a Multicomponent System. *Anal. Chim. Acta* **2017**, *962*, 41–51. [CrossRef]
27. Thomas, S.L.; Myers, C.; Herrington, J.S.; Schug, K.A. Investigation of Operational Fundamentals for Vacuum-Assisted Headspace High-Capacity Solid-Phase Microextraction and Gas Chromatographic Analysis of Semivolatile Compounds from a Model Solid Sample. *J. SeSci.* **2024**, *47*, 2300779. [CrossRef]

28. Sýkora, M.; Vítová, E.; Jeleń, H.H. Application of Vacuum Solid-Phase Microextraction for the Analysis of Semi-Hard Cheese Volatiles. *Eur. Food Res. Technol.* **2020**, *246*, 573–580. [CrossRef]
29. Autelitano, F.; Bianchi, F.; Giuliani, F. Airborne Emissions of Asphalt/Wax Blends for Warm Mix Asphalt Production. *J. Clean. Prod.* **2017**, *164*, 749–756. [CrossRef]
30. Autelitano, F.; Giuliani, F. Analytical Assessment of Asphalt Odor Patterns in Hot Mix Asphalt Production. *J. Clean. Prod.* **2018**, *172*, 1212–1223. [CrossRef]
31. Chlebnikovas, A.; Marčiulaitienė, E.; Šernas, O.; Škultecké, J.; Januševičius, T. Research on Air Pollutants and Odour Emissions from Paving Hot-Mix Asphalt with End-of-Life Tyre Rubber. *Environ. Int.* **2023**, *181*, 108281. [CrossRef] [PubMed]
32. Li, J.; Qin, Y.; Zhang, X.; Shan, B.; Liu, C. Emission Characteristics, Environmental Impacts, and Health Risks of Volatile Organic Compounds from Asphalt Materials: A State-of-the-Art Review. *Energy Fuels* **2024**, *38*, 4787–4802. [CrossRef]
33. International Labour Office. International Chemical Control Toolkit-Draft Guidelines. Available online: https://webapps.ilo.org/static/english/protection/safework/ctrl_banding/toolkit/icct/guide.pdf (accessed on 3 July 2024).
34. Yoko, Y.; Eiichiro, N. Measurement of Odor Threshold by Triangular Odor Bag Method. 2003. Available online: https://www.env.go.jp/en/air/odor/measure/02_3_2.pdf (accessed on 3 July 2024).
35. Royal Society of Chemistry. ChemSpider—Search and Share Chemistry. Available online: https://www.chemspider.com/ (accessed on 3 July 2024).
36. Giri, A.; Osako, K.; Okamoto, A.; Ohshima, T. Olfactometric Characterization of Aroma Active Compounds in Fermented Fish Paste in Comparison with Fish Sauce, Fermented Soy Paste and Sauce Products. *Food Res. Int.* **2010**, *43*, 1027–1040. [CrossRef]
37. Jeleń, H.; Gracka, A. Characterization of Aroma Compounds: Structure, Physico-Chemical and Sensory Properties. In *Flavour*, 1st ed.; Guichard, E., Salles, C., Morzel, M., Le Bon, A., Eds.; Wiley: Hoboken, NJ, USA, 2016; pp. 126–153. [CrossRef]
38. O'Brien, P.J.; Siraki, A.G.; Shangari, N. Aldehyde Sources, Metabolism, Molecular Toxicity Mechanisms, and Possible Effects on Human Health. *Crit. Rev. Toxicol.* **2005**, *35*, 609–662. [CrossRef]
39. LoPachin, R.M.; Gavin, T. Molecular Mechanisms of Aldehyde Toxicity: A Chemical Perspective. *Chem. Res. Toxicol.* **2014**, *27*, 1081–1091. [CrossRef] [PubMed]
40. Carslaw, H.S.; Jaeger, J.C. *Conduction of Heat in Solids*; Clarendon Press: Oxford, UK, 1959.
41. Incropera, F.P.; Dewitt, D.P.; Bergman, T.L.; Lavine, A.S. *Fundamentals of Heat and Mass Transfer*, 6th ed.; John Wiley & Sons: Hoboken, NJ, USA, 2007. Available online: https://hyominsite.wordpress.com/wp-content/uploads/2015/03/fundamentals-of-heat-and-mass-transfer-6th-edition.pdf (accessed on 1 July 2024).
42. Koziel, J.; Jia, M.; Pawliszyn, J. Air Sampling with Porous Solid-Phase Microextraction Fibers. *Anal. Chem.* **2000**, *72*, 5178–5186. [CrossRef] [PubMed]
43. Marrubini, G.; Dugheri, S.; Cappelli, G.; Arcangeli, G.; Mucci, N.; Appelblad, P.; Melzi, C.; Speltini, A. Experimental Designs for Solid-Phase Microextraction Method Development in Bioanalysis: A Review. *Anal. Chim. Acta* **2020**, *1119*, 77–100. [CrossRef]
44. Leardi, R. Experimental Design in Chemistry: A Tutorial. *Anal. Chim. Acta* **2009**, *652*, 161–172. [CrossRef]
45. Marrubini, G.; Melzi, C. *Trattamento dei Dati e Progettazione degli Esperimenti per le Scienze Chimiche e Farmaceutiche*; McGraw-Hill: New York, NY, USA, 2024; Volume 1. Available online: https://www.mheducation.it/trattamento-dei-dati-e-progettazione-degli-esperimenti-per-le-scienze-chimiche-e-farmaceutiche-9788838613463-italy (accessed on 1 July 2024).
46. Psillakis, E. Vacuum-Assisted Headspace Solid-Phase Microextraction: A Tutorial Review. *Anal. Chim. Acta* **2017**, *986*, 12–24. [CrossRef]
47. Psillakis, E.; Yiantzi, E.; Kalogerakis, N. Downsizing Vacuum-Assisted Headspace Solid Phase Microextraction. *J. Chromatogr. A* **2013**, *1300*, 119–126. [CrossRef] [PubMed]
48. Vakinti, M.; Mela, S.-M.; Fernández, E.; Psillakis, E. Room Temperature and Sensitive Determination of Haloanisoles in Wine Using Vacuum-Assisted Headspace Solid-Phase Microextraction. *J. Chromatogr. A* **2019**, *1602*, 142–149. [CrossRef] [PubMed]
49. Glykioti, M.-L.; Yiantzi, E.; Psillakis, E. Room Temperature Determination of Earthy-Musty Odor Compounds in Water Using Vacuum-Assisted Headspace Solid-Phase Microextraction. *Anal. Methods* **2016**, *8*, 8065–8071. [CrossRef]
50. Fatima, S.; Govardhan, B.; Kalyani, S.; Sridhar, S. Extraction of Volatile Organic Compounds from Water and Wastewater by Vacuum-Driven Membrane Process: A Comprehensive Review. *Chem. Eng. J.* **2022**, *434*, 134664. [CrossRef]
51. Jury, W.A.; Russo, D.; Streile, G.; El Abd, H. Evaluation of Volatilization by Organic Chemicals Residing Below the Soil Surface. *Water Resour. Res.* **1990**, *26*, 13–20. [CrossRef]
52. Yiantzi, E.; Kalogerakis, N.; Psillakis, E. Vacuum-Assisted Headspace Solid Phase Microextraction of Polycyclic Aromatic Hydrocarbons in Solid Samples. *Anal. Chim. Acta* **2015**, *890*, 108–116. [CrossRef]
53. Solomou, N.; Bicchi, C.; Sgorbini, B.; Psillakis, E. Vacuum-Assisted Headspace Sorptive Extraction: Theoretical Considerations and Proof-of-Concept Extraction of Polycyclic Aromatic Hydrocarbons from Water Samples. *Anal. Chim. Acta* **2020**, *1096*, 100–107. [CrossRef]
54. Pollien, P.; Ott, A.; Montigon, F.; Baumgartner, M.; Muñoz-Box, R.; Chaintreau, A. Hyphenated Headspace-Gas Chromatography-Sniffing Technique: Screening of Impact Odorants and Quantitative Aromagram Comparisons. *J. Agric. Food Chem.* **1997**, *45*, 2630–2637. [CrossRef]
55. Leardi, R.; Melzi, C.; Polotti, G. CAT (Chemometric Agile Tool). Available online: http://gruppochemiometria.it/index.php/software (accessed on 20 June 2024).

56. Kováts, E. Gas-Chromatographische Charakterisierung Organischer Verbindungen. Teil 1: Retentionsindices Aliphatischer Halogenide, Alkohole, Aldehyde und Ketone. *Helv. Chim. Acta* **1958**, *41*, 1915–1932. [CrossRef]
57. Van Den Dool, H.; Kratz, P.D. A Generalization of the Retention Index System Including Linear Temperature Programmed Gas—Liquid Partition Chromatography. *J. Chromatogr. A* **1963**, *11*, 463–471. [CrossRef]
58. Pacenti, M.; Dugheri, S.; Gagliano-Candela, R.; Strisciullo, G.; Franchi, E.; Degli Esposti, F.; Perchiazzi, N.; Boccalon, P.; Arcangeli, G.; Cupelli, V. Analysis of 2-Chloroacetophenone in Air by Multi-Fiber Solid-Phase Microextraction and Fast Gas Chromatography-Mass Spectrometry. *Acta Chromatogr.* **2009**, *21*, 379–397. [CrossRef]
59. Ji, S.; Gu, S.; Wang, X.; Wu, N. Comparison of Olfactometrically Detected Compounds and Aroma Properties of Four Different Edible Parts of Chinese Mitten Crab. *Fish. Sci.* **2015**, *81*, 1157–1167. [CrossRef]
60. Wu, N.; Gu, S.; Tao, N.; Wang, X.; Ji, S. Characterization of Important Odorants in Steamed Male Chinese Mitten Crab (*Eriocheir sinensis*) Using Gas Chromatography-Mass Spectrometry-Olfactometry. *J. Food Sci.* **2014**, *79*, C1250–C1259. [CrossRef] [PubMed]
61. Pacenti, M.; Dugheri, S.; Traldi, P.; Degli Esposti, F.; Perchiazzi, N.; Franchi, E.; Calamante, M.; Kikic, I.; Alessi, P.; Bonacchi, A.; et al. New Automated and High-Throughput Quantitative Analysis of Urinary Ketones by Multifiber Exchange-Solid Phase Microextraction Coupled to Fast Gas Chromatography/Negative Chemical-Electron Ionization/Mass Spectrometry. *J. Autom. Methods Manag. Chem.* **2010**, *2010*, 972926. [CrossRef]
62. Lee, M.L.; Vassilaros, D.L.; White, C.M. Retention Indices for Programmed-Temperature Capillary-Column Gas Chromatography of Polycyclic Aromatic Hydrocarbons. *Anal. Chem.* **1979**, *51*, 768–773. [CrossRef]

Disclaimer/Publisher's Note: The statements, opinions and data contained in all publications are solely those of the individual author(s) and contributor(s) and not of MDPI and/or the editor(s). MDPI and/or the editor(s) disclaim responsibility for any injury to people or property resulting from any ideas, methods, instructions or products referred to in the content.

Article

The Difference of Volatile Compounds in Female and Male Buds of *Trichosanthes anguina* L. Based on HS-SPME-GC-MS and Multivariate Statistical Analysis

Pingping Song, Bo Xu, Zhenying Liu, Yunxia Cheng and Zhimao Chao *

Institute of Chinese Materia Medica, China Academy of Chinese Medical Sciences, Beijing 100700, China
* Correspondence: chaozhimao@163.com or zmchao@icmm.ac.cn; Tel.: +86-135-2270-5161

Abstract: *Trichosanthes anguina* L. (family Cucurbitaceae) is a monoecious and diclinous plant that can be consumed as a vegetable and has anti-inflammatory and antioxidant effects. The chemical composition and content of volatile compounds in female and male buds of *T. anguina* were explored by headspace solid-phase microextraction-gas chromatography-mass spectrometry (HS-SPME-GC-MS) technology combined with multivariate statistical analysis. The results showed that the content of the volatile compounds was different between female and male buds. 2,2,6-trimethyl-6-vinyltetrahydro-2H-pyran-3-ol and 2,2,6-trimethyl-6-vinyldihydro-2H-pyran-3(4H)-one were the main volatile compounds in both female and male buds. Based on the multivariate statistical analysis of orthogonal projections to latent structures discriminant analysis (OPLS-DA) and *t*-test, the content of seven compounds was significantly different between female and male buds. The content of three compounds in male buds was higher than that in female, i.e., (E)-4,8-dimethyl-1,3,7-nonatriene, 1,5,9,9-tetramethyl-1,4,7-cycloundecatriene, and (E)-caryophyllene. Conversely, the content of (Z)-4-hexen-1-ol, (Z)-3-hexenyl benzoate, (Z)-3-hexenyl salicylate, and 2-hexen-1-ol in female buds was higher than that in male buds. This is the first report on the difference in the volatile compounds between female and male buds of *T. anguina*, which enriches the basic research on the monoecious and diclinous plant and provides a reference for the study of plant sex differentiation.

Keywords: *Trichosanthes anguina*; HS-SPME-GC-MS; monoecious and diclinous plant; bud; volatile compound; OPLS-DA

Citation: Song, P.; Xu, B.; Liu, Z.; Cheng, Y.; Chao, Z. The Difference of Volatile Compounds in Female and Male Buds of *Trichosanthes anguina* L. Based on HS-SPME-GC-MS and Multivariate Statistical Analysis. *Molecules* 2022, 27, 7021. https://doi.org/10.3390/molecules27207021

Academic Editor: Hiroyuki Kataoka

Received: 15 September 2022
Accepted: 14 October 2022
Published: 18 October 2022

Publisher's Note: MDPI stays neutral with regard to jurisdictional claims in published maps and institutional affiliations.

Copyright: © 2022 by the authors. Licensee MDPI, Basel, Switzerland. This article is an open access article distributed under the terms and conditions of the Creative Commons Attribution (CC BY) license (https://creativecommons.org/licenses/by/4.0/).

1. Introduction

Trichosanthes anguina L. (family Cucurbitaceae), an annual climbing herb, is known as snake gourd or snake bean for its slender (up to 200 cm in length and 3 cm in diameter), twisted, and snake-like shape fruit. It is originated in India and Malaysia and commonly cultivated in tropical and subtropical areas and northern China [1]. Its tender fruit is a popular vegetable, which contains proteins, carbohydrates, cellulose, fat, and a variety of minerals [2]. The fruit, seed, and root of *T. anguina* can be used in traditional Chinese medicine for some effects in clearing heat and generating fluid, moisturizing the lung, eliminating dampness, and destroying parasites [3]. It has anti-inflammatory and antioxidant activities and can be used for the treatment of malaria and bronchitis [4]. In addition, two proteins of TR3 from the root and of TS3 from the seed have cytotoxic activity on cancer cell lines [5].

The seed of *T. anguina* contains 3% free sugar, 7% starch, 20% protein, and 43% fat oil, which consists of punicic acid, palmitic acid, stearic acid, oleic acid, and linoleic acid [6,7]. The fruit contains chlorogenic acid, isochlorogenic acid, *p*-coumaric acid, vanillic acid, ferulic acid, protocatechuic acid, caffeic acid, salicylic acid, phloretic acid, 3-indole acetic acid, phloroglucinol, quercetin, calcium, iron, phosphorus, folic acid, carotene, and vitamins B1, B2, and C [8,9]. The leaf contains kaempferol-3-*O*-β-galactoside and kaempferol-3-*O*-β-sophoroside [10].

Most flowering plants are bisexual plants whose flowers have both pistils and stamens. In order to avoid the decrease in the survival adaptability of offspring caused by self-inbreeding or self-mating [11], some plants evolved to unisexual; that is, there are only pistils or stamens in each flower, called female flower or male flower. The plant of which female and male flowers grow in different plants is called a dioecious plant, but in the same plant is known as a monoecious and diclinous plant. There are many differences in apparent structure, physiological function, and chemical components between female and male flowers of dioecious plants, such as *Populus tomentosa* (family Salicaceae) [12], *Herpetospermum pedunculosum* (family Cucurbitaceas), and *T. kirilowii* [13,14]. For monoecious and diclinous plants, the difference between female and male flowers also exists; for example, the chemical compounds from flowers of *Cucurbita moschata* (family Cucurbitaceas) [15] and the flower morphology of *Croton sarcopetalus* (family Euphorbiaceae) [16]. However, there are few comparative reports on the difference between female and male flowers of some monoecious and diclinous plants up to now.

The plant of *T. anguina* is one of the monoecious and diclinous plants. Its calyxes are green, five-lobed, and villous. Both female and male flowers bloom at night, but the blooming time of females is later than that of male flowers. Their corollas are pale yellowish green at the initial opening and turn white at full opening, five-lobed, and with branched and curly terminals. Female flowers are solitary and male flowers are raceme. Female flowers are 5–6 cm long and located at the top of juvenile fruit with a green pistil, two-lobed stigma, and inferior ovary. Male flowers are 3–5 cm long with three stamens and connate anthers. The female and male flowers and tender fruit are shown in Figure 1. The chemical constituents of female and male buds of *T. anguina* were studied, and the difference in sex was compared to make up for the deficiency of sex study of the monoecious and diclinous plants.

Figure 1. The female flower (**a**), male flower (**b**), and tender fruit (**c**) of *T. anguina*.

Solid-phase microextraction (SPME) is a sample treatment method invented by Pawliszyn and Arthur in the 1990s [17], which needs no solvent to extract the sample components. The SPME fiber is used to extract samples directly whose weight is usually less than 5 mg [18]. Headspace (HS) technology combined with SPME is a convenient extraction method. The sample is directly placed in a sealed headspace bottle, and an SPME fiber is put into the headspace bottle and located above the sample. During the heating and extraction process, volatile substances diffuse from the sample to the SPME fiber and are enriched in the fiber. This extraction method is so quick, less than one hour, and does not destroy the sample or require a liquid solvent. HS-SPME is often combined with gas chromatography-mass spectrometry (GC-MS) for widespread use in some volatile substances analysis [19,20].

The material of SPME fiber coating is crucial to the extraction, which determines the fiber's affinity to compounds in the sample. Since the introduction of SPME, many coating materials have been employed in the analysis. Polydimethylsiloxane (PDMS) fiber could be used to determine the content of acrylamide from coffee beans [21]. Polyacrylate (PA) fiber was suitable for the quantification of sesquiterpenes in *Zingiber zerumbet* L. (family Zingiberaceae) volatiles [22]. Polydimethylsiloxane/divinylbenzene (PDMS/DVB) could be applied for the determination of the esters of carboxylic acids in insect lipids [23]. Carboxen/polydimethylsiloxane (CAR/PDMS) fiber was used for the analysis of volatile compounds from European ciders [24]. Divinylbenzene/carboxen/polydimethylsiloxane (DVB/CAR/PDMS) fiber was able to extract more chemical diverse volatile compounds, including ketones, aldehydes, alcohols, terpenoids, and others from *Medicago sativa* L. (family Leguminosae) compared with PDMS and CAR/PDMS fibers [25] and to analyze a variety of plants and their oils such as sweet potato, tomato, and olive oils [26–28]. In this study, DVB/CAR/PDMS fiber was selected due to its better ability to provide a more comprehensive chemical profile of plants and its great extraction effect in the pre-experiment.

The buds, just before blooming, were picked up as experimental samples for the detection of volatile compounds in order to avoid any contamination of external substances. The HS-SPME-GC-MS method was used to detect and identify the volatile compounds from the female and male buds of *T. anguina*. The multivariate statistical analysis of OPLS-DA and *t*-test were carried out to screen out the differential compounds between female and male buds for revealing the difference in the chemical composition of genders of monoecious and diclinous plants.

2. Results

2.1. Volatile Compounds of GC-MS Analysis

Both female and male fresh buds of *T. anguina* just before blooming were individually picked up at night. The samples were weighed as f1 0.1063 g, f2 0.1057 g, f3 0.0975 g, m1 0.0842 g, m2 0.0826 g, and m3 0.0849 g. The volatile compounds were analyzed by HS-SPME-GC-MS successfully. The total ion chromatography is shown in Figure 2. A total of 53 compounds were identified on the base of mass spectrum and retention index (RI). The relative content of these volatile compounds was calculated with peak area normalization and shown in Table 1. The total ion chromatograms of volatile compounds for three female and three male bud samples are shown in Figure 3.

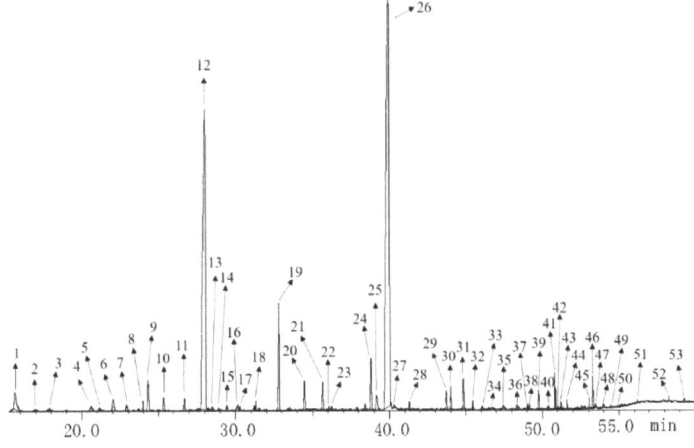

Figure 2. GC-MS total ion chromatogram of volatile compounds of sample f2.

Table 1. Volatile compounds of GC-MS analysis of female and male *T. anguina* buds.

No.	RI	Compound	MF	Fragment (m/z)	CAS	Female			Male			Reports
						f1	f2	f3	m1	m2	m3	
1	1214	3-Methyl-1-butanol	$C_5H_{12}O$	87/70/57/55	123-51-3	2.00	2.67	2.29	2.58	2.41	2.30	-
2	1239	(E)-3,7-Dimethyl-1,3,6-octatriene	$C_{10}H_{16}$	136/121/93/77	3779-61-1	0.21	0.18	0.12	0.09	0.13	0.19	Jasmine tea [29]
3	1256	1-Chloro-3-methylbutane	$C_5H_{11}Cl$	106/91/42/27	107-84-6	0.23	0.30	0.32	0.21	0.19	0.23	-
4	1311	(E)-4,8-Dimethyl-1,3,7-nonatriene	$C_{11}H_{18}$	150/135/69/53	19945-61-0	0.58	0.62	0.50	2.24	4.63	3.57	*Grevillea robusta* [30]
5	1322	(Z)-3-Hexenyl acetate	$C_8H_{14}O_2$	82/67/54	3681-71-8	0.23	0.41	0.27	0.21	0.17	0.21	*Passiflora mollissima* [31]
6	1341	(Z)-2-Hexenyl acetate	$C_8H_{14}O_2$	142/100/67/55	56922-75-9	0.41	0.69	0.40	0.43	0.37	0.49	-
7	1359	1-Hexanol	$C_6H_{14}O$	101/84/69/56	111-27-3	0.41	0.54	0.61	0.56	0.47	0.59	-
8	1375	(E,Z)-2,6-Dimethylocta-2,4,6-triene	$C_{10}H_{16}$	136/121/105/79	7216-56-0	0.13	0.11	0.05	0.07	0.09	0.14	*Pistacia atlantica* [32]
9	1389	(Z)-4-Hexen-1-ol	$C_6H_{12}O$	100/82/67/55	928-91-6	1.90	2.19	2.11	1.29	1.31	1.40	Olive oil [28]
10	1412	2-Hexen-1-ol	$C_6H_{12}O$	100/82/57	928-95-0	1.06	1.13	1.03	0.81	0.75	0.86	
11	1443	(E)-2-Methyl-2-vinyl-(1-hydroxy-1-methylethyl)tetrahydrofuran	$C_{10}H_{18}O_2$	111/94/68/59	34995-77-2	0.50	0.51	0.43	0.41	0.37	0.40	Pu-erh tea [33]
12	1473	2,2,6-Trimethyl-6-vinyldihydro-2H-pyran-3(4H)-one	$C_{12}H_{20}O_3$	168/110/82/68	33933-72-1	30.28	35.82	27.44	28.91	28.34	32.95	-
13	1480	(E)-2-Hexenyl butyrate	$C_{10}H_{18}O_2$	170/155/71/55	53398-83-7	0.07	0.08	0.07	0.06	0.06	0.05	Longjing tea [34]
14	1484	8-Isopropyl-1,3-dimethyltricyclo[4.4.0.0(2,7)]dec-3-ene	$C_{15}H_{24}$	204/161/119/105/81/55	138874-68-7	0.07	0.07	0.05	0.07	0.10	0.10	*Artemisia ordosica* [35]
15	1493	(Z)-3-Hexenyl pentanoate	$C_{11}H_{20}O_2$	103/82/67/55	35852-46-1	0.09	0.13	0.24	0.09	0.05	0.06	*Lysimachia paridiformis* [36]
16	1506	(E)-2-Hexenyl pentanoate	$C_{11}H_{20}O_2$	184/169/85/57	56922-74-8	0.19	0.22	0.41	0.20	0.16	0.16	-
17	1526	8,8-Dimethoxyoct-2-yl 2-formyl-4,6-dimethoxybenzoate	$C_{20}H_{30}O_7$	209/193/165/71	312305-58-1	0.18	0.18	0.25	0.20	0.23	0.18	-
18	1556	3,7-Dimethyl-1,6-octadien-3-ol	$C_{10}H_{18}O$	136/121/93/71/55	78-70-6	0.03	0.02	0.02	0.02	0.01	0.01	La Rioja grape [37]
19	1588	(E)-4,11,11-Trimethyl-8-methylene-bicyclo[7.2.0]undec-4-ene	$C_{15}H_{24}$	205/189/161/133/107/93/69	87-44-5	1.79	1.92	1.73	2.10	2.52	2.49	*Aquilegia japonica* [38]
20	1630	3-Acetoxy-2,2,6-trimethyl-6-vinyltetrahydropyran	$C_{12}H_{20}O_3$	197/179/155/137/94/68/55	67674-42-4	0.38	0.73	0.34	0.53	0.36	0.58	-
21	1660	1,5,9,9-Tetramethyl-1,4,7-cycloundecatriene	$C_{15}H_{24}$	204/189/161/147/121/93/80	515812-15-4	1.10	1.15	0.98	1.23	1.54	1.49	*Artemisia dracunculus* [39]
22	1670	(Z)-3-Hexenyl tiglate	$C_{11}H_{18}O_2$	101/83/71/55	67883-79-8	0.08	0.10	0.04	0.16	0.29	0.29	*Gardenia jasminoides* [40]
23	1675	(E)-2-Hexenyl hexanoate	$C_{12}H_{22}O_2$	198/169/99/71	53398-86-0	0.13	0.13	0.14	0.08	0.06	0.06	Longjing tea [34]
24	1733	Benzyl acetate	$C_9H_{10}O_2$	150/128/91/79	140-11-4	0.26	0.16	0.28	0.24	0.21	0.20	*Prunus mume* [41]

Table 1. Cont.

No.	RI	Compound	MF	Fragment (m/z)	CAS	Female f1	Female f2	Female f3	Male m1	Male m2	Male m3	Reports
25	1753	1-Isopropyl-7-methyl-4-methylene-1,2,3,4,4a,5,6,8a-octahydronaphthalene	$C_{15}H_{24}$	204/161/105	24268-39-1	0.45	0.45	0.35	0.45	0.56	0.60	*Baccharis tridentata* [42]
26	1775	2,2,6-Trimethyl-6-vinyltetrahydro-2H-pyran-3-ol	$C_{10}H_{18}O_2$	170/152/137/109/94/68	14049-11-7	42.71	38.71	41.71	42.64	40.59	39.62	Pu-erh tea [33]
27	1783	Methyl N-hydroxybenzenecarboximidate	$C_8H_9NO_2$	151/133/105/73	67160-14-9	0.45	0.36	0.91	0.58	0.48	0.38	Endophytic fungi from *Baliospermum montanum* [43]
28	1797	1-(2-Butoxyethoxy)ethanol	$C_8H_{18}O_3$	132/100/75/57	54446-78-5	0.09	0.07	0.07	0.07	0.05	0.05	
29	1812	(3E,7E)-4,8,12-Trimethyltrideca-1,3,7,11-tetraene	$C_{16}H_{26}$	218/203/175/137/95/69/53	62235-06-7	0.35	0.34	0.34	0.23	0.16	0.23	Invasive alligatorweed [44]
30	1880	Phenylmethanol	C_7H_8O	108/91/79/65	100-51-6	1.33	0.68	1.36	0.94	1.06	0.74	*Prunus mume* [41]
31	1913	2-Phenylethanol	$C_8H_{10}O$	122/91/77/65	60-12-8	2.81	2.18	4.87	2.33	2.83	1.64	*Populus trichocarpa* [45]
32	1931	7,11,15-Trimethyl-3-methylenehexadec-1-ene	$C_{20}H_{38}$	278/263/137/	504-96-1	0.17	0.11	0.40	0.28	0.33	0.25	*Herpetospermum pedunculosum* [13]
33	1949	Benzothiazole	C_7H_5NS	123/95/68/57	95-16-9	0.25	0.23	0.26	0.24	0.29	0.24	Tomato [27]
34	2022	1,3,3-Trimethyl-2-oxabicyclo[2.2.2]octan-6-ol	$C_{10}H_{18}O_2$	135/108/95/69 108/93/71	60761-00-4	0.24	0.09	0.19	0.23	0.14	0.17	*Pleurotus ostreatus* and *Favolus tenuiculus* [46]
35	2026	3,7-Dimethylocta-1,6-dien-3-yl formate	$C_{11}H_{18}O_2$	182/165/136/121/93/69	115-99-1	0.12	0.09	0.13	0.10	0.06	0.06	-
36	2048	Isopentyl 2-hydroxybenzoate	$C_{12}H_{16}O_3$	208/193/165/138/120/92	87-20-7	0.20	0.23	0.39	0.27	0.11	0.15	-
37	2053	(E)-3,7,11-Trimethyldodeca-1,6,10-trien-3-ol	$C_{15}H_{26}O$	189/161/107/93/69	40716-66-3	0.14	0.10	0.15	0.14	0.13	0.11	Echinacea flower [47]
38	2078	Hexyl benzoate	$C_{13}H_{18}O_2$	206/123/105/77	6789-88-4	0.60	0.58	0.71	0.67	0.44	0.42	*Salvia reuterana* [48]
39	2128	(Z)-3-Hexenyl benzoate	$C_{13}H_{16}O_2$	105/82/67	25152-85-6	1.28	1.03	1.48	0.82	0.58	0.52	Apple [49]
40	2131	6,10,14-Trimethyl pentadecan-2-one	$C_{18}H_{36}O$	250/165/137/109/95/71/58	502-69-2	0.30	0.26	0.17	0.26	0.22	0.53	
41	2141	4-(3,3-Dimethyloxiran-2-yl)-2-(oxiran-2-yl)butan-2-ol	$C_{10}H_{18}O_3$	143/102/84/69/55	1365-19-1	0.15	0.05	0.11	0.12	0.06	0.10	Lemon grass oil [50]
42	2151	(E)-2-Hexenyl benzoate	$C_{13}H_{16}O_2$	204/105/77	76841-70-8	0.64	0.32	0.56	0.45	0.43	0.33	
43	2173	1,6-Dimethyl-4-propan-2-yl-3,4,4a,7,8,8a-hexahydro-2H-naphthalen-1-ol	$C_{15}H_{26}O$	204/189/161/147/119/105	5937-11-1	0.20	0.14	0.20	0.27	0.36	0.35	-
44	2213	Hexyl 2-hydroxybenzoate	$C_{13}H_{18}O_3$	222/138/120/92	259-76-3	0.09	0.08	0.11	0.19	0.11	0.12	-
45	2236	4-Isopropyl-1,6-dimethyl-1,2,3,4,4a,7,8,8a-octahydronaphthalen-1-ol	$C_{15}H_{26}O$	204/161/121/95	81-34-5	0.05	0.02	0.06	0.08	0.10	0.11	*Schisandra chinensis* [51]
46	2288	(Z)-Hex-3-en-1-yl 2-hydroxybenzoate	$C_{13}H_{16}O_3$	220/138/120/82/67/55	65405-77-8	1.06	0.69	1.00	0.61	0.42	0.37	*Ulmus pumila* [52]

Table 1. Cont.

No.	RI	Compound	MF	Fragment (m/z)	CAS	Relative Content/%						Reports
						Female			Male			
						f1	f2	f3	m1	m2	m3	
47	2296	2,3-Dihydroxypropyl acetate	$C_5H_{10}O_4$	134/103/74	106-61-6	0.06	0.11	0.03	0.06	0.13	0.06	-
48	2349	4,4,7a-Trimethyl-5,6,7,7a-tetrahydrobenzofuran-2(4H)-one	$C_{11}H_{16}O_2$	180/137/111/67	15356-74-8	0.06	0.05	0.07	0.09	0.06	0.07	-
49	2384	1-Methyl-4-[(2Z)-6-methylhepta-2,5-dien-2-yl]-7-oxabicyclo[4.1.0]heptane	$C_{15}H_{24}O$	107/93/79/55	121467-35-4	0.01	0.01	0.01	0.01	0.01	0.01	-
50	2446	Benzoic acid	$C_7H_6O_2$	122/105/77/51	65-85-0	0.06	0.06	0.07	0.06	0.06	0.05	*Telfairia occidentalis* [53]
51	2556	Diisobutyl phthalate	$C_{16}H_{22}O_4$	281/167/149/57	84-69-5	0.09	0.11	0.09	0.09	0.11	0.10	-
52	2743	Di(phenethyl) diglycolate	$C_{20}H_{22}O_5$	342/104/77	84-69-5	0.79	0.33	1.50	0.98	1.75	0.56	-
53	2817	Benzyl 2-hydroxybenzoate	$C_{14}H_{12}O_3$	228/109/91/65	118-58-1	0.71	0.25	0.54	1.39	1.38	0.66	-

RI: retention index; MF: molecular formula.

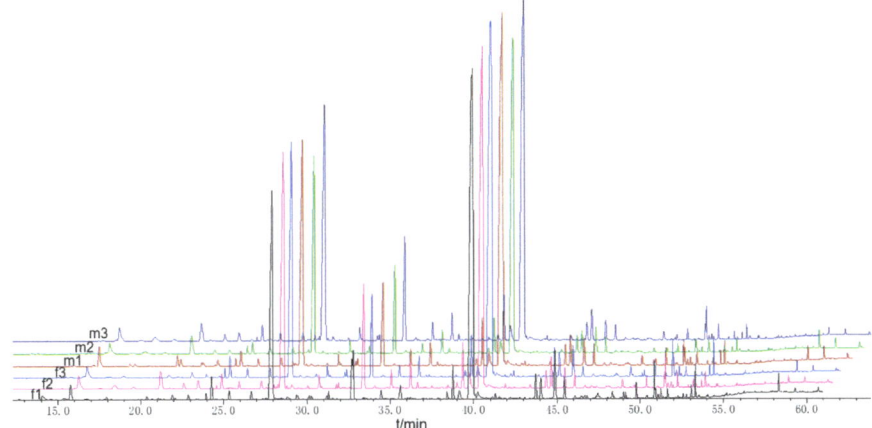

Figure 3. GC-MS total ion chromatograms of volatile compounds from female and male buds of *T. anguina*.

2,2,6-Trimethyl-6-vinyltetrahydro-2H-pyran-3-ol (no. 26) (female 38.71–42.71% and male 39.62–42.64%) and 2,2,6-trimethyl-6-vinyldihydro-2H-pyran-3(4H)-one (no. 12) (female 27.44–35.82% and male 28.34–32.95%) were two compounds with the highest content of *T. anguina* buds.

The volatile components of female and male buds of *T. anguina* included alcohols, ketones, aromatic esters, non-aromatic esters, monoterpenes, sesquiterpenes, diterpenes, alkenes, oximes, heterocycles, alkanes, and acids. The detailed results are shown in Table 2. Among them, the type of heterocycles was the highest content component, with a total relative content of 69.75–75.49% for female and of 69.58–73.39% for male buds. The total content of alcohols was the second highest, with 9.48–12.36% for females and 7.59–8.89% for male buds. Interestingly, there were seven sesquiterpenes whose content in male buds (average 3.56%) was significantly higher than that in female buds (average 2.66%) with $p < 0.05$ in the t-test. In addition, the content of four alkenes in male buds (average 5.34%) was significantly higher than that in female buds (average 2.25%), with $p < 0.05$ in the t-test. These results showed that there was a significant difference between female and male buds in terms of the content of sesquiterpenes and alkenes.

Table 2. The relative content of different types of volatile compounds in female and male *T. anguina* buds.

Type of Compounds	Number of Compounds	Relative Content/%								p-Value
		f1	f2	f3	$\bar{f}$	m1	m2	m3	$\bar{m}$	
Alcohols	8	9.63	9.48	12.36	10.49	8.60	8.89	7.59	8.36	0.10
Ketones	1	0.30	0.26	0.17	0.24	0.26	0.22	0.53	0.34	0.42
Non-aromatic esters	9	1.38	1.96	1.73	1.69	1.39	1.35	1.44	1.39	0.16
Aromatic esters	11	5.90	3.96	6.91	5.59	5.91	5.77	3.61	5.10	0.69
Monoterpenes	4	0.95	0.7	0.8	0.82	0.85	0.63	0.74	0.74	0.47
Sesquiterpenes	7	2.71	2.71	2.55	2.66	3.12	3.78	3.77	3.56	0.02
Diterpenes	1	0.17	0.11	0.4	0.23	0.28	0.33	0.25	0.29	0.55
Alkenes	5	2.37	2.4	1.99	2.25	3.86	6.55	5.62	5.34	0.02
Oximes	1	0.45	0.36	0.91	0.57	0.58	0.48	0.38	0.48	0.63
Heterocycles	4	73.62	75.49	69.75	72.95	72.32	69.58	73.39	71.76	0.59
Alkanes	1	0.23	0.30	0.32	0.28	0.21	0.19	0.23	0.21	0.07
Acids	1	0.06	0.06	0.07	0.06	0.06	0.06	0.05	0.06	0.23

2.2. Multivariate Statistical Analysis

An OPLS-DA model of multivariate statistical analysis was carried out in order to further explore the difference in volatile compounds between female and male buds. This model was verified by 200 times permutation tests. Figure 4a showed that the R^2 and Q^2 values generated by any random arrangement on the left end were smaller than those on the right end, the slope of the regression line was large, and the lower regression line intersected the negative half-axis of the Y-axis, indicating that the model was not overfitting and could be used to find differential compounds. The OPLS-DA score diagram (Figure 4b) showed that the points of f1–f3 of female samples and m1–m3 of male samples were separated along the t1 axis. The value of R^2 was 0.944, and that of Q^2 was 0.838, both of which were greater than 0.5, indicating that the model had suitable interpretation and prediction ability.

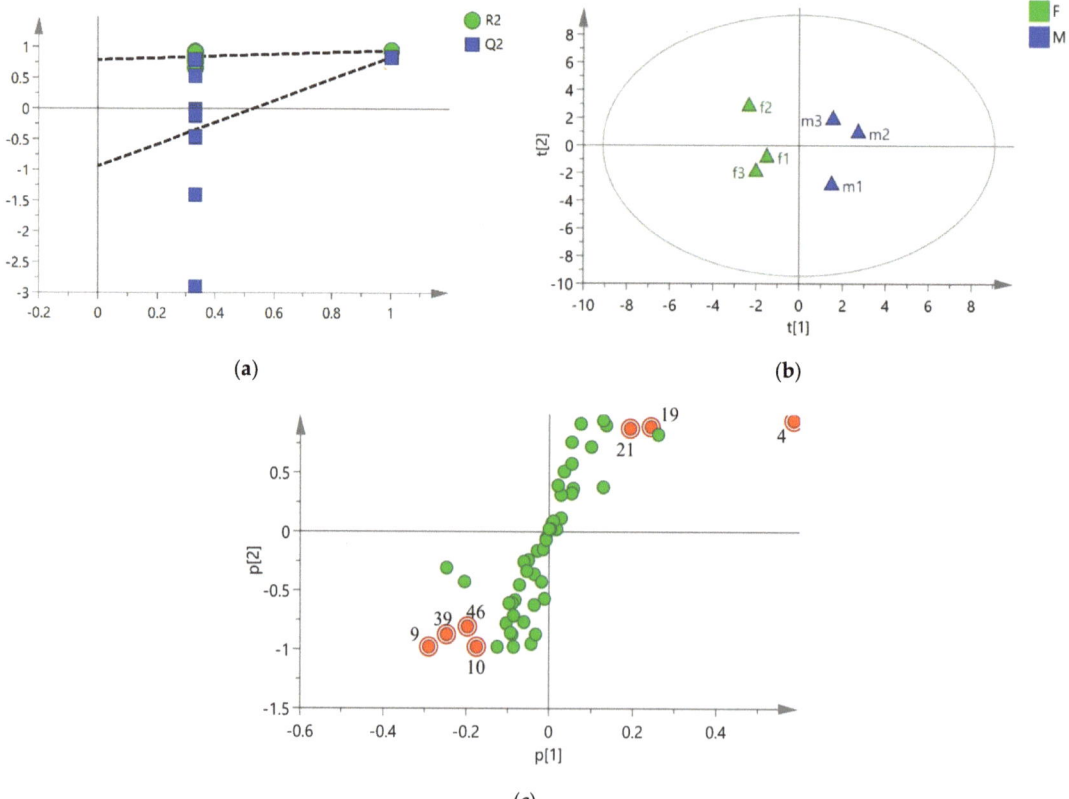

Figure 4. OPLS–DA arranges the verification diagram (a), scores (R^2 = 0.944; Q^2 = 0.838) (b), and S–plot (c) between female and male buds of *T. anguina*. The numbers of the S–plot in Figure (c) were consistent with the no. in Table 1.

Seven compounds were screened out as differential compounds between female and male buds of *T. anguina*, whose VIP values were greater than 1 (Table 3), dots were far away from the origin in the S-plot (Figure 4c), and *p* values of *t*-test were less than 0.05. The content of three compounds in females was significantly lower than that in male buds, i.e., (E)-4,8-dimethyl-1,3,7-nonatriene (no. 4), 1,5,9,9-tetramethyl-1,4,7-cycloundecatriene (no. 21), and (E)-caryophyllene (no. 19). The content of (Z)-4-hexen-1-ol (no. 9), (Z)-3-hexenyl benzoate (no. 39), (Z)-3-hexenyl salicylate (no. 46), and 2-hexen-1-ol (no. 10) in female was higher than that in male buds. These seven differential compounds were marked in red in the S-plot (Figure 4c).

Table 3. Differential volatile compounds between female and male buds of *T. anguina*.

Section	No.	Compound	VIP
A	4	(*E*)-4,8-Dimethyl-1,3,7-nonatriene	3.68
A	19	(*E*)-Caryophyllene	1.57
A	21	1,5,9,9-Tetramethyl-1,4,7-cycloundecatriene	1.25
B	9	(*Z*)-4-Hexen-1-ol	1.81
B	39	(*Z*)-3-Hexenyl benzoate	1.60
B	46	(*Z*)-3-Hexenyl salicylate	1.30
B	10	2-Hexen-1-ol	1.10

A: higher in males than in females; B: higher in females than in males.

2.3. Heat Map

The content distribution of seven differential compounds in the female and male buds of *T. anguina* was visualized in the form of a heat map. Horizontal columns represented different samples (f1–3 and m1–3), and vertical columns represented different compounds (no. 4, 19, 21, 39, 46, 9, and 10). If the color of the block was red, the deeper the red was, the higher the content of the compound in the sample was; if the color was blue, the deeper the blue was, the lower the content of the compound was. As shown in Figure 5, six samples were clearly separated into two groups, i.e., all three female samples were classified into a group on the left, and three male samples were classified into a group on the right. Therefore, the different content levels of seven differential compounds in the female and male buds could be intuitively and clearly observed in Figure 5.

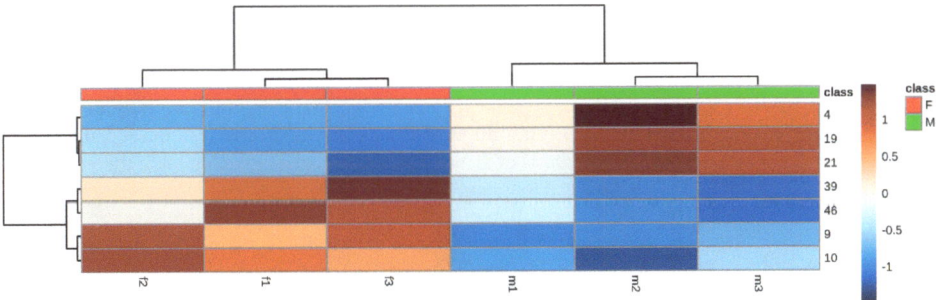

Figure 5. The heatmap of the differential compounds in female and male buds of *T. anguina*. The numbers of the heat map were consistent with the no. in Table 1.

3. Discussion

As a monoecious and diclinous plant, the structure and function of female and male buds and flowers of *T. anguina* are quite different. The buds just before opening were collected for our experiment, which confirmed some differences in the volatile compounds between female and male buds. The results must be more accurate and reliable than those of opening flowers, which can avoid the loss of bud volatile compounds and the pollution of external substances [13]. This is the first report to study the volatile compounds from the buds and to find the differential compounds between female and male buds of *T. anguina*.

Diisobutyl phthalate (no. 51) is usually used as a plasticizer, which is often detected in soil, water, and air, besides in plastics. It is a detection index component of environmental pollution [54]. In this experiment, the fresh buds were treated by HS-SPME without any solvent extraction or plastics exposure. Therefore, it was suspected to be derived from the soil during growth or from the air.

(E)-4,8-Dimethyl-1,3,7-nonatriene (DMNT) is one of the floral compounds of some night-flowering plants [55]. DMNT can attract insects to pollinate [56] and has the function of attracting natural enemies of pests to avoid pest invasion [57]. The average content of DMNT in males (3.48%) was 6.14 times that in female buds (0.57%), whose content gap between genders was the largest. The results pointed out that high DMNT content in male buds was more conducive to attracting insects to pollinate and improving the reproductive capacity of *T. anguina*.

(E)-Caryophyllene, a bicyclic sesquiterpene, has many pharmacological effects such as anti-inflammation, antidepression, and anti-convulsion [58]. Its average content in male (2.37%) was higher than that in female buds (1.81%), which is a significant difference with $p < 0.05$ in the *t*-test between female and male buds. The result indicated that the male buds should have stronger pharmacological activities than the female buds of *T. anguina*.

1,5,9,9-Tetramethyl-1,4,7-cycloundecatriene was detected, whose average content in male (1.42%) was higher than that in female (1.08%) buds and had a significant difference ($p < 0.05$ in *t*-test) in genders. It was often found in volatile components from some plants, such as *Zanthoxylum dissitum* (family Rutaceae) and *Artemisia dracunculus* (family Compositae) [39,59].

(Z)-3-Hexenyl benzoate and (Z)-3-hexenyl salicylate are two (Z)-3-hexenol esters of aromatic acids. The average content of the former was 1.26% in females and 0.64% in males, and of the latter was 0.92% in female and 0.47% in male buds. Both of their content in females was about two times that in male buds. (Z)-3-Hexenyl benzoate showed specific binding to odorant-binding proteins of *Halyomorpha halys* and *Plautia stali*, which was similar to alarm pheromones of both two pests with the function of repelling [60]. The higher content of (Z)-3-hexenylbenzoate in female buds indicated that the compound might help female flowers avoid pests and protect female flowers' pollination and development. (Z)-3-Hexenyl salicylate is a useful compound for people, often used as fragrance ingredients for fine fragrances, shampoos, toilet soaps, and household cleaners [61].

2,2,6-Trimethyl-6-vinyltetrahydro-2H-pyran-3-ol was the most abundant compound of all the volatile compounds from *T. anguina* buds, whose content was 38.71−42.71% and 39.62−42.64% in female and male buds, respectively. It was also the main aroma compound from pu-erh teas, a popular fermented tea that originated in Yunnan [33]. The content of 2,2,6-Trimethyl-6-vinyldihydro-2H-pyran-3(4H)-one was the second highest volatile compound in both buds of *T. anguina*. It also had a high content in the volatile compounds of *Camellia sasanqua* 'Dongxing' flowers (family Theaceae) [62].

The chemical components of some dioecious plants were reported to be different between females and males. The essential oils of the female and male aerial parts of *Baccharis tridentata* Vahl. (family Asteraceae) were explored with GC-MS analysis. α-Pinene was the main compound in the essential oil of both genders, which was presented at higher content in males (1173 ± 60 μg/L) than that in females (794 ± 40 μg/L). Conversely, the concentrations of α-phellandrene, α-terpinene, and *trans*-verbenol in female essential oil were higher than those in males [42]. The chemical components from female and male flowers of *Schisandra chinensis* (family Magnoliaceae) were differentiated with GC-MS analysis. The results showed that 16 compounds were found only in female flowers (including α-farnesene, α-pinene, and 3-carene), and 19 compounds (including *p*-xylene, 3-pyridinecarboxaldehyde, and 1,2-epoxydodecane) were detected only in male flowers. A number of compounds detected both in female and male flowers were quantitatively different; for example, the content of β-pinene in females was 6.36 times that in male flowers (0.70 ± 0.06% and 0.11 ± 0.01%, respectively) [51].

The plant *T. kirilowii*, a different species of the same genus of *T. anguina*, was a dioecious plant. The highest content component of *T. kirilowii* flowers was alcohol, the same as *T. anguina* buds. Some same compounds of linalool, benzyl alcohol, and (E)-linalool oxide were found both in *T. anguina* and *T. kirilowii*. However, the highest content compound in females was linalool, but in male flowers of *T. kirilowii* was benzyl alcohol. Some compounds, such as β-myrcene and α-ocimene, were only detected in females, and benzaldehyde was only detected in male flowers of *T. kirilowii* [14]. The difference between female and male flowers of *T. kirilowii* was larger than that between female and male buds of *T. anguina*. In other words, the difference in volatile compounds between the two genders from the dioecious plant was larger than that from the monoecious and diclinous plants. Dioecious plants have more advanced evolution than monoecious and diclinous plants in plant sexology [63]. From the perspective of the flowers and buds of sexual organs, the higher inequality was found based on their volatile compounds.

4. Materials and Methods

4.1. Apparatus and Materials

A Shimadzu GC-MS-QP 2010 plus gas chromatography-mass spectrometer and a Swiss CTC Combi-xt PAL three-in-one multifunctional automatic sampler were purchased from Shimadzu (Tokyo, Japan). A polyethylene glycol capillary chromatographic column INNOWAX (30 m × 0.25 mm, 0.25 μm) was purchased from Shanghai Troody Analysis Instrument Co., Ltd. (Shanghai, China). The SPME fiber 50/30 μm carboxen/polydimethylsiloxane/divinylbenzene (CAR/PDMS/DVB) was purchased from Supelco (Bellefonte, PA, USA). An ME 204/02 electronic balance was purchased from Mettler Toledo Instruments Co., Ltd. (Shanghai, China). An *n*-alkanes standard (C_{11}–C_{32}) was purchased from Sigma-Aldrich (St. Louis, MO, USA).

4.2. Sample Collection

The plant of *T. anguina* was planted in the courtyard of 16 Dongzhimen South Street, Dongcheng District, Beijing (39°56′18.45″ N, 116°25′41.06″ E, and 48 m altitude). On 6th August, just before their buds bloomed, male buds were picked from pedicels at 8 p.m., and female buds were taken from the top of their juvenile fruits at 9 p.m. The original plant was identified as *Trichosanthes anguina* L. (family Cucurbitaceae) by Prof. Zhimao Chao (Institute of Chinese Materia Medica, China Academy of Chinese Medical Sciences) according to the description in Flora of China (Editorial Board of Flora of China, 1984). The voucher specimens (TAF 1–3 and TAM 1–3) were deposited at the 1022 laboratory of the Institute of Chinese Materia Medica, China Academy of Chinese Medical Sciences, Beijing, China.

4.3. Sample Preparation

The fresh bud samples were collected and placed in 15 mL glass headspace bottles. Each headspace bottle was placed in one of the bud samples, tightly covered, weighed, and subtracted from the bottle's body weight to obtain the sample weight. Moreover, the material of the headspace bottle gasket was polytetrafluoroethylene (PTFE). The 50/30 μm CAR/PDMS/DVB SPME fiber was aged at 260 °C for 30 min, extended through the needle, exposed into the headspace bottle to adsorb volatile compounds at 50 °C for 30 min, and immediately injected into the gas chromatography injection port at 250 °C for 3 min to desorb volatile compounds.

4.4. Chromatographic Conditions

The volatile compounds of the samples were analyzed by the GC-MS method. A GC-MS-QP 2010 plus gas chromatography-mass spectrometer was used coupled to polyethylene glycol capillary chromatographic column Agilent HP-INNOWAX (30 m × 0.25 mm, 0.25 μm). The splitless injection mode was used. The carrier gas was high-purity helium, which was used at a constant flow rate of 1.01 mL·min^{-1}. The temperature of the injection port was set at 250 °C. The heating program was as follows: the initial temperature was 40 °C maintained for 8 min, raised to 160 °C at a rate of 3 °C·min^{-1}, and subsequently raised to 240 °C at a rate of 10 °C·min^{-1} and held for 5 min.

4.5. MS Conditions

The electron ionization (EI) source was used and operated at 70 eV. The ion source temperature was 200 °C. The interface temperature was 220 °C. Moreover, the scanning range was m/z 29–350.

4.6. Data Processing

The volatile compounds from samples were identified with the mass spectrum and RI. The mass spectra obtained from the GC-MS experiments were compared with the National Institute of Standards and Technology (NIST) 14 spectrum library. The RIs were calculated according to the peak retention time of these volatile compounds and of the series of *n*-alkanes (C_{11}–C_{32}) under the same temperament condition and were compared with the values in the previous reports. The peak area normalization was carried out for semi-quantitative analysis and for comparison of the difference of the volatile compounds between female and male samples.

4.7. Statistical Analysis

OPLS-DA is a supervised discriminant analysis that can effectively distinguish the difference between groups. The experimental data were imported into SIMCA-P software (version 14.1, Umetrics, Malmö, Sweden) to establish an OPLS-DA model to distinguish the female and male buds. The experimental data were also imported into SPSS (version 19.0, IBM, America), analyzed with a *t*-test, and then combined with OPLS-DA results to identify the significantly different compounds of the volatile components from female and male buds of *T. anguina*.

4.8. Heat Map

After data were imported into the Metabo Analyst 5.0 website, the heat map was generated to visualize the distribution of different compounds in different samples so that readers could intuitively observe the content gap of these compounds in female and male samples.

5. Conclusions

In this study, HS-SPME-GC-MS combined with multivariate statistical analysis was used to explore some differences in volatile compounds of *T. anguina* buds, and it was found that there was a significant difference between female and male buds. A total of 53 volatile compounds were identified by GC-MS. There were the same volatile compounds from female and male buds, but their content was different. Based on multivariate statistical analysis, seven different compounds were screened out. Among them, the content of (E)-4,8-dimethyl-1,3,7-nonatriene, (E)-caryophyllene, and 1,5,9,9-tetramethyl-1,4,7-cycloundecatriene from male was higher than that from female buds, and the content of (Z)-4-hexen-1-ol, (Z)-3-hexenyl benzoate, (Z)-3-hexenyl salicylate, and 2-hexen-1-ol from male was lower than that from female buds. Further comparison between the monoecious and diclinous plant of *T. anguina* and the dioecious plant of *T. kirilowii*, an opinion suggested that the difference of volatile compounds between female and male buds of the monoecious and diclinous plant was smaller than that of the dioecious plant. This opinion was consistent with the evolutionary view of plant sex [63].

This is the first report that the difference in volatile compounds between female and male buds of *T. anguina* was analyzed. Furthermore, a comparison between monoecious and diclinous plants and dioecious plants was first carried out.

Author Contributions: Data curation, P.S.; Formal analysis, B.X.; Funding acquisition, Z.C.; Investigation, P.S.; Methodology, Z.L.; Resources, Z.C.; Validation, B.X. and Y.C.; Visualization, Y.C.; Writing—original draft, P.S.; Writing—review and editing, Z.C. All authors have read and agreed to the published version of the manuscript.

Funding: Thanks for the support of China Agriculture Research System of the Ministry of Finance and the Ministry of Agriculture and Rural Areas (CARS-21).

Institutional Review Board Statement: Not applicable.

Informed Consent Statement: Not applicable.

Data Availability Statement: All the relevant data have been provided in the manuscript. The authors will provide additional details if required.

Conflicts of Interest: The authors declare no conflict of interest.

References

1. Raj, M.N.; Prasanna, K.P.; Peter, K.V. Snake gourd. *Genet. Improv. Veg. Crop* **1993**, *223*, 259–264.
2. Ojiako, O.A.; Igwe, C.U. The Nutritive, Anti-Nutritive and Hepatotoxic Properties of *Trichosanthes anguina* (Snake Tomato) Fruits from Nigeria. *Pak. J. Nutr.* **2008**, *7*, 85–89. [CrossRef]
3. Zong, J.; Cao, H.B. *Nutrition and Scientific Consumption of 68 Special Vegetables*; Jindun Press: Beijing, China, 2013; p. 36.
4. Sandhya, S.; Vinod, K.R.; Chandra Sekhar, J.; Aradhana, R.; Nath, V.S. An updated review on *Tricosanthes cucumerina* L. *Int. J. Pharm. Sci. Rev. Res.* **2010**, *1*, 56–60.
5. Churiyah, W.S. Antiproliferative protein from *Trichosanthes cucumerina* L. var *anguina* (L.) Haines. *Biotropia Southeast Asian J. Trop. Biol.* **2010**, *17*, 8–16.
6. Takagi, T.; Itabashi, Y. Occurrence of mixtures of geometrical isomers of conjugated octadecatrienoic acids in some seed oils: Analysis by open-tubular gas liquid chromatography and high performance liquid chromatography. *Lipids* **1981**, *16*, 546–551. [CrossRef]
7. Ali, M.A.; Sayeed, M.A.; Islam, M.S.; Yeasmin, M.S.; Khan, G.; Muhamad, I.I. Physicochemical and antimicrobial properties of Trichosanthes anguina and Swietenia mahagoni seeds. *Bull. Chem. Soc. Ethiop.* **2011**, *25*, 427–436. [CrossRef]
8. Chao, Z.M. Review on chemical constituents of Trichosanthes. *China J. Chin. Mater. Med.* **1990**, *15*, 49–51.
9. Venkataramaiah, C.; Rao, K.N. Studies on Indolyl-3-acetic Acid Oxidase and Phenolic Acid Pattern in Cucurbitaceous Fruits. *Z. Pflanzenphysiol.* **1983**, *111*, 459–463. [CrossRef]
10. Yoshizaki, M.; Fujino, H.; Masuyama, M.; Arisawa, M.; Morita, N. A chemotaxonomic study of flavonoids in the leaves of six Trichosanthes species. *Phytochemistry* **1987**, *26*, 2557–2558. [CrossRef]
11. Benesh, D.P.; Weinreich, F.; Kalbe, M.; Milinski, M. Lifetime inbreeding depression, purging, and mating system evolution in a simultaneous hermaphrodite tapeworm. *Evolution* **2014**, *68*, 1762–1774. [CrossRef]
12. Xu, B.; Wu, C.; Li, Z.; Song, P.; Chao, Z. ^{1}H NMR Combined with Multivariate Statistics for Discrimination of Female and Male Flower Buds of *Populus tomentosa*. *Molecules* **2021**, *26*, 6458. [CrossRef]
13. Liu, Z.; Fang, Y.; Wu, C.; Hai, X.; Xu, B.; Li, Z.; Song, P.; Wang, H.; Chao, Z. The Difference of Volatile Compounds in Female and Male Buds of *Herpetospermum pedunculosum* Based on HS-SPME-GC-MS and Multivariate Statistical Analysis. *Molecules* **2022**, *27*, 1288. [CrossRef]
14. Sun, W.; Chao, Z.; Wang, C.; Wu, X.; Tan, Z. [Difference of volatile constituents contained in female and male flowers of Trichosanthes kirilowii by HS-SPME-GC-MS]. *China J. Chin. Mater. Medica* **2012**, *37*, 1570–1574.
15. Li, C.Q.; Lu, Y.; Li, X.Z.; Xing, H.; Kang, W.Y. Volatile constituents from flower of Tianmian *Cucurbita moschata* Duch. by head-space solid micro-extraction coupled with GC-MS. *Sci. Technol. Food Ind.* **2012**, *33*, 151–156.
16. Freitas, L.; Bernardello, G.; Galetto, L.; Paoli, A.A.S. Nectaries and reproductive biology of *Croton sarcopetalus* (Euphorbiaceae). *Bot. J. Linn. Soc.* **2001**, *136*, 267–277. [CrossRef]
17. Hawthorne, S.B.; Miller, D.J.; Pawliszyn, J.; Arthur, C.L. Solventless determination of caffeine in beverages using solid-phase microextraction with fused-silica fibers. *J. Chromatogr. A* **1992**, *603*, 185–191. [CrossRef]
18. Zaitsev, V.N.; Zui, M.F. Preconcentration by solid-phase microextraction. *J. Anal. Chem.* **2014**, *69*, 715–727. [CrossRef]
19. Sven, U. Solid-phase microextraction in biomedical analysis. *J. Chromatogr. A* **2000**, *902*, 167–194.
20. Xu, L.; Liu, H.; Ma, Y.; Wu, C.; Li, R.; Chao, Z. Comparative study of volatile components from male and female flower buds of *Populus × tomentosa* by HS-SPME-GC-MS. *Nat. Prod. Res.* **2019**, *33*, 2105–2108. [CrossRef]
21. Wawrzyniak, R.; Jasiewicz, B. Straightforward and rapid determination of acrylamide in coffee beans by means of HS-SPME/GC-MS. *Food Chem.* **2019**, *301*, 125264. [CrossRef]
22. Bhavya, M.L.; Ravi, R.; Naidu, M.M. Development and validation of headspace Solid-Phase microextraction coupled with gas chromatography (HS-SPME-GC) method for the analysis of *Zingiber zerumbet* L. *Nat. Prod. Res.* **2021**, *35*, 1221–1225. [CrossRef] [PubMed]
23. Cerkowniak, M.; Boguś, M.I.; Włóka, E.; Stepnowski, P.; Gołębiowski, M. Application of headspace solid-phase microextraction followed by gas chromatography coupled with mass spectrometry to determine esters of carboxylic acids and other volatile compounds in *Dermestes maculatus* and *Dermestes ater* lipids. *Biomed. Chromatogr.* **2018**, *32*, e4051. [CrossRef] [PubMed]
24. Nešpor, J.; Karabín, M.; Štulíková, K.; Dostálek, P. An HS-SPME-GC-MS Method for Profiling Volatile Compounds as Related to Technology Used in Cider Production. *Molecules* **2019**, *24*, 2117. [CrossRef] [PubMed]

25. Yang, D.-S.; Lei, Z.; Bedair, M.; Sumner, L.W. An Optimized SPME-GC-MS Method for Volatile Metabolite Profiling of Different Alfalfa (*Medicago sativa* L.) Tissues. *Molecules* **2021**, *26*, 6473. [CrossRef]
26. Zhang, R.; Tang, C.; Jiang, B.; Mo, X.; Wang, Z. Optimization of HS-SPME for GC-MS Analysis and Its Application in Characterization of Volatile Compounds in Sweet Potato. *Molecules* **2021**, *26*, 5808. [CrossRef]
27. Lee, J.H.J.; Jayaprakasha, G.K.; Rush, C.M.; Crosby, K.M.; Patil, B.S. Production system influences volatile biomarkers in tomato. *Metabolomics* **2018**, *14*, 99. [CrossRef]
28. Cecchi, L.; Migliorini, M.; Giambanelli, E.; Cane, A.; Mulinacci, N.; Zanoni, B. Volatile Profile of Two-Phase Olive Pomace (Alperujo) by HS-SPME-GC–MS as a Key to Defining Volatile Markers of Sensory Defects Caused by Biological Phenomena in Virgin Olive Oil. *J. Agric. Food Chem.* **2021**, *69*, 5155–5166. [CrossRef]
29. An, H.M.; Ou, X.C.; Xiong, Y.F.; Zhang, Y.B.; Li, J.; Li, Q.; Li, Q.; Li, S.; Huang, J.A. Study on the characteristic aroma components of jasmine tea. *J. Tea Sci.* **2020**, *40*, 225–237.
30. Carriero, G.; Brunetti, C.; Fares, S.; Hayes, F.; Hoshika, Y.; Mills, G.; Tattini, M.; Paoletti, E. BVOC responses to realistic nitrogen fertilization and ozone exposure in silver birch. *Environ. Pollut.* **2016**, *213*, 988–995. [CrossRef]
31. Conde-Martínez, N.; Sinuco, D.C.; Osorio, C. Chemical studies on curuba (Passiflora mollissima (Kunth) L. H. Bailey) fruit flavour. *Food Chem.* **2014**, *157*, 356–363. [CrossRef]
32. Falahati, M.; Sepahvand, A.; Mahmoudvand, H.; Baharvand, P.; Jabbarnia, S.; Ghojoghi, A.; Yarahmadi, M. Evaluation of the antifungal activities of various extracts from Pistacia atlantica Desf. *Curr. Med. Mycol.* **2015**, *1*, 25–32. [CrossRef] [PubMed]
33. Nian, B.; Jiao, W.W.; He, M.Z.; Liu, Q.T.; Zhou, L.X.; Jiang, B.; Zhang, Z.Y.; Liu, M.L.; Ma, Y.; Chen, L.J.; et al. Determination and comparison of biochemical components and aroma substances in the pu-erh teas with mellow flavor and floral-fruity aroma. *Mod. Food Sci. Technol.* **2020**, *36*, 241–248.
34. Wang, G.-C.; Sun, X.-L.; Cai, X.-M.; Chen, Z.-M. Effects of tea plant volatiles on foraging behavior of Xysticus ephippiatus Simon. *Chin. J. Eco-Agric.* **2012**, *20*, 612–618. [CrossRef]
35. Zhang, H.; Zhou, D.; Luo, Y.; Wang, J.; Zong, S. Identification of volatile compounds emitted by Artemisia ordosica (Artemisia, Asteraceae) and changes due to mechanical damage and weevil infestation. *Z. Naturforsch. C J. Biosci.* **2013**, *68*, 313. [CrossRef] [PubMed]
36. Wei, J.-F.; Yin, Z.-H.; Kang, W.-Y. Volatiles of Lysimachia Paridiformis Var. Stenophylla, Lysimachia Fortumei and Lysimachia Chikungensis by HS-SPME-GC-MS. *Afr. J. Tradit. Complement. Altern. Med.* **2014**, *11*, 70–75. [CrossRef] [PubMed]
37. Marín-San Román, S.; Carot, J.M.; Sáenz de Urturi, I.; Rubio-Bretón, P.; Pérez-Álvarez, E.P.; Garde-Cerdán, T. Optimization of thin film-microextraction (TF-SPME) method in order to determine musts volatile compounds. *Anal. Chim. Acta.* **2022**, *1226*, 340254. [CrossRef] [PubMed]
38. Wang, H.Y.; Zhang, W.; Dong, J.H.; Wu, H.; Wang, Y.H.; Xiao, H.X. Optimization of SPME-GC-MS and characterization of floral scents from Aquilegia japonica and *A. amurensis* flowers. *BMC Chem.* **2021**, *15*, 26. [CrossRef]
39. Qader, K.O.; Salah, T.F.M.; Rasul, A.A. GC-MS analysis of essential oil Extract from leaves and stems of tarragon (*Artemisia dracunculus* L.). *J. Biol. Agric. Healthc.* **2017**, *7*, 50–57.
40. Cao, Y.; Wang, J.; Germinara, G.S.; Wang, L.; Yang, H.; Gao, Y.; Li, C. Behavioral Responses of *Thrips hawaiiensis* (Thysanoptera: Thripidae) to Volatile Compounds Identified from *Gardenia jasminoides* Ellis (Gentianales: Rubiaceae). *Insects* **2020**, *11*, 408. [CrossRef]
41. Zhang, T.; Bao, F.; Yang, Y.; Hu, L.; Ding, A.; Ding, A.; Wang, J.; Cheng, T.; Zhang, Q. A Comparative Analysis of Floral Scent Compounds in Intraspecific Cultivars of Prunus mume with Different Corolla Colours. *Molecules* **2019**, *25*, 145. [CrossRef]
42. Minteguiaga, M.; Fariña, L.; Cassel, E.; Fiedler, S.; Catalán, C.A.N.; Dellacassa, E. Chemical compositions of essential oil from the aerial parts of male and female plants of Baccharis tridentata Vahl. (Asteraceae). *J. Essent. Oil Res.* **2021**, *33*, 299–307. [CrossRef]
43. Jagannath, S.; Konappa, N.; Lokesh, A.; Bhuvaneshwari; Dasegowda, T.; Udayashankar, A.C.; Chowdappa, S.; Cheluviah, M.; Satapute, P.; Jogaiah, S. Bioactive compounds guided diversity of endophytic fungi from *Baliospermum montanum* and their potential extracellular enzymes. *Anal. Biochem.* **2021**, *614*, 114024. [CrossRef] [PubMed]
44. Shi, M.Z.; Li, J.Y.; Chen, Y.T.; Fang, L.; Wei, H.; Fu, J.W. Plant volatile compounds of the Invasive Alligatorweed, *Alternanthera philoxeroides* (Mart.) Griseb, infested by *Agasicles hygrophila* Selman and Vogt (Coleoptera: Chrysomelidae). *Life* **2022**, *12*, 1257. [CrossRef]
45. Günther, J.; Lackus, N.D.; Schmidt, A.; Huber, M.; Stödtler, H.-J.; Reichelt, M.; Gershenzon, J.; Köllner, T.G. Separate Pathways Contribute to the Herbivore-Induced Formation of 2-Phenylethanol in Poplar. *Plant Physiol.* **2019**, *180*, 767–782. [CrossRef] [PubMed]
46. Omarini, A.; Dambolena, J.S.; Lucini, E.; Mejía, S.J.; Albertó, E.; Zygadlo, J.A. Biotransformation of 1,8-cineole by solid-state fermentation of Eucalyptus waste from the essential oil industry using Pleurotus ostreatus and Favolus tenuiculus. *Folia Microbiol.* **2016**, *61*, 149–157. [CrossRef]
47. Kaya, M.; Merdivan, M.; Tashakkori, P.; Erdem, P.; Anderson, J.L. Analysis of *Echinacea* flower volatile constituents by HS-SPME-GC/MS using laboratory-prepared and commercial SPME fibers. *J. Essent. Oil Res.* **2019**, *31*, 91–98. [CrossRef]
48. Panahi, Y.; Ghanei, M.; Hadjiakhoondi, A.; Ahmadi-Koulaei, S.; Delnavazi, M.-R. Free Radical Scavenging Principles of Salvia reuterana Boiss. Aerial Parts. *Iran J. Pharm. Res.* **2020**, *19*, 283–290. [CrossRef]

49. Badra, Z.; Herrera, S.L.; Cappellin, L.; Biasioli, F.; Dekker, T.; Angeli, S.; Tasin, M. Species-Specific Induction of Plant Volatiles by Two Aphid Species in Apple: Real Time Measurement of Plant Emission and Attraction of Lacewings in the Wind Tunnel. *J. Chem. Ecol.* **2021**, *47*, 653–663. [CrossRef]
50. Akinkunmia, E.O.; Oladeleb, A.; Eshoa, O.; Oduseguna, I. Effects of storage time on the antimicrobial activities and composition of lemon grass oil. *J. Appl. Res. Med. Aromat. Plants* **2016**, *3*, 105–111. [CrossRef]
51. Sowndhararajan, K.; Kim, J.-H.; Song, J.E.; Kim, M.; Kim, S. Chemical components of male and female flowers of Schisandra chinensis. *Biochem. Syst. Ecol.* **2020**, *92*, 104121. [CrossRef]
52. Wei, H.S.; Qin, J.H.; Cao, Y.Z.; Li, K.B.; Yin, J. Two classic OBPs modulate the responses of female *Holotrichia oblita* to three major ester host plant volatiles. *Insect Mol. Biol.* **2021**, *30*, 390–399. [CrossRef] [PubMed]
53. Eseyin, O.A.; Sattar, M.A.; Rathore, H.; Aigbe, F.; Afzal, S.; Ahmad, A.; Lazhari, M.; Akthar, S. GC-MS and HPLC profiles of phenolic fractions of the leaf of Telfairia occidentalis. *Pak. J. Pharm. Sci.* **2018**, *31*, 45–50. [PubMed]
54. Gao, D.-W.; Wen, Z.-D. Phthalate esters in the environment: A critical review of their occurrence, biodegradation, and removal during wastewater treatment processes. *Sci. Total Environ.* **2016**, *541*, 986–1001. [CrossRef]
55. Azuma, H.; Toyota, M.; Asakawa, Y. Intraspecific Variation of Floral Scent Chemistry in Magnolia kobus DC. (Magnoliaceae). *J. Plant Res.* **2001**, *114*, 411–422. [CrossRef]
56. Svensson, G.P.; Hickman, M.O., Jr.; Bartram, S.; Boland, W.; Pellmyr, O.; Raguso, R.A. Chemistry and geographic variation of floral scent in *Yucca filamentosa* (Agavaceae). *Am. J. Bot.* **2005**, *92*, 1624–1631. [CrossRef] [PubMed]
57. Li, W.; Lin, Y.J.; Zhou, F. The recent research progress on DMNT and TMTT in plants. *J. Plant Prot.* **2018**, *45*, 946–953.
58. Francomano, F.; Caruso, A.; Barbarossa, A.; Fazio, A.; La Torre, C.; Ceramella, J.; Mallamaci, R.; Saturnino, C.; Iacopetta, D.; Sinicropi, M.S. β-(E)-Caryophyllene: A sesquiterpene with countless biological properties. *Appl. Sci.* **2019**, *9*, 5420. [CrossRef]
59. Wang, C.-F.; Yang, K.; You, C.-X.; Zhang, W.-J.; Guo, S.-S.; Geng, Z.-F.; Du, S.-S.; Wang, Y.-Y. Chemical Composition and Insecticidal Activity of Essential Oils from Zanthoxylum dissitum Leaves and Roots against Three Species of Storage Pests. *Molecules* **2015**, *20*, 7990–7999. [CrossRef]
60. Wang, Z.; Yang, F.; Sun, A.; Song, J.; Shan, S.; Zhang, Y.; Wang, S. Expressional and functional comparisons of five clustered odorant binding proteins in the brown marmorated stink bug Halyomorpha halys. *Int. J. Biol. Macromol.* **2022**, *206*, 759–767. [CrossRef]
61. Lapczynski, A.; McGinty, D.; Jones, L.; Letizia, C.S.; Api, A.M. Fragrance material review on cis-3-hexenyl salicylate. *Food Chem. Toxicol.* **2007**, *45*, S402–S405. [CrossRef]
62. Wang, J.; Li, X.L.; Yin, H.F.; Fan, Z.Q.; Li, J.Y. Volatile components in different floral organs and flowering stages of *Camellia sasanqua* 'Dongxing'. *J. Yunnan Agric. Univ. Nat. Sci.* **2018**, *33*, 904–910.
63. Husband, B.C.; Schemske, D.W. Evolution of the magnitude and timing of inbreeding. *Evolution* **1996**, *50*, 54. [CrossRef] [PubMed]

Article

Volatile Organic Compounds (VOCs) Produced by *Levilactobacillus brevis* WLP672 Fermentation in Defined Media Supplemented with Different Amino Acids

Sarathadevi Rajendran [1,2], Patrick Silcock [1] and Phil Bremer [1,*]

1 Department of Food Science, University of Otago, Dunedin 9054, New Zealand; sarathadevi.rajendran@postgrad.otago.ac.nz (S.R.); pat.silcock@otago.ac.nz (P.S.)
2 Department of Agricultural Chemistry, Faculty of Agriculture, University of Jaffna, Kilinochchi 44000, Sri Lanka
* Correspondence: phil.bremer@otago.ac.nz

Abstract: Fermentation by lactic acid bacteria (LAB) is a promising approach to meet the increasing demand for meat or dairy plant-based analogues with realistic flavours. However, a detailed understanding of the impact of the substrate, fermentation conditions, and bacterial strains on the volatile organic compounds (VOCs) produced during fermentation is lacking. As a first step, the current study used a defined medium (DM) supplemented with the amino acids L-leucine (Leu), L-isoleucine (Ile), L-phenylalanine (Phe), L-threonine (Thr), L-methionine (Met), or L-glutamic acid (Glu) separately or combined to determine their impact on the VOCs produced by *Levilactobacillus brevis* WLP672 (LB672). VOCs were measured using headspace solid-phase microextraction (HS-SPME) gas chromatography–mass spectrometry (GC-MS). VOCs associated with the specific amino acids added included: benzaldehyde, phenylethyl alcohol, and benzyl alcohol with added Phe; methanethiol, methional, and dimethyl disulphide with added Met; 3-methyl butanol with added Leu; and 2-methyl butanol with added Ile. This research demonstrated that fermentation by LB672 of a DM supplemented with different amino acids separately or combined resulted in the formation of a range of dairy- and meat-related VOCs and provides information on how plant-based fermentations could be manipulated to generate desirable flavours.

Keywords: amino acids; defined medium (DM); lactic acid bacteria (LAB); volatile organic compounds (VOCs)

Citation: Rajendran, S.; Silcock, P.; Bremer, P. Volatile Organic Compounds (VOCs) Produced by *Levilactobacillus brevis* WLP672 Fermentation in Defined Media Supplemented with Different Amino Acids. *Molecules* **2024**, *29*, 753. https://doi.org/10.3390/molecules29040753

Academic Editor: Hiroyuki Kataoka

Received: 20 December 2023
Revised: 25 January 2024
Accepted: 2 February 2024
Published: 6 February 2024

Copyright: © 2024 by the authors. Licensee MDPI, Basel, Switzerland. This article is an open access article distributed under the terms and conditions of the Creative Commons Attribution (CC BY) license (https://creativecommons.org/licenses/by/4.0/).

1. Introduction

Plant-based foods have gained popularity as consumers choose to reduce their meat and dairy intake due to concerns about their health, the environment, and/or animal welfare [1–4]. This desire has resulted in an increase in sales of meat or dairy analogues worldwide as consumers seek out alternative forms and flavours of products they are familiar with [5–12]. A challenge with producing plant based-analogues is obtaining realistic meat- or dairy-like flavours.

Flavour is a complex sensory modality that encompasses volatile organic compounds (VOCs) sensed in the nose at the olfactory epithelium retronasally, non-volatile organic compounds sensed on the tongue (taste attributes: sweet, salt, sour, bitter, and umami), and chemesthetic responses (hot, spicy, and pungent) sensed in the oral cavity. Generally, studies on flavour focus on the analysis of VOCs owing to the importance of aroma/odour in overall flavour perception [13–15]. VOCs are a low-molecular-weight (<400 Da) compound with a relatively high vapour pressure at room temperature, which means that they can be easily transferred into the gaseous phase [16] and subsequently to olfactory receptors [13,17].

In meat or dairy analogues, the addition of flavour compounds extracted from meat or dairy products is generally not acceptable. In addition, the chemical generation of flavour compounds can be environmentally unfriendly owing to the requirement to use solvents, and such processes lack selectivity, which leads to the formation of unwanted compounds, reduced process efficiency, and increased downstream costs [18]. Although desirable flavours can be extracted directly from plants, this is only economically feasible for a small number of VOCs, as plants contain complex mixtures of VOCs [19]. A promising route for the generation of flavour VOCs that imitate meat or dairy flavour compounds is via the microbial biosynthesis or fermentation of plant material [20]. For example, lactic acid bacteria (LAB) can use plant substrates for energy and nutrition and produce a range of volatile secondary metabolites. In fermented foods, these volatile secondary metabolites are responsible for the production of flavour compounds or flavour precursors [21,22]. The composition of the plant substrates and the LAB strains used have been reported to have the greatest impacts on the resulting fermentation flavours produced [23,24]. However, as the substrate composition of plants can vary widely, understanding how best to generate targeted flavour VOCs through plant-based fermentation is challenging.

LAB are a fastidious microorganism that require a rich cultivation medium for growth, as the majority of them are auxotrophic for a wide range of amino acids and vitamins [25]. Carbohydrates (simple sugars), protein (peptone, yeast extract, beef extract, or whey protein), minerals, vitamins, and buffering agents are common ingredients in LAB cultivation media [26]. A rich cultivation medium, however, is not suitable for determining the role of substrates on VOC formation by LAB fermentation owing to the difficulty in determining which substrates and metabolic pathways are responsible for the VOCs detected. To reduce the complexity in the system, a defined medium with only sufficient nutrients to support LAB growth can be used [27–32].

Amino acids are important not only for the growth of LAB but also for the production of flavour compounds. Most amino acids do not have a direct impact on flavour, but they do contribute indirectly because they are precursors to key flavour compounds [33]. Enzymes found in the LAB, such as deaminases, decarboxylases, transaminases (aminotransferases), and lyases, can convert amino acids in different ways. Amino acid transamination is the primary initiator of the conversion of amino acids to flavour compounds. Transamination of amino acids results in α-keto acid formation, which can subsequently be decarboxylated into aldehydes, which in turn can be dehydrogenated into alcohols or carboxylic acids by alcohol dehydrogenases and aldehyde dehydrogenases, respectively [34–36].

For LAB strains to be used in commercial production of plant-based flavours, the bacteria need to be readily available, food grade, and not very fastidious in terms of nutrient requirements or growth temperatures. An initial trial with *Levilactobacillus brevis* WLP672 (*Lev. brevis* WLP672) and *Lactobacillus delbrueckii* WLP677 strains identified the LAB brewing strain *Lev. brevis* WLP672 as a good candidate for use in fermentation trials. *Lev. brevis* WLP672 is an obligatory heterofermentative LAB that is used in the production of a wide range of fermented products worldwide. The bacterium uses the phosphoketolase pathway (PK) to ferment hexoses to produce a mixture of lactic acid, ethanol, acetic acid, CO_2, and an array of volatile secondary metabolites [37].

To date, most studies on the amino acid-derived VOCs produced by yeasts [38–42], fungi [43], and LAB [44,45] have used complex media (natural or synthetic), with only a few studies using LAB in the defined medium [33,46]. This is the first study to use headspace solid-phase microextraction (HS-SPME) gas chromatography–mass spectrometry (GC-MS) to provide a comprehensive analysis of VOCs produced by *Lev. brevis* WLP672 in response to the addition of single or combined amino acids in a defined medium.

2. Results and Discussion

2.1. Development of Defined Medium

Based on past literature [27,47–52] and a series of trials, a defined medium (DM) was developed (Table 1), which supported the growth of *Lev. brevis* WLP672 (thereafter referred to as LB672). In addition to the other components in the DM, growth of LB672 did not occur in the absence of sodium acetate, as previously reported [48]. Sodium acetate is postulated to stimulate the growth of several LAB species, owing to it being both a buffering agent and an energy source [47,48]. Sodium acetate concentrations of 0.1 and 1.2% were the lowest and highest concentrations previously used in nutrient-rich media designed to support the growth of LAB [47,49]. In the current study, sodium acetate stimulated the growth of LB672 in the DM when glucose was present, but it did not enable growth (function as an energy source) when glucose was absent. LB672 also did not grow in the DM in the absence of peptone (enzymatic protein digest, Bacto peptone), even in the presence of individual amino acid. *Lev. brevis* strains have previously been reported to lack genes for amino acid biosynthesis [53,54]. Bacto peptone was used as its composition was defined [55,56]. Further, LB672 did not grow in the absence of the added vitamins or minerals.

Table 1. Overview of the composition of the different media used.

Media	Glucose	Peptone	Vitamins	Salt	Sodium Acetate	Glu	Leu	Ile	Phe	Thr	Met
DML0.1	2%	0.5%	✓	✓	0.1%	0.2%	0.2%	-	-	-	-
DMI0.1	2%	0.5%	✓	✓	0.1%	0.2%	-	0.2%	-	-	-
DMP0.1	2%	0.5%	✓	✓	0.1%	0.2%	-	-	0.2%	-	-
DMT0.1	2%	0.5%	✓	✓	0.1%	0.2%	-	-	-	0.2%	-
DMM0.1	2%	0.5%	✓	✓	0.1%	0.2%	-	-	-	-	0.2%
DMG0.1	2%	0.5%	✓	✓	0.1%	0.2%	-	-	-	-	-
DM0.1	2%	0.5%	✓	✓	0.1%	-	-	-	-	-	-
DMAa0.1	2%	0.5%	✓	✓	0.1%	0.2%	0.04%	0.04%	0.04%	0.04%	0.04%
DML1.2	2%	0.5%	✓	✓	1.2%	0.2%	0.2%	-	-	-	-
DMI1.2	2%	0.5%	✓	✓	1.2%	0.2%	-	0.2%	-	-	-
DMP1.2	2%	0.5%	✓	✓	1.2%	0.2%	-	-	0.2%	-	-
DMT1.2	2%	0.5%	✓	✓	1.2%	0.2%	-	-	-	0.2%	-
DMM1.2	2%	0.5%	✓	✓	1.2%	0.2%	-	-	-	-	0.2%
DMG1.2	2%	0.5%	✓	✓	1.2%	0.2%	-	-	-	-	-
DM1.2	2%	0.5%	✓	✓	1.2%	-	-	-	-	-	-
DMAa1.2	2%	0.5%	✓	✓	1.2%	0.2%	0.04%	0.04%	0.04%	0.04%	0.04%

Amino acids: glutamic acid (Glu), leucine (Leu), isoleucine (Ile), phenylalanine (Phe), threonine (Thr), methionine (Met).

To investigate the role of amino acid supplementation on the VOCs produced by LB672 during fermentation, the DM was supplemented with amino acids, either separately or in combination (Table 1). Note that amino acid supplementation in the DM resulted in an amino acid concentration 4 to 30 times higher than the amino acid concentration in the DM (from the peptone) [55,56]. Almost all LAB require glutamic acid (Glu) because of their inability to synthesise its precursor α-ketoglutarate de novo, owing to their lack of a complete TCA cycle [25]. This means aminotransferases, which initiate transamination reactions of amino acid, utilise Glu as the donor substrate of the amino groups and convert them into α-ketoglutarate [36]. To ensure that sufficient Glu was present to enable transamination reactions, Glu was added to all media except the original DM (DM0.1 and DM1.2, Table 1).

2.2. Physiochemical Properties

The pH of the medium impacts on the growth of bacteria, and during LAB fermentation, the pH of the medium can decrease owing to the production of lactic acid [57]. In the current study, pH values decreased over the 16 days of fermentation (Table 2) from 5.2 to 4.6 in media containing 1.2% acetate and from 6.6 to 6.1 in media containing 0.1% acetate. During fermentation, the turbidity (OD_{600}) of the media increased over time (Table 2) by about 0.5 to 0.7 units in media containing 1.2% acetate and in the range of 0.2 to 0.4 in media containing 0.1% acetate. It is obvious that there were differences obtained in OD_{600} between the two acetate concentrations used in the DM. However, due to the variation in the initial pH of the DM, it is not possible to confirm the impact of acetate on the growth. Further, after 16 days of fermentation in either DM0.1 or DM1.2 (Table 2), OD_{600} values were higher in DM containing 0.1 and 1.2% acetate compared to in DM supplemented with amino acids. The reason for the decreased LB672 growth in the presence of amino acids is unclear. However, as the addition of amino acids decreased the initial pH of the medium, it is speculated that this could be one of the reasons for the decreased growth (lower final OD_{600} values).

Table 2. The pH and OD_{600} of samples after 16 days of fermentation by LB672 in different medium compositions.

Media	Initial pH	After 16 Days of Fermentation	
		pH	OD_{600}
DML0.1	6.66	6.24	0.269 [de]
DMI0.1	6.64	6.30	0.271 [de]
DMP0.1	6.65	6.17	0.223 [e]
DMT0.1	6.68	6.10	0.297 [de]
DMM0.1	6.67	6.09	0.249 [de]
DMG0.1	6.83	6.20	0.286 [de]
DM0.1	7.85	7.28	0.391 [cd]
DMAa0.1	6.63	6.45	0.233 [e]
DML1.2	5.27	4.63	0.511 [bc]
DMI1.2	5.28	4.64	0.536 [bc]
DMP1.2	5.29	4.72	0.542 [bc]
DMT1.2	5.26	4.66	0.517 [bc]
DMM1.2	5.3	4.66	0.515 [bc]
DMG1.2	5.4	4.89	0.605 [ab]
DM1.2	7.01	5.0	0.711 [a]
DMAa1.2	5.31	4.87	0.492 [bc]

Results are the mean value of duplicate samples. Values with different superscript lowercase letters ([a–e]) in the column (OD_{600}) are significantly different according to Tukey's test at $p < 0.05$. DML0.1 to DMAa1.2: media supplemented with different amino acids (see Table 1).

2.3. Volatile Organic Compounds (VOCs) after Fermentation

A total of 49 VOCs were detected after 16 days of LB672 fermentation, which were attributed to supplementation of the DM with different amino acids separately or combined. The VOCs detected included alcohols (19 compounds), acids (9), esters (6), sulphur compounds (5), ketones (4), aldehydes (1), and unknown compounds (5) (Table 3). In order to distinguish differences across different medium compositions in the relative abundance of the VOCs detected, hierarchical clustering analysis and heatmap visualisation were carried out (Figures 1 and 2). In the dendrogram, VOCs on nearby branches tend to show positive correlations, whereas distant branches tend to show negative correlations.

Table 3. LB672 fermentation VOCs identified in different medium compositions via HS-SPME-GC-MS analysis.

No.	Compound Name	RI (Calc.)	RI (Lit.)	R Match	Identification Method
	Alcohols				
1	Ethanol	930	932	942	MS, RI
2	2-Methyl propanol	1086	1092	827	MS, RI
3	2-Pentanol	1116	1119	881	MS, RI
4	2-Methyl butanol	1199	1208	905	MS, RI
5	3-Methyl butanol (isoamyl alcohol)	1200	1209	962	MS, RI
6	3-Heptanol	1287	1290	959	MS, RI
7	2-Heptanol	1310	1320	953	MS, RI
8	3-Methyl-2-buten-1-ol (prenol)	1312	1320	815	MS, RI

Table 3. *Cont.*

No.	Compound Name	RI (Calc.)	RI (Lit.)	R Match	Identification Method
9	1-Hexanol	1343	1355	832	MS, RI
10	4-Methyl-2-heptanol	1349	1372	900	MS, RI
11	2-Octanol	1410	1412	849	MS, RI
12	2-Nonanol	1508	1521	952	MS, RI
13	1-Octanol	1545	1557	912	MS, RI
14	1-Nonanol	1647	1660	883	MS, RI
15	1-Decanol	1751	1760	824	MS, RI
16	Citronellol	1752	1765	801	MS, RI
17	Geraniol	1833	1847	879	MS, RI
18	Benzyl alcohol	1868	1870	924	MS, RI
19	Phenylethyl alcohol	1904	1906	914	MS, RI
	Acids				
20	Acetic acid	1429	1449	946	MS, RI
21	Butanoic acid	1620	1625	905	MS, RI
22	2-Methyl butanoic acid	1659	1662	929	MS, RI
23	3-Methyl butanoic acid (isovaleric acid)	1658	1666	956	MS, RI
24	Hexanoic acid	1836	1846	903	MS, RI
25	Heptanoic acid	1944	1950	864	MS, RI
26	Octanoic acid	2049	2060	920	MS, RI
27	Nonanoic acid	2154	2171	912	MS, RI
28	n-Decanoic acid	2261	2276	801	MS, RI
	Esters				
29	Ethyl acetate	889	888	957	MS, RI
30	Butyl acetate	1069	1074	959	MS, RI
31	3-Methylbutyl acetate	1120	1122	867	MS, RI
32	Ethyl hexanoate	1230	1233	860	MS, RI
33	Ethyl heptanoate *	1331	1331	771	MS, RI
34	2-Phenylethyl acetate	1813	1813	800	MS, RI
	Sulphur compounds				
35	Methanethiol *	690	692	734	MS, RI
36	Dimethyl disulphide	1072	1077	965	MS, RI
37	Dimethyl trisulphide	1386	1377	935	MS, RI
38	Methional	1455	1454	809	MS, RI
39	5-Ethenyl-4-methyl thiazole	1527	1520	917	MS, RI
	Ketones				
40	4-Methyl-4-penten-2-one	1069	1110	804	MS, RI
41	2-Heptanone	1183	1182	921	MS, RI
42	2-Nonanone	1390	1390	954	MS, RI
43	2-Undecanone	1600	1598	939	MS, RI
	Aldehydes				
44	Benzaldehyde	1530	1520	961	MS, RI
	Unknown compounds				
45	Unknown 1	1027	NA		
46	Unknown 2	1139	NA		
47	Unknown 3	1181	NA		
48	Unknown 4	1597	NA		
49	Unknown 5	1803	NA		

Identification method: MS—mass spectrum, RI—retention indices. *: co-eluted with air peaks/background compounds.

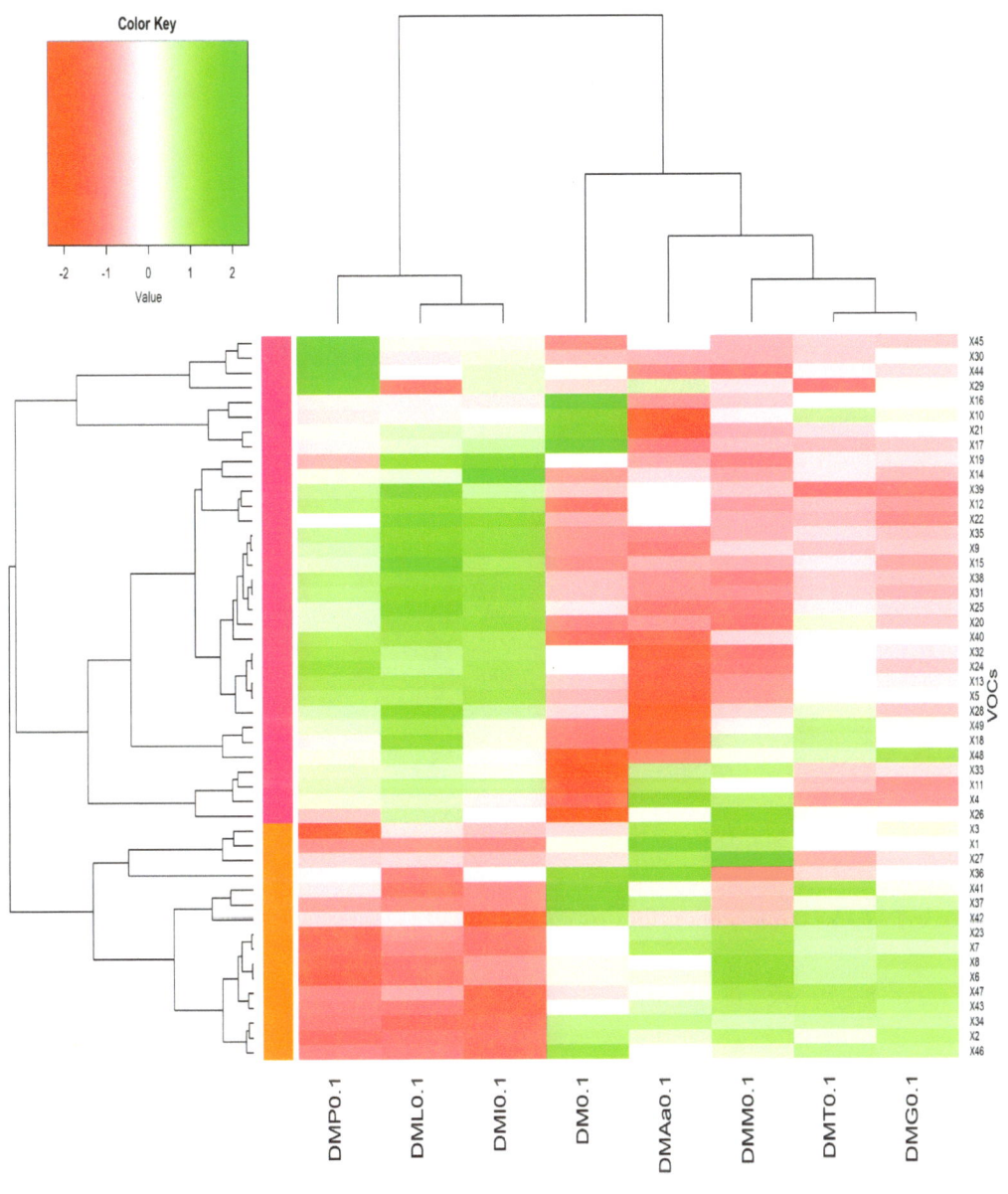

Figure 1. Heatmap visualisation and hierarchical clustering analysis of VOCs produced by LB672 based on the log 2 transformed average peak area of each VOC. Fermentation was carried out in the defined medium (DM) with 0.1% acetate added and supplemented with different amino acids separately or in combination (DMP: Phe added, DML: Leu added, DMI: Ile added, DM: defined medium, DMAa: Aamix added, DMM: Met added, DMT: Thr added, and DMG: Glu added). The green colour represents a higher abundance, whereas the red colour indicates a lower abundance. The VOCs represented in the heatmap are numbered according to the peak numbers (Table S1).

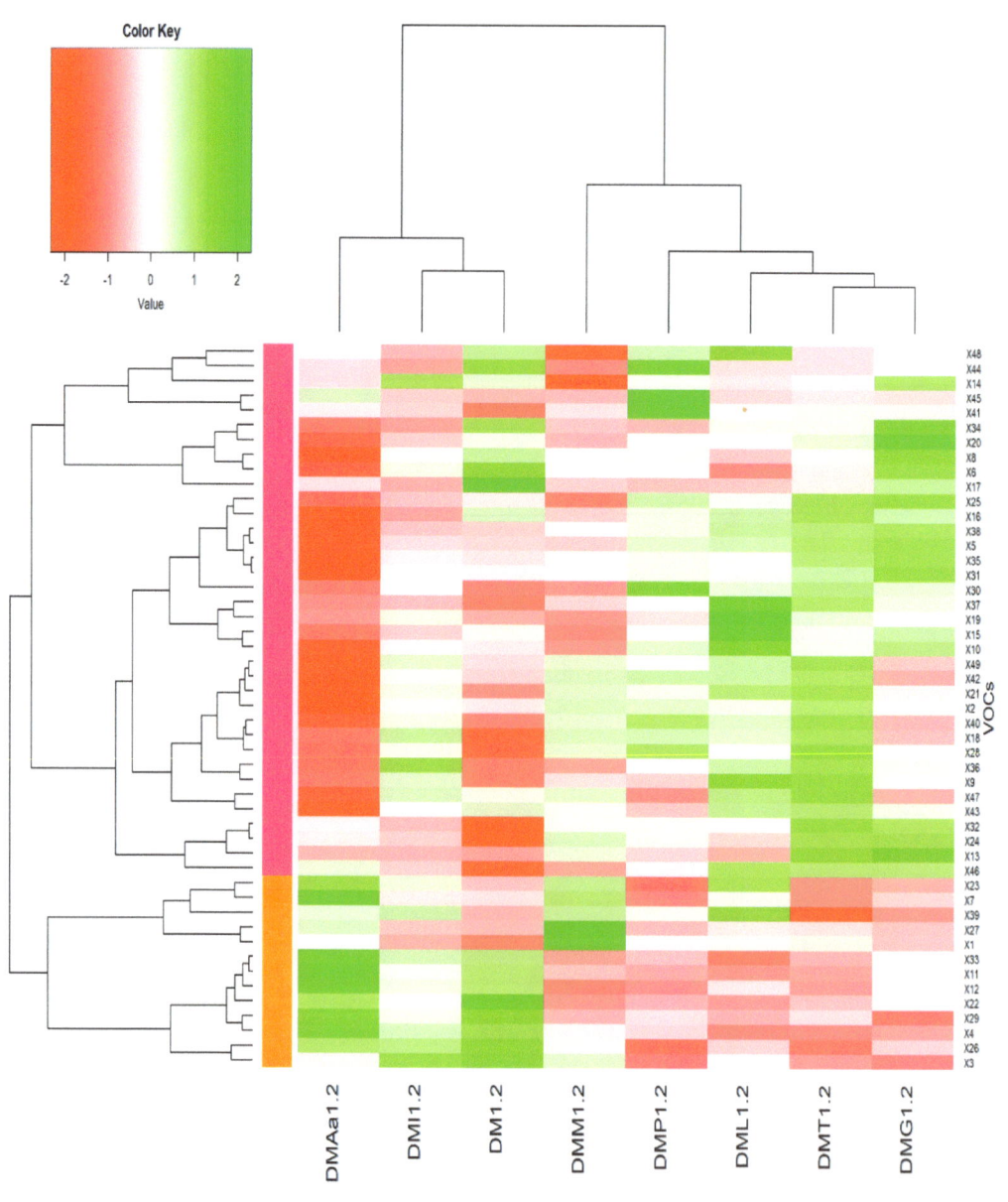

Figure 2. Heatmap visualisation and hierarchical clustering analysis of VOCs produced by LB672 based on the log 2 transformed average peak area of each VOC. Fermentation was carried out in the defined medium (DM) with 1.2% acetate added and supplemented with different amino acids separately or in combination (DMAa: Aamix added, DMI: Ile added, DM: defined medium, DMM: Met added, DMP: Phe added, DML: Leu added, DMT: Thr added, and DMG: Glu added). The green colour represents a higher abundance, whereas the red colour indicates a lower abundance. The VOCs represented in the heatmap are numbered according to the peak numbers (Table S2).

In heatmap 1 (Figure 1), VOCs detected after LB672 fermentation in DM containing 0.1% acetate were primarily grouped (column-wise) into two clusters based on the different medium compositions used: cluster 1—phenylalanine (Phe) added (DMP0.1), leucine (Leu) added (DML0.1), and isoleucine (Ile) added (DMI0.1), and cluster 2—DM (DM0.1), amino acid mixture (Aamix) added (DMAa0.1), methionine (Met) added (DMM0.1), threonine (Thr) added (DMT0.1), and Glu only added (DMG0.1). The alcohols (2-methyl propanol, 2-pentanol, 1-hexanol, 1-octanol, 2-octanol, 1-nonanol, 2-nonanol, 1-decanol, and citronellol), Ile/Leu-derived alcohols (2-methyl butanol, and 3-methyl butanol), and Phe-derived compound (phenylethyl alcohol) were present in high proportions in the cluster of Phe, Leu, and Ile, where acids (butanoic acid, hexanoic acid, heptanoic acid, and octanoic acid), Ile/Leu-derived acids (2-methyl butanoic acid, and 3-methyl butanoic acid), and Met-derived compounds (methanethiol, methional, dimethyl disulphide, and dimethyl trisulphide) were in low proportions. In contrast, the alcohols (2-pentanol, 4-methyl-2-heptanol, 1-octanol, 2-octanol, 1-nonanol, 1-decanol, and citronellol) were present in low proportions in the cluster of DM, Aamix, Met, Thr, and Glu, where Met-derived compounds (dimethyl disulphide, and dimethyl trisulphide) and acids (butanoic acid, hexanoic acid, and heptanoic acid) were in high proportions. Further, VOCs were row-wise clustered, mainly into two clusters (Figure 1 and Table S1); cluster 1 (pink) contains higher proportions of alcohols, ketones, Phe-derived compounds, and Leu/Ile-derived alcohols, while cluster 2 (orange) is characterized by acids, Met-derived compounds and Leu/Ile-derived acids.

In DM containing 1.2% acetate, VOCs formed two main clusters (column-wise), as shown by heatmap 2 (Figure 2): an Aamix (DMAa1.2), Ile (DMI1.2), and DM (DM1.2) cluster and a Met (DMM1.2), Phe (DMP1.2), Leu (DML1.2), Thr (DMT1.2), and Glu (DMG1.2) cluster. The alcohols (2-methyl propanol, 1-octanol, 2-octanol, 1-nonanol, 1-decanol) and ketones (2-heptanone, 2-nonanone, and 2-undecanone) were present in low proportions in the cluster of Aamix, Ile, and DM. In the subcluster of Ile and DM, Met-derived compounds (methanethiol, methional, and dimethyl disulphide) and Phe-derived compounds (phenylethyl alcohol, and 2-phenylethyl acetate) were in low proportions. As the Aamix medium contained all of the amino acids, the possibility of forming high proportions of all amino acid-derived VOCs was increased. In contrast to this cluster, the alcohols (2-methyl propanol, 1-octanol, 2-octanol, 1-nonanol, and 1-decanol) were present in high proportions in the subcluster of Phe, Thr, Leu, and Glu. Further, VOCs were row-wise clustered into two clusters (Figure 2 and Table S2); cluster 1 (pink) is characterized by higher proportions of acids, alcohols, ketones, esters, and Leu/Ile/Phe-derived compounds, while cluster 2 (orange) is presented by higher proportions of Met-derived compounds.

Since the heat map analysis highlighted differences between the VOC profile across different medium compositions, the specific amino acid-derived VOCs produced by LB672 across different medium compositions are discussed in the following sections separately.

2.4. Phe-Derived VOCs

Phe, an aromatic amino acid, can be converted by LAB into phenyl pyruvate via the action of aromatic aminotransferases (ArAT) in the presence of α-ketoglutarate. The ArAT is active on all aromatic amino acids (tryptophan (Try), Phe, and tyrosine (Tyr)) and also on Leu and Met [36]. Further, phenyl pyruvate can subsequently be converted into benzeneacetaldehyde (phenylacetaldehyde), phenyl acetic acid, phenylethyl alcohol (phenyl ethanol), benzyl alcohol (phenyl methanol), and benzaldehyde by LAB via enzymatic, non-enzymatic, and unknown mechanisms [36,58].

Benzaldehyde, which has characteristic almond and burnt sugar notes [57], is produced chemically (non-enzymatic reactions) from phenyl pyruvate in the presence of oxygen and manganese [34,58]. DM with added Phe and 0.1% acetate had the highest benzaldehyde peak area (Figure 3A, medium DMP). Benzaldehyde was detected in all media possibly due to the presence of peptone, which is speculated to be the reason why many of the compounds discussed in the following sections were also detected in many of the ferments at low levels.

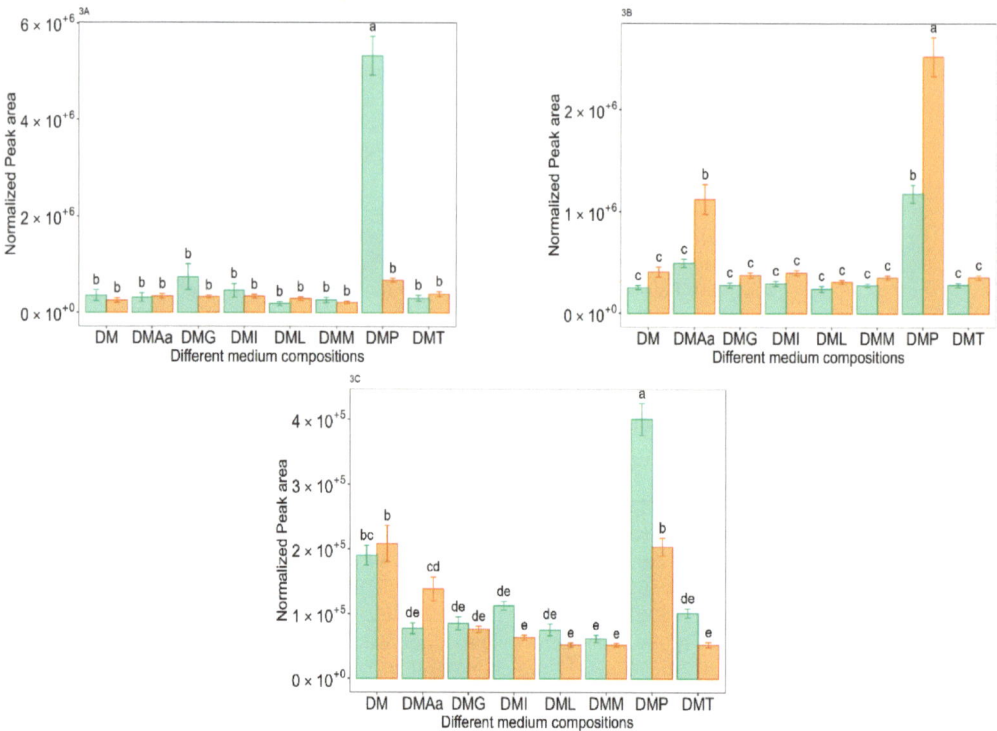

Figure 3. Normalized peak area of benzaldehyde (**3A**), phenylethyl alcohol (**3B**), and benzyl alcohol (**3C**) across different medium compositions (see Table 1) at 0.1% ▇ and 1.2% ▇ acetate. Values are presented as mean ± standard error ($n = 6$). Different letters represent significant difference between the different medium compositions according to Tukey's test at $p < 0.05$.

Phenylethyl alcohol, which has a characteristic floral note [59], can be produced from phenyl pyruvate, which is generated by transamination of Phe. Phenylpyruvate is then decarboxylated to phenylacetaldehyde (benzeneacetaldehyde) via the action of decarboxylase. Further, it can be dehydrogenated to phenylethyl alcohol via the action of the enzyme alcohol dehydrogenase (AlcDH) [36]. The highest peak area for phenylethyl alcohol was obtained in media with added Phe, where a significantly ($p < 0.05$) higher peak area was obtained with 1.2% acetate compared to 0.1% acetate (Figure 3B, medium DMP). Phe addition has previously been reported to increase phenylethyl alcohol production during the fermentation of lychee wine using *Saccharomyces cerevisiae* var. *cerevisiae* MERIT.ferm [38], in synthetic grape must fermentation by commercial wine yeast strains [39], in papaya juice fermented by *Williopsis saturnus* var. *mrakii* [41], and in synthetic grape juice fermented by *Saccharomyces cerevisiae* var. *bayanus* [42]. The addition of an amino acid mixture (valine (Val), Leu, Ile, and Phe) to soy (tofu) whey has also been reported to increase the amount of phenylethyl alcohol following *Torulaspora delbrueckii* Biodiva fermentation [40].

Benzyl alcohol, which imparts floral notes, can be produced from benzaldehyde via oxidative and non-oxidative pathways [60]. It was detected at the highest peak area in media supplemented with Phe (Figure 3C, medium DMP), where a significantly ($p < 0.05$) higher amount was obtained with 0.1% acetate compared to 1.2% acetate.

A genomic study by Liu et al. [53] reported that *Lev. Brevis* ATCC 367 lacked the gene *araT* encoding ArAT (putative) enzyme, which initiates the transamination reaction that converts Phe into phenyl pyruvate. However, in the current study, LB672 produced benzaldehyde, phenylethyl alcohol, and benzyl alcohol presumably via the intermediate phenyl pyruvate. Yvon and Rijnen [61] reported that the presence of aspartate aminotransferase (AspAT) in *Brevibacterium linens* is responsible for aspartate (Asp) transamination and is also active on aromatic amino acids. The gene *aspAT* encoding for AspAT enzyme has been detected in several *Lev. Brevis* strains, including *Lev. Brevis* ATCC 367 [53] and *Lev. Brevis* CGMCC 1306 [62]. Hence, it seems more likely that, in the current study, AspAT in LB672 carried out the transamination reaction of Phe rather than the ArAT enzyme.

2.5. Met-Derived VOCs

The catabolism of Met, a sulphur-containing amino acid, is initiated by a transamination step involving ArAT or the branched-chain aminotransferase (BcAT) in the presence of α-ketoglutarate, yielding 4-methylthio-2-ketobutyric acid (KMBA). KMBA can be biochemically converted via a decarboxylation reaction into methional and subsequently converted into methanethiol and α-ketobutyrate via an unknown pathway. KMBA can also be directly converted into 2-hydroxyl-4-methylthiobutyric acid and methanethiol via dehydrogenation. Further, demethiolation of Met produces methanethiol, α-ketobutyrate, and ammonia via two pyridoxal phosphate-dependent lyases (cystathionine β-lyase (CBL), cystathionine γ-lyase (CGL)). Lastly, KMBA can be chemically converted into methanethiol [34,36,58,63]. Methanethiol produced through these pathways can be further converted into dimethyl sulphide, dimethyl disulphide, and dimethyl trisulphide by auto-oxidation [36,63]. In addition, methanethiol can also react with carboxylic acids to produce thioesters [58]. The α-ketobutyrate formed from the Met catabolism can be converted into propanoic acid and yields ATP via substrate-level phosphorylation [63].

In the current study, the highest peak area for methanethiol was in DM at 0.1% acetate supplemented with the Aamix (Figure 4A, medium DMAa), which contained Met along with other amino acids. The highest peak area for methional was in medium supplemented with Met (Figure 4B, medium DMM), where a significantly ($p < 0.05$) higher peak area was obtained for 1.2% acetate compared to 0.1% acetate. For dimethyl disulphide, the highest peak area was observed in the medium with Aamix added (Figure 4C, medium DMAa) at 1.2% acetate, suggesting that it was converted from methanethiol.

According to a study by Liu et al. [53], the genes *cblA* and *cglA*, which encode CBL and CGL lyases, are present in *Lev. Brevis* ATCC 367, and catalyse the demethiolation reaction of Met. Lyases, play an important role in the biosynthesis of sulphur-containing amino acids but not a major role in Met catabolism by LAB [36]. Therefore, aminotransferases ArAT or BcAT, which might be present in LB672 and involved in the transamination reaction, would appear to be the major source of the degradation products. As the peak area for methanethiol was high and the peak area for methional was low, the results suggest that either methional was produced through transamination (either by ArAT or by BcAT) and converted into methanethiol or that methanethiol was produced directly from Met via the activity of CBL/CGL.

2.6. Leu/Ile-Derived VOCs

The catabolism of Leu and Ile (branched-chain amino acids) is initiated by an aminotransferase enzyme, BcAT, that catalyses the hydrolysis of Leu and Ile to the α-keto acids 4-methyl-2-oxopentanoic acid (α-ketoisocaproate) and 3-methyl-2-oxopentanoic acid (α-keto-β-methylvalerate), respectively. In many LAB, ArAT is also involved in the transamination of Leu. After the transamination reaction, α-keto acids undergo the following biochemical reactions: 1. Oxidative decarboxylation of 4-methyl-2-oxopentanoic acid and 3-methyl-2-oxopentanoic acid to 3-methyl butanoic acid and 2-methyl butanoic acid, respectively (enzyme: ketoacid dehydrogenase (KaDH)); 2. Decarboxylation of 4-methyl-2-oxopentanoic acid and 3-methyl-2-oxopentanoic acid to 3-methyl butanal and 2-methyl butanal, respectively (enzyme: keto acid decarboxylase (KdcA)); and 3. Reduction of 4-methyl-2-oxopentanoic acid and 3-methyl-2-oxopentanoic acid to 2-hydroxy-4-methylpentanoic acid and 2-hydroxymethylvalerate, respectively (enzyme: hydroxy acid dehydrogenase (HycDH)). Further, 3-methyl butanal and 2-methyl butanal can be either reduced to alcohols by AlcDH (3-methyl butanol and 2-methyl butanol, respectively) or oxidised to acids by an aldehyde dehydrogenase (AldDH) (3-methyl butanoic acid and 2-methyl butanoic acid, respectively) [34–36].

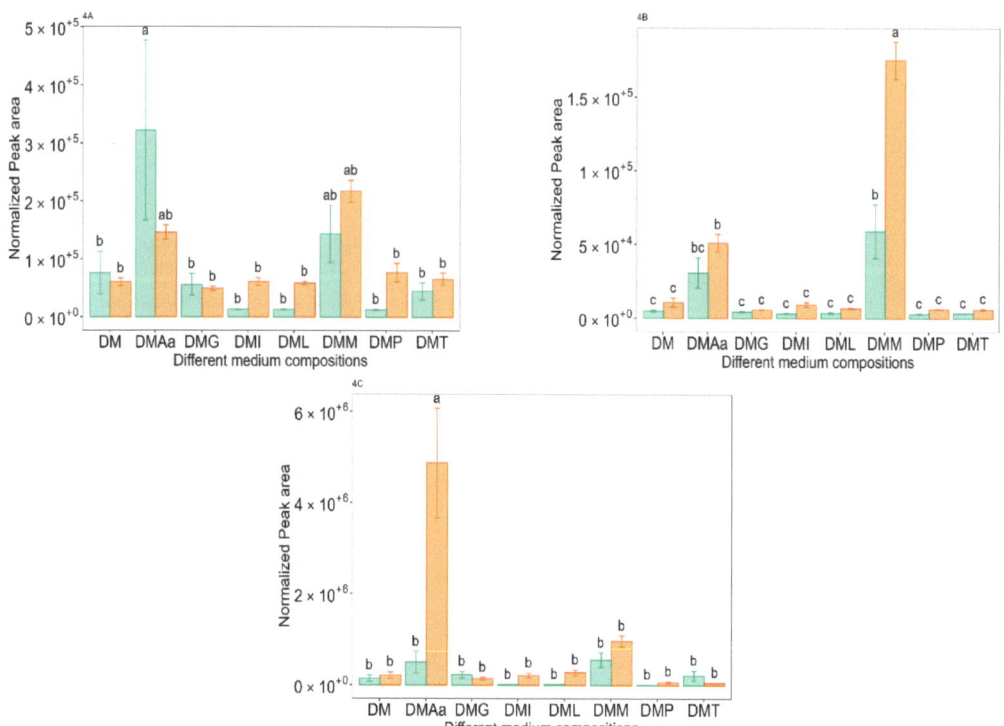

Figure 4. Normalized peak area of methanethiol (**4A**), methional (**4B**), and dimethyl disulphide (**4C**) across different medium compositions (see Table 1) at 0.1% ▢ and 1.2% ▢ acetate. Values are presented as mean ± standard error (n = 6). Different letters represent significant difference between the different medium compositions according to Tukey's test at $p < 0.05$.

In the current study, the highest peak area for 2-methyl butanol was observed in DM supplemented with Ile (Figure 5A, DMI) at 0.1% acetate. Similarly, the highest peak area for 3-methyl butanol was obtained in DM supplemented with Leu (Figure 5B, DML) at 0.1% acetate. The addition of Leu has previously been reported to increase 3-methyl butanol production: in synthetic grape must fermented by commercial wine yeast strains [39], in synthetic cassava medium fermented by *Ceratocystis fimbriata* [43], and in papaya juice fermented by *Williopsis saturnus* var. *mrakii* [41]. The addition of an amino acid mixture (Val, Leu, Ile, and Phe) to soy (tofu) whey has also been reported to increase the peak area of 3-methyl butanol fermented by *Torulaspora delbrueckii* Biodiva [40]. Further, the addition of Ile has also been reported to increase 2-methyl butanol production during the fermentation of papaya juice by *Williopsis saturnus* var. *mrakii* [41].

The gene *bcaT* encodes the BcAT enzyme which initiates the transamination reaction on branched-chain amino acids, and the gene *araT* encodes the ArAT (putative) enzyme that initiates the transamination reaction on Leu, were both absent in a genome study of *Lev. Brevis* ATCC 367 [53]. In the current study, LB672 fermentation resulted in the production of 2- and 3-methyl butanol from Ile and Leu, respectively. This result suggests that LB672 possesses BcAT or ArAT enzymes, which catalysed the transamination of Ile and Leu because 2- and 3-methyl butanol could not have been produced without a transamination step.

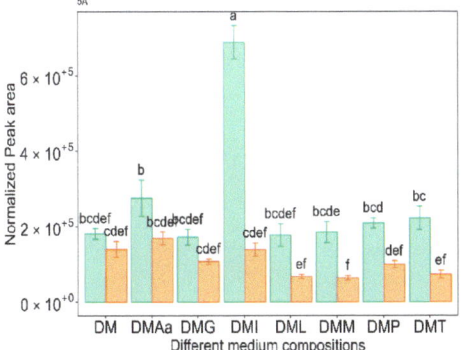
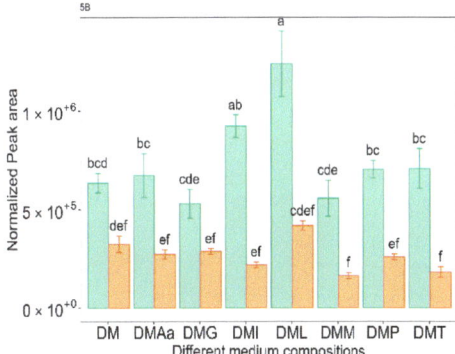

Figure 5. Normalized peak area of 2-methyl butanol (**5A**) and 3-methyl butanol (**5B**) across different medium compositions (see Table 1) at 0.1% ▇ and 1.2% ▇ acetate. Values are presented as mean ± standard error ($n = 6$). Different letters represent significant difference between different medium compositions according to Tukey's test at $p < 0.05$.

Ethanol, which is a marker compound in fermentation studies, can be produced from sugars via an intermediate acetaldehyde using the PK pathway [21], from the degradation of Thr [64], or from acetate through acetyl-CoA [65]. Overall, the ethanol peak area was higher in DM containing 1.2% acetate compared to 0.1% (Figure 6A). This may have been a result of the acetate being converted into ethanol. Interestingly, in DM supplemented with either Leu, Ile, or Phe at 0.1% acetate, the peak area for ethanol was significantly ($p < 0.05$) lower than in other media with either 0.1 or 1.2% acetate. This finding suggests that LB672 in the presence of added Leu, Ile, or Phe at the lower acetate concentrations is using a pathway other than an ethanol-producing pathway.

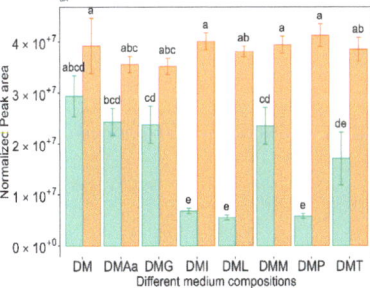
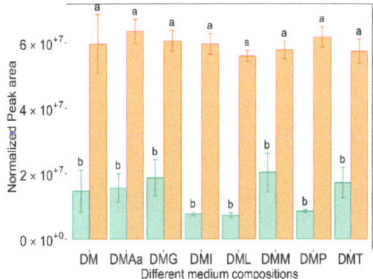

Figure 6. Normalized peak area of ethanol (**6A**) and acetic acid (**6B**) across different medium compositions (see Table 1) at 0.1% ▇ and 1.2% ▇ acetate. Values are presented as mean ± standard error ($n = 6$). Different letters represent significant difference between the different medium compositions according to Tukey's test at $p < 0.05$.

Acetic acid was detected in all media, with its concentration being significantly ($p < 0.05$) higher at 1.2% acetate-containing DM compared to 0.1% acetate-containing DM (Figure 6B). Though *Lev. Brevis* strains have been shown to produce acetic acid in addition to lactic acid [66], owing to the fact that acetate was present in the DM, it is not possible to confirm that LB672 produced additional acetate. Acetic acid, however, is the building block for fatty acids consisting of an even number of carbons [32,67]. In the current study, butanoic, hexanoic, octanoic, and decanoic acids were present in a higher abundance in media containing 1.2% acetate compared to 0.1%. This result demonstrates in a concentration-dependent manner that acetate could be converted into fatty acids with even carbon chain-length numbers.

In the current study, the addition of some individual amino acids (Phe, Met, Leu, and Ile) or the addition of the amino acid mixture (Aamix) had an impact on the relative proportions of VOCs formed by LB672 fermentation. In contrast, no particular VOCs were detected in the medium supplemented with Thr. Thr catabolism by LAB is initiated by threonine aldolase (TA) or serine hydroxymethyltransferase (SHMT), which converts Thr to acetaldehyde and glycine [35,36]. Acetaldehyde was not detected after LB672 fermentation in the current study. As acetaldehyde is an intermediate compound in the ethanol production pathway [64], it was considered that it was most likely not detected rather than not being present. The inability to detect acetaldehyde was due to the fact that the HS-SPME method used was not optimised for its detection—rather, it was designed to detect a broad range of compounds.

The addition of different concentrations of acetate (0.1, or 1.2%) in the DM significantly influenced the relative abundance of VOCs generated after LB672 fermentation. The mechanism(s) underpinning this observation were, however, unclear, as acetate's role in LAB growth and VOC generation is complicated. In addition to serving as a buffering agent, acetate has also been reported to demonstrate growth-stimulating activities for several LAB [47], and omitting acetate from the media can result in diminished growth [48]. Acetate can also serve as a precursor to a number of lipids/fatty acids [32,68] and ethanol [65].

The presence of aminotransferases such as BcAT, ArAT, and AspAT and other enzymes involved in amino acid catabolism, including KdcA, AlcDH, AldDH, and CBL/CGL, in LB672 is likely in the current study because of the production of VOCs derived from branched-chain, aromatic, or sulphur (Met) amino acids. This could be confirmed by applying a genomic analysis of LB672.

In the current study, the VOCs produced after 16 days of fermentation were analysed using GC-MS. Although GC-MS is a commonly used reference method for VOC analysis, it cannot be used to track changes in VOCs in real time. However, over time, the concentration of VOC produced can fluctuate dramatically. It is therefore possible that compounds were produced but were not detected in the analysis performed at the end of the fermentation (16 days). Proton transfer reaction-time of flight-mass spectrometry (PTR-ToF-MS) is a quick, direct, non-invasive, and highly sensitive (parts per trillion by volume) online method that can be used to monitor the production of VOCs [69]. Such an approach would be useful in determining how VOCs change over time in the DM supplemented with different amino acids.

Among the fermentation-derived VOCs detected in the present study, dimethyl trisulphide, methional, 2-nonanone, 2-undecanone, 1-hexanol, and 1-octanol were characteristic odour-active VOCs in cooked meat [70]; acetic acid, 3-methyl butanoic acid, methional, dimethyl trisulphide, and hexanoic acid were odour-active VOCs in dairy yoghurt [71]; and acetic acid, butanoic acid, 3-methyl butanoic acid, 3-methyl butanol, methional, methanethiol, dimethyl disulphide, dimethyl trisulphide, 2-phenylethyl acetate, benzaldehyde, ethyl hexanoate, 2-heptanone, 2-nonanone, and 2-undecanone were odour-active VOCs in dairy cheese [61,72]. Hence, the current study has reinforced the critical role that substrate composition plays in VOC production, and the knowledge gained will help researchers to develop plant-based fermentations designed to generate specific VOCs that, when added to a product, will contribute to meat or dairy flavours. For example, using a plant-based substrate high in methionine or cysteine is likely to result in the production of higher concentrations of sulphur-derived VOCs such as methional, methanethiol, methionol, dimethyl sulphide, dimethyl disulphide, and dimethyl trisulphide, which are important contributors to meat or dairy flavours.

3. Materials and Methods

3.1. LAB Strain

The LB672 culture was obtained from White Labs, USA. A stock culture was maintained at 4 °C until use. For working cultures, LB672 was initially cultivated in de Man, Rogosa, and Sharpe (MRS) broth at 25 °C for 3 days in sealed containers using anaerobic packs (Mitsubishi Gas Chemical (MGC) Company, Tokyo, Japan). An aliquot of the resulting culture was streaked onto MRS agar plates, which were incubated at 25 °C in sealed containers containing MGC anaerobic packs for 3 days or until large colonies were visible on the plates. An inoculating suspension was created by picking colonies from the streak plate and adding them to 10 mL of MRS broth, which was incubated at 25 °C for 3 days in sealed containers using MGC anaerobic packs. To prevent carryover of medium components to fermentation experiments, the cells were added to an Eppendorf tube (1 mL) and centrifugated (5000× g for 5 min at 20 °C) using a microcentrifuge (IEC Micromax, Milford, MA, USA). Cells were then washed twice with sterilised phosphate-buffered saline (PBS) (100 mL; 0.8 g NaCl, 0.02 g KCl, 0.144 g Na_2HPO_4, and 0.0245 g KH_2PO_4, pH of 7.4) and then resuspended to a final concentration of 1×10^9 CFU/mL. The resulting cell suspension was used for the fermentation studies.

3.2. Medium Compositions

The base for the DM contained D-glucose (20 g/L); peptone (enzymatic protein digest, Bacto peptone) (5 g/L), sodium acetate (either 1 or 12 g/L), mineral salts ($MgSO_4 \cdot 7H_2O$ (0.2 g/L), NaCl (0.01 g/L), $FeSO_4 \cdot 7H_2O$ (0.01 g/L), and $MnSO_4 \cdot 5H_2O$ (0.04 g/L)), and vitamins (calcium pantothenate (B5) (0.4 mg/L), nicotinic acid (B3) (0.2 mg/L), riboflavin (B2) (0.4 mg/L), and thiamine HCl (B1) (0.2 mg/L)). The DM was supplemented with amino acids L-leucine (Leu) (media DML0.1 and DML1.2), L-isoleucine (Ile) (media DMI0.1 and DMI1.2), L-phenylalanine (Phe) (media DMP0.1 and DMP1.2), L-threonine (Thr) (media DMT0.1 and DMT1.2), L-methionine (Met) (media DMM0.1 and DMM1.2), or L-glutamic acid (Glu) (media DMG0.1 and DMG1.2) separately (2 g/L) or combined (0.4 g/L each) (Aamix) (media DMAa0.1 and DMAa1.2) (Table 1). Glu was added to all media (2 g/L) except the two-base DM without added amino acid (media DM0.1 and DM1.2). The pH of the DM containing 0.1% acetate was adjusted to 6.6 using K_2HPO_4 (1M) and a pH meter (pH 213 microprocessor, HANNA, Cluj-Napoca, Romania). All the chemicals used were analytical grade unless otherwise stated. Most components were prepared as concentrated stock solutions and stored at 4 °C until used. The amino acids were dissolved in HCl solution (50 mM). All the stock solutions were prepared using reverse osmosis (RO) water unless otherwise stated. During the media preparation, the glucose and vitamin solutions were filter (nylon membrane: 0.22 µm, BIOFIL, Kowloon, Hong Kong) sterilised, with the other components being sterilised via autoclaving (121 °C, 15 min) (Astell Scientific, Sidcup, UK). In a class II biological safety cabinet (NUAIRE/NU-425-400, Plymouth, MN, USA), the sterilised media components were dispensed into sterile Schott bottles (100 mL) to achieve a final volume of 50 mL.

3.3. Fermentation

To confirm their sterility, prepared media were incubated at 25 °C for at least three days prior and checked for an absence of turbidity prior to inoculation. Each medium was inoculated with 100 µL of the LB672 cell suspension (1×10^9 CFU/mL) and incubated at 25 °C for 16 days under anaerobic conditions, as mentioned previously. Uninoculated sterile medium was used as control. All the fermentations were carried out in duplicate. At the end of the fermentation, growth was confirmed by measuring the pH and optical density (OD_{600}, Ultrospec 3300 pro, Amersham Biosciences, Amersham, UK) of an aseptically removed sub-sample.

3.4. Determination of Volatile Organic Compounds

The volatile organic compounds (VOCs) present at the end of fermentation were measured using headspace–solid phase microextraction (HS-SPME) gas chromatography–mass spectrometry (GC-MS) (Agilent 6890N GC system, Beijing, China; coupled to an Agilent 5975B VL mass spectrometer with triple axis detector, Wilmington, DE, USA). Samples were prepared by centrifuging (2000 rpm, 4 °C for 10 min, Sorvall RT 6000, Wilmington, DE, USA) a sub-sample (35 mL) of the ferment and adding the resulting supernatant (10 mL) to a 20 mL headspace vial (Phenomenex, Auckland, New Zealand) containing 3 g of NaCl along with 20 µL of the internal standard (3-heptanone (Sigma-Aldrich, Saint Louis, MO, USA) (0.008 mg/mL, prepared with methanol)). The vials were quickly capped with a Teflon-faced septa and stored at 4 °C in a refrigeration block (Peltier cooling tray) in the autosampler until analysis. Three analytical replicates were prepared from each supernatant to provide a total of six replicates for each fermentation medium (biological replicates: 2 and analytical replicates: 3). Samples were analysed in a randomised order, blocked by replicates. At the beginning of each replicate and at the end of each replicate, two blank samples consisting solely of RO water were analysed.

Prior to GC-MS analysis, each headspace vial was incubated at 40 °C for 5 min, after which the SPME fibre (divinylbenzene/carboxen/polydimethylsiloxane (DVB/CAR/PDMS), 50/30 µm Stableflex, Supelco, Bellefonte, PA, USA) was inserted through the septum of the vial and exposed to the headspace for 40 min at 40 °C. Upon completion of the extraction procedure, for thermal desorption, the fibre was inserted into the front injector port of the GC (set at 230 °C) for 2 min in splitless mode followed by 3 min in split mode (purge flow: 50 mL/min). Between samples, the SPME fibre was conditioned in the back inlet at 270 °C with a purge flow of 50 mL/min for 2 min. VOCs were separated through a Zebron ZB-Wax capillary column (60 m × 0.32 mm inner diameter × 0.5 µm film thickness: Phenomenex, Torrance, CA, USA) using helium as the carrier gas at a constant flow rate of 1 mL/min. The oven temperature programme was set at 50 °C for 5 min initially, followed by heating at a rate of 5 °C/min to 210 °C, followed by 10 °C/min until 240 °C, and finally held for 5 min, for a total run time of 45 min. Mass ions were measured between 29 and 300 m/z via electron ionisation (EI mode 70 eV). The ion source temperatures of mass ion traps were sustained at 230 °C with a quadrupole temperature of 150 °C.

3.5. Data Analysis

3.5.1. GC-MS Data Extraction

The GC-MS raw data files were exported in CDF format and imported into PARADISe (Version V6.0.0.14) software. PARADISe is based on PARAFAC2 modelling, which allowed simultaneous deconvolution of pure mass spectra of peaks and integration of areas of deconvoluted peaks for all samples. Resolved peaks were identified by matching their deconvoluted pure mass spectra and retention index (RI) [73,74] against the NIST spectral library (NIST14, version 2.2, National Institute of Standards and Technology). The compounds' RI (NIST spectral library) was compared using the calculated RI. A cubic spline interpolation using the n-alkane series (C7–C25), which was run under the same conditions in GC-MS, was used to calculate the RI of each compound [75]. The compounds were further selected based on their R match greater than 800. On a few occasions, compounds with R matches below 800 were chosen because they co-eluted with background or air peaks. If a compound only had a calculated RI or if its identification was not confident, VOCs were considered "unknown". Finally, the data matrix of the peak area for each compound (including not-identified peaks (unknown)) for each sample replicate was exported from PARADISe.

3.5.2. Statistical Analysis

The data exported from PARADISe were examined and compounds that had a significant difference in peak area between controls and treatments were identified using t-tests. Significant differences across treatments were assessed via one-way analysis of variance (ANOVA) with a generalised linear model (significance level at $p < 0.05$) using SPSS (IBM SPSS statistics, version 29.0.0.0 (241)).

Statistical analysis for selected compounds was carried out in R (version R 4.2.1) (R Foundation for Statistical Computing, Vienna, Austria). Heatmaps were developed using "cluster", "gplots", "factoextra", and "tidyverse" external packages as a non-targeted approach to encompass variations between medium compositions and selected VOCs. The peak area average of each VOC was log 2 transformed and used to create heatmaps. The VOCs were clustered using a distant matrix computed with the Pearson correlation coefficient.

One-way ANOVA was performed for each selected VOC across different medium compositions for the targeted approach. The mean separations were calculated using Tukey's HSD test at $p < 0.05$. Graphs were plotted using the "ggplot2" external package in R.

4. Conclusions

The use of a defined medium (DM) helped to understand the impact of amino acid addition on the generation of specific fermentation VOCs. When an Aamix or either Phe, Met, Leu, or Ile were added to the DM, the VOC profile produced was noticeably affected after LB672 fermentation; however, no specific VOCs were detected in Thr-supplemented DM. The results provide a foundation for understanding LB672's role in amino acid catabolism by indicating which amino acid catabolic enzymes may be present and by highlighting the VOC they produce. To gain a better understanding of how specific VOCs could be produced during LAB fermentation, different amino acid combinations and LAB strains of the same or different species should be studied using an online VOC-tracking method (PTR-ToF-MS).

Supplementary Materials: The following supporting information can be downloaded at: https://www.mdpi.com/article/10.3390/molecules29040753/s1, Table S1: VOCs detected in 0.1% acetate media; Table S2: VOCs detected in 1.2% acetate media.

Author Contributions: Conceptualisation, P.S. and P.B.; methodology, S.R., P.S. and P.B.; investigation, S.R.; formal analysis, S.R.; data curation, S.R.; writing—original draft, S.R.; writing—review and editing, S.R., P.S. and P.B.; supervision, P.S and P.B. All authors have read and agreed to the published version of the manuscript.

Funding: This work was supported by the Accelerating Higher Education Expansion and Development (AHEAD) operation (AHEAD/PhD/R3/Agri/394), a World Bank-funded project, Ministry of Education, Sri Lanka, University of Otago doctoral scholarship, and Catalyst: Seeding funding was provided by the New Zealand Ministry of Business, Innovation and Employment and administered by the Royal Society Te Apārangi.

Institutional Review Board Statement: Not applicable.

Informed Consent Statement: Not applicable.

Data Availability Statement: Data will be made available on request.

Acknowledgments: S.R. would like to thank Michelle Petrie and Ian Ross for the microbiology laboratory assistance, Michelle Leus for the technical assistance in GC-MS, and Andrea Warburton for the GC-MS data extraction using PARADISe.

Conflicts of Interest: The authors declare no conflicts of interest.

References

1. Lea, E.J.; Crawford, D.; Worsley, A. Consumers' readiness to eat a plant-based diet. *Eur. J. Clin. Nutr.* **2006**, *60*, 342–351. [CrossRef] [PubMed]
2. Szejda, K.; Urbanovich, T.; Wilks, M. *Accelerating Consumer Adoption of Plant-Based Meat: An Evidence-Based Guide for Effective Practice*; The Good Food Institute: Washington, DC, USA, 2020.
3. Austgulen, M.; Skuland, S.; Schjøll, A.; Alfnes, F. Consumer readiness to reduce meat consumption for the purpose of environmental sustainability: Insights from Norway. *Sustainability* **2018**, *10*, 3058. [CrossRef]
4. Clem, J.; Barthel, B. A look at plant-based diets. *Mo. Med.* **2021**, *118*, 233–238. [PubMed]
5. Alcorta, A.; Porta, A.; Tarrega, A.; Alvarez, M.D.; Vaquero, M.P. Foods for plant-based diets: Challenges and innovations. *Foods* **2021**, *10*, 293. [CrossRef] [PubMed]
6. Lima, M.; Costa, R.; Rodrigues, I.; Lameiras, J.; Botelho, G. A narrative review of alternative protein sources: Highlights on meat, fish, egg and dairy analogues. *Foods* **2022**, *11*, 2053. [CrossRef] [PubMed]
7. Bryant, C.J. Plant-based animal product alternatives are healthier and more environmentally sustainable than animal products. *Future Foods* **2022**, *6*, 100174–100186. [CrossRef]
8. Ishaq, A.; Irfan, S.; Sameen, A.; Khalid, N. Plant-based meat analogs: A review with reference to formulation and gastrointestinal fate. *Curr. Res. Food Sci.* **2022**, *5*, 973–983. [CrossRef]

9. Michel, F.; Hartmann, C.; Siegrist, M. Consumers' associations, perceptions and acceptance of meat and plant-based meat alternatives. *Food Qual. Prefer.* **2021**, *87*, 104063–104073. [CrossRef]
10. Aschemann-Witzel, J.; Gantriis, R.F.; Fraga, P.; Perez-Cueto, F.J.A. Plant-based food and protein trend from a business perspective: Markets, consumers, and the challenges and opportunities in the future. *Crit. Rev. Food Sci. Nutr.* **2021**, *61*, 3119–3128. [CrossRef]
11. Pointke, M.; Pawelzik, E. Plant-based alternative products: Are they healthy alternatives? Micro- and macronutrients and nutritional scoring. *Nutrients* **2022**, *14*, 601. [CrossRef]
12. Szenderak, J.; Frona, D.; Rakos, M. Consumer acceptance of plant-based meat substitutes: A narrative review. *Foods* **2022**, *11*, 1274. [CrossRef]
13. Reineccius, G. *Flavor Chemistry and Technology*, 2nd ed.; Taylor & Francis Group: Boca Raton, FL, USA, 2006.
14. Lawless, H. The sense of smell in food quality and sensory evaluation. *J. Food Qual.* **1991**, *14*, 33–60. [CrossRef]
15. Astray, G.; García-Río, L.; Mejuto, J.C.; Pastrana, L. Chemistry in food: Flavours. *Electron. J. Environ. Agric. Food Chem.* **2007**, *6*, 1742–1763.
16. Paravisini, L.; Guichard, E. Interactions between aroma compounds and food matrix. In *Flavour: From Food to Perception*; Wiley: New York, NY, USA, 2016; pp. 208–234.
17. Small, D.M.; Gerber, J.C.; Mak, Y.E.; Hummel, T. Differential neural responses evoked by orthonasal versus retronasal odorant perception in humans. *Neuron* **2005**, *47*, 593–605. [CrossRef]
18. Dastager, S.G. *Aroma Compounds*; Springer: Berlin/Heidelberg, Germany, 2009.
19. Janssens, L.; De Pooter, H.L.; Schamp, N.M.; Vandamme, E.J. Production of flavours by microorganisms. *Process Biochem.* **1992**, *27*, 195–215. [CrossRef]
20. Longo, M.A.; Sanromán, M.A. Production of food aroma compounds: Microbial and enzymatic methodologies. *Food Technol. Biotechnol.* **2006**, *44*, 335–353. [CrossRef]
21. Bamforth, C.W.; Cook, D.J. *Food, Fermentation, and Micro-Organisms*, 2nd ed.; Wiley: New York, NY, USA, 2019. [CrossRef]
22. Petrovici, A.R.; Ciolacu, D.E. Natural flavours obtained by microbiological pathway. In *Generation of Aromas and Flavours*; InTech: London, UK, 2018; pp. 33–52.
23. Rajendran, S.; Silcock, P.; Bremer, P. Flavour volatiles of fermented vegetable and fruit substrates: A review. *Molecules* **2023**, *28*, 3236. [CrossRef]
24. Szutowska, J. Functional properties of lactic acid bacteria in fermented fruit and vegetable juices: A systematic literature review. *Eur. Food Res. Technol.* **2020**, *246*, 357–372. [CrossRef]
25. Teusink, B.; Molenaar, D. Systems biology of lactic acid bacteria: For food and thought. *Curr. Opin. Syst. Biol.* **2017**, *6*, 7–13. [CrossRef]
26. Hayek, S.A.; Gyawali, R.; Aljaloud, S.O.; Krastanov, A.; Ibrahim, S.A. Cultivation media for lactic acid bacteria used in dairy products. *J. Dairy Res.* **2019**, *86*, 490–502. [CrossRef]
27. Wegkamp, A.; Teusink, B.; de Vos, W.M.; Smid, E.J. Development of a minimal growth medium for *Lactobacillus plantarum*. *Lett. Appl. Microbiol.* **2010**, *50*, 57–64. [CrossRef] [PubMed]
28. Kwoji, I.D.; Okpeku, M.; Adeleke, M.A.; Aiyegoro, O.A. Formulation of chemically defined media and growth evaluation of *Ligilactobacillus salivarius* ZJ614 and *Limosilactobacillus reuteri* ZJ625. *Front. Microbiol.* **2022**, *13*, 1450. [CrossRef] [PubMed]
29. Jensen, P.R.; Hammer, K. Minimal requirements for exponential growth of *Lactococcus lactis*. *Appl. Environ. Microbiol.* **1993**, *59*, 4363–4366. [CrossRef] [PubMed]
30. Niven, C.F. Nutrition of *Streptococcus lactis*. *J. Bacteriol.* **1944**, *47*, 343–350. [CrossRef]
31. Cocaign-Bousquet, M.; Garrigues, C.; Novak, L.; Lindley, N.D.; Loublere, P. Rational development of a simple synthetic medium for the sustained growth of *Lactococcus lactis*. *J. Appl. Bacteriol.* **1995**, *79*, 108–116. [CrossRef]
32. van Niel, E.W.J.; Hahn-Hägerdal, B. Nutrient requirements of lactococci in defined growth media. *Appl. Microbiol. Biotechnol.* **1999**, *52*, 617–627. [CrossRef]
33. Pastink, M.I.; Teusink, B.; Hols, P.; Visser, S.; de Vos, W.M.; Hugenholtz, J. Genome-scale model of *Streptococcus thermophilus* LMG18311 for metabolic comparison of lactic acid bacteria. *Appl. Environ. Microbiol.* **2009**, *75*, 3627–3633. [CrossRef]
34. Kranenburg, R.V.; Kleerebezem, M.; van Hylckama Vlieg, J.; Ursing, B.M.; Boekhorst, J.; Smit, B.A.; Ayad, E.H.E.; Smit, G.; Siezen, R.J. Flavour formation from amino acids by lactic acid bacteria: Predictions from genome sequence analysis. *Int. Dairy J.* **2002**, *12*, 111–121. [CrossRef]
35. Christensen, J.E.; Dudley, E.G.; Pederson, J.A.; Steele, J.L. Peptidases and amino acid catabolism in lactic acid bacteria. *Antonie Van Leeuwenhoek* **1999**, *76*, 217–246. [CrossRef]
36. Fernandez, M.; Zuniga, M. Amino acid catabolic pathways of lactic acid bacteria. *Crit. Rev. Microbiol.* **2006**, *32*, 155–183. [CrossRef]
37. Teixeira, P. *Lactobacillus, Lactobacillus brevis*. In *Encyclopedia of Food Microbiology*; Elsevier: Amsterdam, The Netherlands, 2014; Volume 2, pp. 418–424. [CrossRef]
38. Chen, D.; Chia, J.Y.; Liu, S.Q. Impact of addition of aromatic amino acids on non-volatile and volatile compounds in lychee wine fermented with *Saccharomyces cerevisiae* MERIT.ferm. *Int. J. Food Microbiol.* **2014**, *170*, 12–20. [CrossRef]
39. Fairbairn, S.; McKinnon, A.; Musarurwa, H.T.; Ferreira, A.C.; Bauer, F.F. The impact of single amino acids on growth and volatile aroma production by *Saccharomyces cerevisiae* strains. *Front. Microbiol.* **2017**, *8*, 2554. [CrossRef]
40. Chua, J.Y.; Tan, S.J.; Liu, S.Q. The impact of mixed amino acids supplementation on *Torulaspora delbrueckii* growth and volatile compound modulation in soy whey alcohol fermentation. *Food Res. Int.* **2021**, *140*, 109901–109913. [CrossRef]

41. Lee, P.-R.; Yu, B.; Curran, P.; Liu, S.-Q. Impact of amino acid addition on aroma compounds in papaya wine fermented with *Williopsis mrakii*. *S. Afr. J. Enol. Vitic.* **2011**, *32*, 220–228. [CrossRef]
42. Wang, Y.-Q.; Ye, D.-Q.; Liu, P.-T.; Duan, L.-L.; Duan, C.-Q.; Yan, G.-L. Synergistic effects of branched-chain amino acids and phenylalanine addition on major volatile compounds in wine during alcoholic fermentation. *S. Afr. J. Enol. Vitic.* **2016**, *37*, 169–175. [CrossRef]
43. Meza, J.C.; Christen, P.; Revah, S. Effect of added amino acids on the production of a fruity aroma by *Ceratocystis fimbriata*. *Sci. Aliment.* **1998**, *18*, 627–636.
44. Gutsche, K.A.; Tran, T.B.; Vogel, R.F. Production of volatile compounds by *Lactobacillus sakei* from branched chain alpha-keto acids. *Food Microbiol.* **2012**, *29*, 224–228. [CrossRef]
45. Tavaria, F.K.; Dahl, S.; Carballo, F.J.; Malcata, F.X. Amino acid catabolism and generation of volatiles by lactic acid bacteria. *J. Dairy Sci.* **2002**, *85*, 2462–2470. [CrossRef]
46. Canon, F.; Maillard, M.B.; Henry, G.; Thierry, A.; Gagnaire, V. Positive interactions between lactic acid bacteria promoted by nitrogen-based nutritional dependencies. *Appl. Environ. Microbiol.* **2021**, *87*, e01055-21. [CrossRef]
47. Henderson, L.M.; Snell, E.E. A uniform medium for determination of amino acids with various microorganisms. *J. Biol. Chem.* **1948**, *172*, 15–29. [CrossRef]
48. De man, J.C.; Rogosa, M.; Sharpe, M.E. A medium for the cultivation of lactobacilli. *J. Appl. Bacteriol.* **1960**, *23*, 130–135. [CrossRef]
49. Russell, C.; Bhandari, R.R.; Walker, T.K. Vitamin requirements of thirty-four lactic acid bacteria associated with brewery products. *J. Gen. Microbiol.* **1954**, *10*, 371–376. [CrossRef] [PubMed]
50. Hébert, E.M.; Raya, R.R.; Savoy de Giori, G. *Evaluation of Minimal Nutritional Requirements of Lactic Acid Bacteria Used in Functional Foods*; Humana Press Inc.: Totowa, NJ, USA, 2004.
51. MacLeod, R.A.; Snell, E.E. Some mineral requirements of the lactic acid bacteria. *J. Biol. Chem.* **1947**, *170*, 351–365. [CrossRef]
52. Zacharof, M.-P.; Lovitt, R.W. Partially chemically defined liquid medium development for intensive propagation of industrial fermentation lactobacilli strains. *Ann. Microbiol.* **2012**, *63*, 1235–1245. [CrossRef]
53. Liu, M.; Nauta, A.; Francke, C.; Siezen, R.J. Comparative genomics of enzymes in flavor-forming pathways from amino acids in lactic acid bacteria. *Appl. Environ. Microbiol.* **2008**, *74*, 4590–4600. [CrossRef] [PubMed]
54. Makarova, K.; Slesarev, A.; Wolf, Y.; Sorokin, A.; Mirkin, B.; Koonin, E.; Pavlov, A.; Pavlov, N.; Karamychev, V.; Polouchine, V.; et al. Comparative genomics of the lactic acid bacteria. *Proc. Natl. Acad. Sci. USA* **2006**, *103*, 15611–15616. [CrossRef]
55. Klompong, V.; Benjakul, S.; Kantachote, D.; Shahidi, F. Characteristics and use of yellow stripe trevally hydrolysate as culture media. *J. Food Sci.* **2009**, *74*, 219–225. [CrossRef]
56. Foudin, A.S.; Wynn, W.K. Growth of *Puccinia graminis* f. sp. *tritici*. *Phytopathology* **1972**, *62*, 1032–1040. [CrossRef]
57. Li, T.; Jiang, T.; Liu, N.; Wu, C.; Xu, H.; Lei, H. Biotransformation of phenolic profiles and improvement of antioxidant capacities in jujube juice by select lactic acid bacteria. *Food Chem.* **2021**, *339*, 127859–127869. [CrossRef]
58. McSweeney, P.L.H.; Sousa, M.J. Biochemical pathways for the production of flavour compounds in cheeses during ripening: A review. *Le Lait* **2000**, *80*, 293–324. [CrossRef]
59. Ricci, A.; Cirlini, M.; Levante, A.; Dall'Asta, C.; Galaverna, G.; Lazzi, C. Volatile profile of elderberry juice: Effect of lactic acid fermentation using *L. plantarum*, *L. rhamnosus* and *L. casei* strains. *Food Res. Int.* **2018**, *105*, 412–422. [CrossRef]
60. Valera, M.J.; Boido, E.; Ramos, J.C.; Manta, E.; Radi, R.; Dellacassa, E.; Carrau, F. The mandelate pathway, an alternative to the phenylalanine ammonia lyase pathway for the synthesis of benzenoids in Ascomycete Yeasts. *Appl. Environ. Microbiol.* **2020**, *86*, e00701-20. [CrossRef]
61. Yvon, M.; Rijnen, L. Cheese flavour formation by amino acid catabolism. *Int. Dairy J.* **2001**, *11*, 185–201. [CrossRef]
62. Hu, S.; Zhang, X.; Lu, Y.; Lin, Y.-C.; Xie, D.-F.; Fang, W.; Huang, J.; Mei, L.-H. Cloning, expression and characterization of an aspartate aminotransferase gene from *Lactobacillus brevis* CGMCC 1306. *Biotechnol. Biotechnol. Equip.* **2017**, *31*, 544–553. [CrossRef]
63. Marilley, L.; Casey, M.G. Flavours of cheese products: Metabolic pathways, analytical tools and identification of producing strains. *Int. J. Food Microbiol.* **2004**, *90*, 139–159. [CrossRef] [PubMed]
64. Ardö, Y. Flavour formation by amino acid catabolism. *Biotechnol. Adv.* **2006**, *24*, 238–242. [CrossRef] [PubMed]
65. Hols, P.; Ramos, A.; Hugenholtz, J.; Delcour, J.; de Vos, W.M.; Santos, H.; Kleerebezem, M. Acetate utilization in *Lactococcus lactis* deficient in lactate dehydrogenase: A rescue pathway for maintaining redox balance. *J. Bacteriol.* **1999**, *181*, 5521–5526. [CrossRef]
66. Feyereisen, M.; Mahony, J.; Kelleher, P.; Roberts, R.J.; O'Sullivan, T.; Geertman, J.A.; van Sinderen, D. Comparative genome analysis of the *Lactobacillus brevis* species. *BMC Genom.* **2019**, *20*, 416–431. [CrossRef]
67. Barker, H.A.; Kamen, M.D.; Bornstein, B.T. The synthesis of butyric and caproic acids from ethanol and acetic acid by *Clostridium kluyveri*. *Proc. Natl. Acad. Sci. USA* **1945**, *31*, 373–381. [CrossRef]
68. Reed, L.J.; DeBusk, B.G.; Johnston, P.M.; Getzendaner, M.E. Acetate-replacing factors for lactic acid bacteria. *J. Biol. Chem.* **1951**, *192*, 851–858. [CrossRef]
69. Lindinger, W.; Hansel, A.; Jordan, A. Proton-transfer-reaction mass spectrometry (PTR-MS): On-line monitoring of volatile organic compounds at pptv levels. *Chem. Soc. Rev.* **1998**, *27*, 347–354. [CrossRef]
70. Ba, V.H.; Hwang, I.; Jeong, D.; Touseef, A. Principle of meat aroma flavors and future prospect. In *Latest Research into Quality Control*; InTech: London, UK, 2012; pp. 145–176.
71. Marsili, R. Flavors and off-flavors in dairy foods. In *Encyclopedia of Dairy Sciences*; Elsevier: Amsterdam, The Netherlands, 2022; pp. 560–578. [CrossRef]

72. Smit, G.; Smit, B.A.; Engels, W.J. Flavour formation by lactic acid bacteria and biochemical flavour profiling of cheese products. *FEMS Microbiol. Rev.* **2005**, *29*, 591–610. [CrossRef] [PubMed]
73. Johnsen, L.G.; Skou, P.B.; Khakimov, B.; Bro, R. Gas chromatography—Mass spectrometry data processing made easy. *J. Chromatogr. A* **2017**, *1503*, 57–64. [CrossRef]
74. Baccolo, G.; Quintanilla-Casas, B.; Vichi, S.; Augustijn, D.; Bro, R. From untargeted chemical profiling to peak tables—A fully automated AI driven approach to untargeted GC-MS. *Trends Anal. Chem.* **2021**, *145*, 116451–116459. [CrossRef]
75. Halang, W.A.; Langlais, R.; Kugler, E. Cubic spline interpolation for the calculation of retention indices in temperature programmed gas-liquid chromatography. *Anal. Chem.* **1978**, *50*, 1829–1832. [CrossRef]

Disclaimer/Publisher's Note: The statements, opinions and data contained in all publications are solely those of the individual author(s) and contributor(s) and not of MDPI and/or the editor(s). MDPI and/or the editor(s) disclaim responsibility for any injury to people or property resulting from any ideas, methods, instructions or products referred to in the content.

Article

A New Perspective on SPME and SPME Arrow: Formaldehyde Determination by On-Sample Derivatization Coupled with Multiple and Cooling-Assisted Extractions

Stefano Dugheri [1,*], Giovanni Cappelli [2], Niccolò Fanfani [3], Jacopo Ceccarelli [2], Giorgio Marrubini [4], Donato Squillaci [2], Veronica Traversini [2], Riccardo Gori [5], Nicola Mucci [2] and Giulio Arcangeli [2]

1 Industrial Hygiene and Toxicology Laboratory, University Hospital Careggi, 50134 Florence, Italy
2 Department of Experimental and Clinical Medicine, University of Florence, 50121 Florence, Italy; donato.squillaci@unifi.it (D.S.)
3 Department of Experimental and Clinical Biomedical Sciences "Mario Serio", University of Florence, 50121 Florence, Italy
4 Department of Drug Sciences, University of Pavia, 27100 Pavia, Italy; giorgio.marrubini@unipv.it
5 Department of Civil and Environmental Engineering, University of Florence, 50121 Florence, Italy
* Correspondence: stefano.dugheri@unifi.it

Abstract: Formaldehyde (FA) is a toxic compound and a human carcinogen. Regulating FA-releasing substances in commercial goods is a growing and interesting topic: worldwide production sectors, like food industries, textiles, wood manufacture, and cosmetics, are involved. Thus, there is a need for sensitive, economical, and specific FA monitoring tools. Solid-phase microextraction (SPME), with O-(2,3,4,5,6-pentafluorobenzyl)-hydroxylamine (PFBHA) on-sample derivatization and gas chromatography, is proposed for FA monitoring of real-life samples. This study reports the use of polydimethylsiloxane (PDMS) as a sorbent phase combined with innovative commercial methods, such as multiple SPME (MSPME) and cooling-assisted SPME, for FA determination. Critical steps, such as extraction and sampling, were evaluated in method development. The derivatization was performed at 60 °C for 30 min, followed by 15 min sampling at 10 °C, in three cycles (SPME Arrow) or six cycles (SPME). The sensitivity was satisfactory for the method's purposes (LOD-LOQ at 11-36 ng L^{-1}, and 8-26 ng L^{-1}, for SPME and SPME Arrow, respectively). The method's linearity ranges from the lower LOQ at trace level (ng L^{-1}) to the upper LOQ at 40 mg L^{-1}. The precision range was 5.7–10.2% and 4.8–9.6% and the accuracy was 97.4% and 96.3% for SPME and SPME Arrow, respectively. The cooling MSPME set-up applied to real commercial goods provided results of quality comparable to previously published data.

Keywords: solid-phase microextraction; environmental analysis; food analysis; formaldehyde; on-sample derivatization; SPME; SPME Arrow; headspace

1. Introduction

Formaldehyde (FA) is globally synthesized on a large scale by catalytic oxidation of methanol in the vapor phase [1]. FA is commonly used to produce industrial chemicals [2–4], and its aqueous solution (known as formalin) has many applications as a sanitizer and preservative [5].

Human FA exposure can result in acute and chronic toxicity. Concerning FA acute toxicity, skin, eyes, and respiratory tract irritation, skin sensitization, and developmental toxicity have been reported [6,7].

Chronic exposure to FA has been recognized to enhance the risk of asthma [8] and miscarriage [9]. Long-term FA exposure has also been related to nasopharyngeal cancer [10] and increased risk of rare neck and head cancers [11].

The complex scenario of FA toxicity, in addition to its mass use, has recently led to the adoption of new regulations in various production sectors. In 2022, the World Trade

Organization (WTO) issued a statement from the European Union (EU) on its plan to regulate both FA and FA-releasing substances in wood-based articles and furniture and the interior of vehicles [12,13]. Moreover, in 2021, the European Commission's Scientific Committee for Consumer Safety (SCCS) proposed lowering the threshold for cosmetics containing FA-releasing substances from 0.05 to 0.001% [14,15]. Vietnam and the European Commission released specific restrictions concerning FA in textiles [16–18]. Recently, it has been reported that formalin is extensively used in several tropical countries as an artificial preservative for food [19], contravening the daily limit of oral exposure to FA from the total diet by the European Food Safety Authority (EFSA), which should not exceed 100 mg FA per day [20]. As for food-contact materials, the current US regulation requires that the amount of FA in melamine resins employed on the surface of food-contact products must not exceed 0.5 milligrams per square inch [21]. Since the Federal Hazardous Substances Act (FHSA) in the US listed FA as a "strong sensitizer" substance, the Consumer Product Safety Commission (CPSC) stated that products containing more than 1% of FA could lead to severe hypersensitivity in humans [22]. Concerning the commercialization of children's products containing FA, the United States' regulatory agencies set restrictions [23–25]. Because of this heterogeneous scenario, evaluating FA content in commercial products is a key factor for compliance with regulations [26]. In this context, green and more efficient approaches to analytical measurements are also aspects to be considered [27]. Also, miniaturization has been implemented in every analytical field, including exposome investigations [28], resulting in cost and time savings throughout the sampling process. The miniaturization of traditional sample preparation devices for liquid chromatography (LC) and gas chromatography (GC) led to the development of new eco-friendly analytical instruments and methods, limiting the impact of chemicals on the environment.

Moreover, applying automated microextraction techniques (METs) [29,30] allows the lowering of time-consuming manual sample preparations. With the introduction of the first commercial device in 1993 by Supelco (Bellefonte, US), solid-phase microextraction (SPME) [31] has proven to be one of the most applied and versatile METs thanks to continuous technological innovations, especially in recent years [29]. In 2015, the Restek Corporation (Bellefonte, US) updated the SPME technology, proposing the SPME Arrow. This tool is a larger-diameter SPME probe with rugged construction, designed to enhance mechanical durability, with a greater phase volume than standard SPME fibers [32,33]. Thus far, polydimethylsiloxane (PDMS) is one of the most widely available coatings employed in the entrapment of volatile analytes via METs; this is mainly due to the ease of control of the extraction process, as well as the absence of competition among analytes during absorption, making it the best choice in the presence of complex matrices [29]. The recent applications involve coupling SPME with on-fiber derivatization [27,34–36]. In particular, several reports described the reaction of carbonyl compounds with o-(2,3,4,5,6-pentafluorobenzyl) hydroxylamine (PFBHA) and subsequent separation of PFB-oxime derivatives by GC coupled with different detectors [37–41], showing that the derivatives had outstanding chromatographic properties. Other derivatizing reagents are available on the market, such as 2,4-dinitrophenylhydrazine (DNPH) and 2,2,2-trifluoroethylhydrazine (TFEH). However, although DNPH derivatization affords low limits of detection and low chemical noise in the blank samples [38], hydrazones must be extracted or preconcentrated before analysis [42–44]. As for TFEH, it could be used to form the TFE–hydrazone [44,45], but the reagent is more expensive than PFBHA. Only a few articles reported the determination of FA using PFBHA on-sample derivatization and SPME sampling in various matrices and the related GC analysis of the PFB-FA-oxime [46–52]. It is, however, of some interest to investigate the coupling of headspace (HS) or direct injection (DI) SPME with PFB-oxime derivatives to analyze carbonyl functional groups.

We propose the use of a commercially available cooling-assisted SPME [53,54] together with a multiple SPME (MSPME) technique [55,56] for FA monitoring in real-life samples.

Cooling-assisted SPME enhances the fiber efficiency caption while maintaining efficient stripping from the matrix through two different temperature layers.

To our knowledge, combining the cooling-assisted SPME with MSPME has not been reported. Using this technique, we could apply multiple extraction cycles on FA-containing samples, thus enabling the fully automated routine monitoring of FA content of real-life samples. To achieve this goal, we investigated the key points of the on-sample PFBHA derivatization using both SPME and SPME Arrow PDMS fibers.

The developed cooling MSPME method was applied and tested on different matrices to prove its general applicability. Several critical aspects of the technique, such as pH, derivatizing agent excess, and FA blank signal, are discussed.

2. Results and Discussion

2.1. Method Optimization

Fundamental aspects regarding the optimization of the method are investigated and presented here; in particular, the derivatization temperature, the concentration of the derivatizing reagent, and the effect of salt are considered.

Firstly, we considered the background solution of FA since both reagent water and derivatizing reagent can represent a potential source of contamination [38,57]. Hudson et al. [58] sparged Milli-Q water with ultrapure Argon for 45 min, lowering the FA levels in blanks by 10%. Other procedures reported in the literature, such as UV irradiation and the addition of hydrogen peroxide, showed opposite results in terms of reduction. As a general consideration, it is impossible to define the outcome of FA lowering from the start, as the effect is strictly dependent on the number of dissolved substances that can be transformed and can cause an increase in its concentration [48,58,59]. We observed the FA depletion of roughly 20% (accounting for 8 ± 3 ng) [60] after irradiating with UV light (254 nm, 7.2 W m^{-2}) Milli-Q water for 60 min.

Moreover, we considered previous works employing PFBHA-based on-sample derivatization methods (see Table 1).

Table 1. The literature conditions for PFBHA-based on-sample derivatization methods.

PFBHA Solution Concentration (mg/mL)	PFBHA Solution Volume Used (µL)	Derivatization Target	Sample Volume (mL)	Sample Matrix	Reference
12	50	Acetone	10	Seawater	[58]
12	100	Aliphatic aldehydes C1 to C9 (from formaldehyde to nonanal), acetone, acrolein, butanone, furfural, benzaldehyde, methylglyoxal, glyoxal, 2,4-pentane dione	20	Seawater	[48]
2	1000	Formaldehyde, acetaldehyde, acetone, propionaldehyde, acrolein, isobutyraldehyde, butyraldehyde, pentanal, crotonaldehyde, isovaleraldehyde, hexanal	9	Alcoholic beverages	[47]
6	40	23 carbonyl compounds, including C1–C10 saturated aliphatic and unsaturated aldehydes, ketones, and dialdehydes	4.0–8.5	Water	[49]

In this framework, we set up the method using 100 µL of a 20 mg mL^{-1} PFBHA water solution for 1.85 mL of liquid sample and 18 mL of HS.

As for derivatization temperature, Güneş et al. suggested that an increase in the incubation temperature positively affects the reaction efficiency, which was therefore set to 60 °C [61]. The authors also indicated that there was no visible effect on the analysis derived from the salt type; for this reason, NaCl was chosen.

A customized CTC PAL3 System xyz Autosampler (CTC Analytics AG Industrie Strasse 20 CH-4222, Zwingen, Switzerland) installed online to the GC instrument was used to achieve a fully automated procedure, from sample preparation to injection into the GC apparatus. The instrument described improves productivity, minimizing dead times between samples and ultimately reducing the costs of the analytical assay. The full automation of the system requires minimal operator supervision and can process more samples during each analytical session.

A scheme of the procedure, and a picture of the autosampler GC system, are reported in Figure 1.

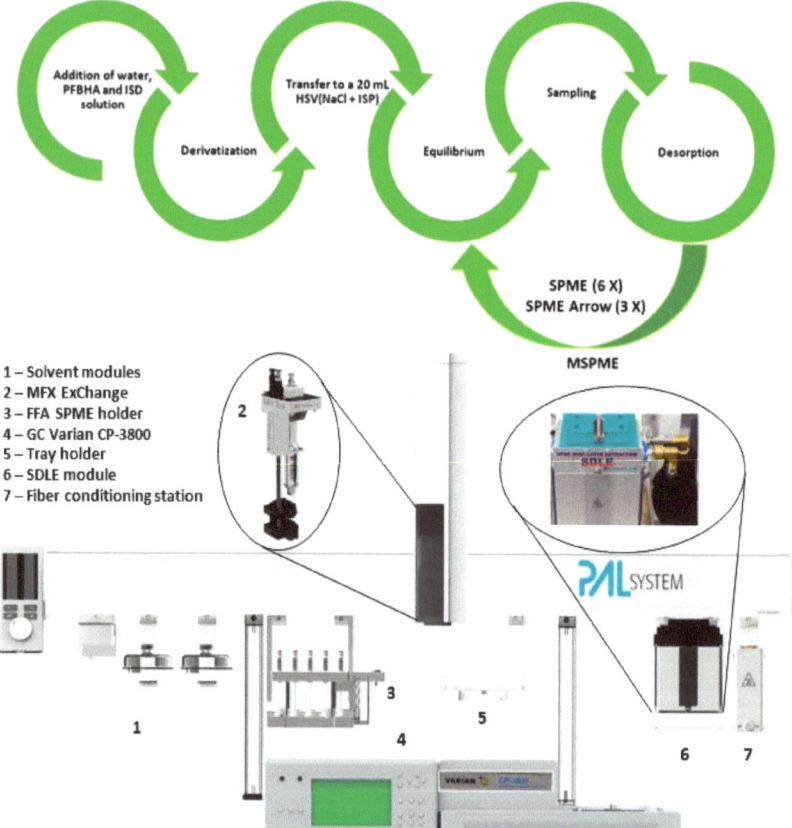

Figure 1. Flow chart and image of autosampler for the fully automated on-sample derivatization and analysis of PFBHA FA derivative with internal standardization (Picture courtesy of Filippo Degli Esposti, Chromline, Prato, Italy, free domain).

2.2. SPME Absorption by PDMS Coating

The PDMS absorptive coating was chosen for sampling complex matrices since analytes do not compete, unlike using porous phases such as divinylbenzene [62]. The theory of the liquid phase SPME technique was described in previous works. Wardencki et al. demonstrated that increasing the PDMS thickness enhances the analytes' recovery [47]. Louch, Motlagh, and Pawliszyn reported that the extraction time is inversely proportional to the square of the coating thickness [63]. Moreover, PDMS is the best sorbent for analytes having molecular mass in the 75–300 Da range [64]. Since the weight of PFB-FA-oxime is 225 Da, PDMS was deemed a suitable sorbent phase. We thus selected 100 μm PDMS SPME fibers (among the commercially available 7, 30, and 100 μm, respectively) and 250 μm PDMS SPME Arrow fibers (among the commercially available 100 and 250 μm, respectively).

Considering the liquid fiber coating's features, the extraction obeys the rules of liquid–liquid partitioning equilibrium. In a three-phase system composed of a liquid polymeric coating, a gaseous headspace, and an aqueous medium, the mass of analyte absorbed (using HS technique) by a coating at equilibrium, n (μg), is calculated by Equation (1) [65]:

$$n = \frac{C_o \times V_1 \times V_2 \times K_1 \times K_2}{K_1 \times K_2 \times V_1 + K_2 V_3 + V_2} \quad (1)$$

where K_1 is the SPME coating/HS partition coefficient, K_2 is the HS/aqueous matrix partition coefficient, C_0 is the initial concentration of the analyte in the aqueous solution (μg mL^{-1}), and V_1, V_2 and V_3 are the coating, the aqueous solution, and the HS volumes (mL), respectively.

In the case of DI-SPME sampling from an aqueous medium, Equation (1) can be simplified using Equation (2):

$$n = \frac{C_o \times V_1 \times V_2 \times K}{KV_1 + V_2} \quad (2)$$

where K, defined as $K_1 \cdot K_2$, is the partition coefficient between the SPME liquid polymeric coating and the sample.

Pacenti et al. [66] indicated K_{ow} as a good estimator of K. Moreover, K_2 can be described by means of Equation (3):

$$K_2 = \frac{K_H}{R \times T} \quad (3)$$

where K_H is Henry's constant (mol atm^{-1} L^{-1}), R is the universal gas constant (L atm K^{-1} mol^{-1}), and T is the sampling temperature (K, in Kelvin scale).

SPME performs extraction at equilibrium, and therefore it is not exhaustive. Hence, the hypothesis of ideal conditions needed by mathematical modeling must be verified. The distribution constants estimated from physico-chemical tables or by the structural unit contribution method can anticipate trends in SPME analysis; in particular, K_1 and K_2 values suggest whether the DI or HS mode is advantageous [52]. To calculate the theoretical mass, using Equations (1) and (2), we considered the same values employed in the method presented, i.e., 20 μg mL^{-1} calibration level (normalized for the final volume and converted in PFB-FA-oxime), V_2 and V_3 as 2.5 mL and 17.5 mL, respectively. Physico-chemical constants of the PFB-FA-oxime were obtained by Performs Automated Reasoning in Chemistry (SPARChem, Danielsville, GA, USA). Theoretic recoveries were calculated considering the reaction between immediate and exhaustive FA and PFBHA [37]. The calculation provided n values of 200 and 197 μg and theoretical recoveries of 68 and 67% for HS and DI, respectively.

To confirm these theoretical results, we propose the comparison between HS- and DI-SPME using SPME and SPME Arrow fibers. The chromatogram is shown in Figure 2 (the comparison calibration curves are shown in the Supplementary Materials in Figure S1).

As shown in Table 2, the two techniques provide comparable sensitivity (i.e., slope), recoveries, and precision, as found in previous works [49,67]. In more detail, Table 2 shows that the method's sensitivity was similar using either DI or HS. Exposure time with DI was lower than using HS, while the recoveries were comparable between the two techniques, as observed previously [52].

With water as the medium, DI causes a reduction of the GC column lifetime in comparison with HS. The mean recovery values obtained depend on the non-exhaustive nature of SPME and the inverse proportionality between temperature and captured analyte amount found using each absorbent phase. Equations (1) and (2) show an expectable overestimation of recovered PFB-FA-oxime compared to the experimental values due to the non-exhaustive inherent nature of the process of absorption of the analyte into the sorbent phase. The methods' precision, expressed by the average CV% of Table 2, was good and found to be similar for the two techniques. Mean recovery is the mean value of the recoveries found for each of the five calibration levels obtained by the PFB-FA-oxime hexane solutions regression curve. The method's precision was computed by the average variation coefficient (Average CV%) at each calibration level of the curves. Considering these results, we selected HS-SPME as the more suitable technique for determining FA.

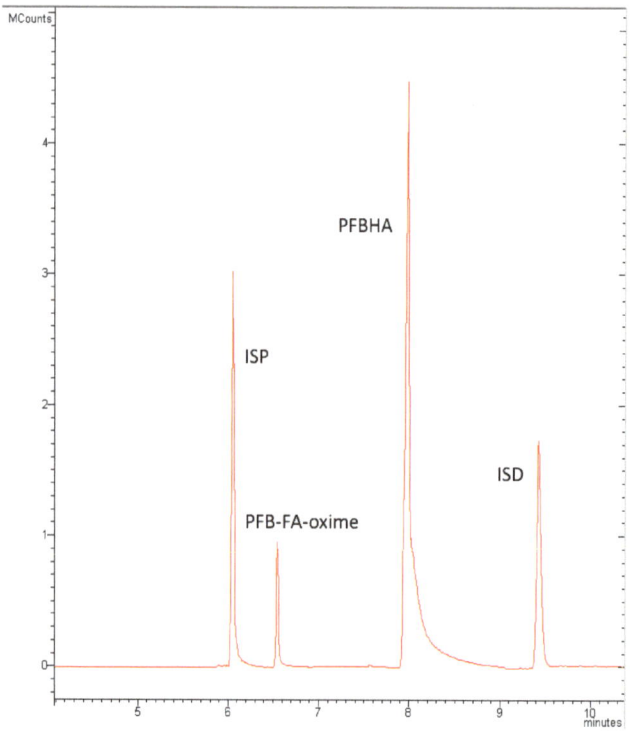

Figure 2. Total ion current chromatogram in EI mode, showing ISP, PFB-FA-oxime, PFBHA, and ISD.

Table 2. Comparison of the slope ± standard error, exposure time, mean recovery, average CV%, and R^2 between HS and DI using SPME and SPME Arrow.

	SPME		SPME Arrow	
	HS	DI	HS	DI
Slope ± Standard Error	$(8.7 \pm 0.1) \times 10^4$	$(7.6 \pm 0.1) \times 10^4$	$(2.37 \pm 0.04) \times 10^5$	$(2.01 \pm 0.08) \times 10^5$
Intercept ± Standard Error	$(1.0 \pm 2.5) \times 10^4$	$(3.4 \pm 3.0) \times 10^4$	$-(1.0 \pm 0.9) \times 10^4$	$(1.7 \pm 1.5) \times 10^5$
Exposure time (min)	30	20	30	20
Mean recovery (%)	14.8	14.2	38.3	37.4
R^2	0.9994	0.9989	0.9990	0.9962
Average CV%	9.1	10.3	8.9	9.7

2.3. Cooling-Assisted SPME

In HS-SPME, heating the sample solution is beneficial since it increases the HS concentration of volatile analytes. However, the increase in sampling temperature decreases the amount of analytes trapped in the fiber coating because the absorption process is exothermic. Given these premises, cooling-assisted SPME is designed to effectively extract the analytes into the HS while improving SPME sensitivity. This technique consists of heating the sample matrix while cooling the sampling fiber, thus simultaneously increasing the sample matrix/HS and HS/fiber coating distribution constants.

Cooling-assisted SPME is very efficient in complex matrices such as sludge, soil, and clay, where the analytes are firmly adsorbed to the active sites of the sample medium [68–70].

The cooling-assisted extraction has been described and classified into three approaches: (i) internally cooled SPME using liquid CO_2 directly on the fiber; (ii) internally cooled fiber based on a thermoelectric cooler (TEC); (iii) externally cooled SPME using circulating fluids (alcohol or cold water), where the HS of the sample vial is cooled from the outside [55,69,71–75].

In the present study, considering the multivariate nature of the underlying phenomena involved in the cooling-assisted HS-SPME, a rapid study of the operating conditions of the SPME and SPME Arrow devices was carried out by applying a 2^3 full factorial experimental design. We planned eight experiments, adding three additional tests to validate the models computed (the experimental plan with responses is given in the Supplementary Materials). The responses selected were the peak area of PFB-FA-oxime, and the factors studied were the exposure temperature studied in the range from 10 to 20 °C, the fiber exposure time in cooling mode in the range from 15 to 30 min, and the sampling temperature of the derivatized analyte studied in the range from 60 to 80 °C. The models computed were validated in the experimental domain center point (15 °C, 22.5 min, and 70 °C) at the 95% and 99% confidence levels with satisfactory relative error percentages in prediction (8% for conventional SPME and 17% for SPME Arrow). The models for the two types of fiber showed that, within the ranges evaluated, the only significant effect on the process is produced by the fiber exposure temperature, which must be kept at the lower level investigated (10 °C). The remaining two factors have non-statistically significant and negligible effects compared to the former. Therefore, the sampling temperature and the fiber exposure time can be set at any convenient value within the variation range investigated. We thus selected the following operating conditions: sampling temperature at 60 °C, 15 min time, and 10 °C fiber exposure temperature.

Calibration curves were analyzed and compared to traditional HS sampling described in the previous paragraph using a fully automated system, operating as described in point (iii) of the above classification. Table 3 shows cooling HS performances on calibration levels for both SPME and SPME Arrow: the sensitivity expanded compared to HS (2.6×10^5 and 3.7×10^5, as opposed to 8.7×10^4 and 2.4×10^5, for SPME and SPME Arrow, respectively), as well as the recovery (41.7% and 60.5%, as opposed to 14.8% and 38.3%, for SPME and SPME Arrow, respectively), while the average CV% does not show a significant difference from traditional HS curves (8.9% and 7.6%, compared to previous 9.1% and 8.9%, for SPME and SPME Arrow, respectively) (the calibration curves are showed in the Supplementary Materials in Figure S4).

Table 3. Slope and intercept ± standard error, exposure temperature, mean recovery, average CV%, and R^2 for cooling HS using SPME and SPME Arrow.

	Cooling SPME	Cooling SPME Arrow
Slope ± Standard Error	$(2.44 \pm 0.03) \times 10^5$	$(3.74 \pm 0.07) \times 10^5$
Intercept ± Standard Error	$(2.7 \pm 7.1) \times 10^4$	$-(1.5 \pm 1.4) \times 10^5$
Exposure temperature (°C)	10	10
Mean recovery (%)	41.7	60.5
R^2	0.9994	0.9991
Average CV%	8.9	7.6

2.4. MSPME Extraction

The time used to reach the partition equilibrium in SPME sampling depends on several parameters, such as sample matrix, sample agitation, temperature, and properties of the coating/analyte [76]. An equilibrium time of 30–60 min is very common for SPME sampling [27,66]; on the contrary, extraction time can be shorter, but this leads to lower extraction yields and, therefore, higher detection limits. Generally, a compromise between extraction time and yield is mandatory; yet, specific adjustments can be introduced to reach a higher throughput or an increased sensitivity.

In this framework, MSPME is a rugged procedure suitable for magnifying the analyte's response in quantitative analyses of complex matrices [77–80]; it is based on calculating the mass extracted using the peak areas of a few consecutive extractions from the same sample [53,54,77]. The extraction can also be performed using multiple fibers in a single chamber containing the sample, sequentially desorbed in the GC injection block, trapping the volatile compounds at the beginning of the GC column using a cryoscopic technique, before performing a 'single shot' chromatographic run.

A practical calculation of the total area can be performed via Equation (4):

$$A_T = \frac{A_1}{1 - \beta} \qquad (4)$$

where A_1 is the peak area in the first extraction and β is calculated from the linear regression of the logarithms of the individual peak areas, as shown in Equation (5) [80]:

$$\ln A_i = (i - 1) \times \ln \ln \beta + \ln \ln A_1 \qquad (5)$$

where A_i is the peak area obtained in the ith extraction. To obtain a linear trend in the logarithm of peak areas, Tena et al. suggested that the analyte loaded on the fiber must be considered in relation to its concentration—β has an influence since it should be below 0.95 to achieve at least a 5% difference in two consecutive areas, whereas values below 0.4 allow for calculating the analyte simply with a sum of the areas since four extractions provide a recovery above 97% [54].

The MSPME approach was implemented in the cooling HS and tested on calibration levels to achieve complete extraction of FA. Figure 3 shows the peak area depletion performing consecutive extractions on the calibration levels. Table 4 reports the mean recovery, mean β, the CV% range, and the R^2 on the five calibration levels (the calibration curves obtained with both SPME and SPME Arrow are shown in the Supplementary Materials in Figure S5). The approach resulted in a mean recovery of 97.4% for SPME and 96.3% for SPME Arrow, confirming the feasibility of the method to reach an exhaustive stripping of FA from the sample and, therefore, high sensitivities. The calculated value of β for SPME Arrow, i.e., 0.41, confirms its higher efficiency in trapping the PFB-FA-oxime in our experimental conditions: in fact, three extractions reduce the concentration of the analyte in the HS to blank levels. The maximum CVs% are 10.2% for SPME and 9.6% for SPME Arrow, corresponding to the sixth and third extraction, respectively. Conversely, the time necessary to perform a complete sample analysis increases from about 100 min to about 410 min for SPME and to 215 min for SPME Arrow.

The results obtained on standard solutions confirm that the combination of HS sampling with cooling MSPME represents the best set-up for sampling PFB-FA-oxime following on-sample derivatization.

Table 4. Slope and intercept ± standard error, mean recovery, mean β values, CV% range, and R^2 for the calibration levels, studied with cooling MSPME and cooling MSPME Arrow techniques.

	Cooling MSPME	Cooling MSPME Arrow
Slope ± Standard Error	$(4.8 \pm 0.1) \times 10^5$	$(6.32 \pm 0.05) \times 10^5$
Intercept ± Standard Error	$(0.1 \pm 2.0) \times 10^5$	$-(2.0 \pm 1.0) \times 10^5$
Mean recovery (%)	97.4	96.3
β	0.56	0.41
CV% range	5.7–10.2	4.8–9.6
R^2	0.9990	0.9998

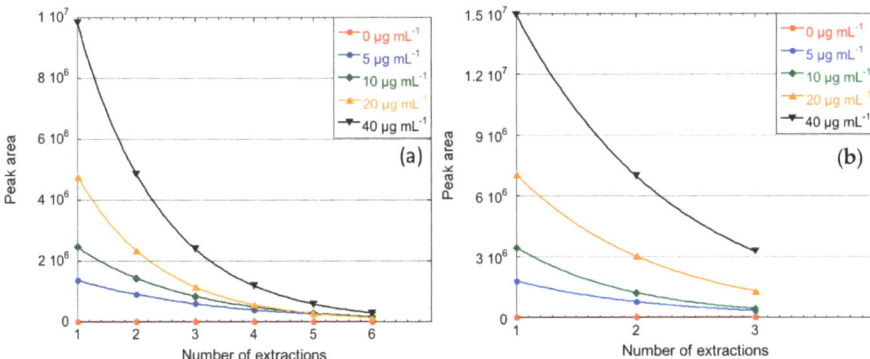

Figure 3. Decrease in peak area with the number of injections on calibration levels for SPME (**a**) and SPME Arrow (**b**).

The method, optimized using cooling and MSPME approaches, is characterized by enhanced performances compared to the on-sample PFBHA derivatization with the initially presented HS sampling approach, as shown in Table 5. LOD was calculated by multiplying 3.3 by the ratio between the standard deviation of blanks and the intercept of the curve, and the LOQ is three times the LOD. The implementation of these two procedures led to a lowering in the LOD and LOQ values from 22 ng L^{-1} (LOD)–73 ng L^{-1} (LOQ) and 14 ng L^{-1} (LOD)–46 ng L^{-1} (LOQ) to 11 ng L^{-1} (LOD)–36 ng L^{-1} (LOQ) and 8 ng L^{-1} (LOD)–26 ng L^{-1} (LOQ), for SPME and SPME Arrow, respectively. The LOD obtained is limited by the FA content in Milli-Q water, which can not be further lowered. The results obtained are comparable with the previous literature studies for the determination of carbonyl compounds [47,49,58,61] using PFBHA on-sample derivatization; nonetheless, the implementation of the newly proposed automated analytical tool leads to a higher sensitivity, albeit requiring more analysis time due to the multiple extraction steps.

Table 5. Comparison between the performances obtained in this work and previous studies.

	Column	Technique	LOD	LOQ
Cooling MSPME	Fused-silica 35% Ph (30 m × 0.25 mm, 0.25 μm film thickness)	GC-MS	11 ng L^{-1}	36 ng L^{-1}
Cooling MSPME Arrow	Fused-silica 35% Ph (30 m × 0.25 mm, 0.25 μm film thickness)	GC-MS	8 ng L^{-1}	26 ng L^{-1}
Bao et al. [49]	Fused-silica capillary column (30 m × 0.25 mm I.D., 0.25 μm film thickness)	SPME-GC-ECD	20 ng L^{-1}	-
Gunes et al. [61]	Wax capillary column (30 m × 0.32 mm ID, 0.25 μm film thickness)	GC-FID	50 μg L^{-1}	167 μg L^{-1}
Hudson et al. [58]	5%-phenyl-methylpolysi-loxane (30 m × 0.25 mm id × 0.25 μm film thickness)	SPME-GC-MS	3.0 ng L^{-1}	-
Wardencki et al. [47]	Fused-silica containing Rtx-5 (30 m × 0.32 i.d., 3 μm film thickness)	SPME-GC-ECD	5 ng L^{-1}	-

Further modifications can be introduced to obtain high analytical throughput, whether or not it is necessary to process a large batch of samples. Louch et al. suggested that the extraction time, i.e., the diffusion time through the watery layer, is proportional to the square migration extent and inversely proportional to the water diffusion coefficient [63]. In particular, by reducing the vial diameter by a factor of three, the authors achieved a decrease in extraction time of an order of magnitude. Furthermore, the higher amount of phase in SPME Arrow allows to perform fewer extraction cycles and can therefore represent a valid choice to improve throughput. Hence, further tests should be conducted to customize the method depending on the analytical requirements. As far as sensitivity is concerned, Cancilla et al. [81] reported that PFB-FA-oxime formation in water differed only in reproducibility, varying the pH, whereas, for higher-molecular-weight aldehydes, yields increased at pH < 3. This observation is justified by the greater availability of the unprotonated hydroxylamine, involved in the first step as pH increases, while the ease of removal of the protonated carbonyl oxygen, i.e., the second step, is enhanced when pH decreases [58,82]. A modification in the pH value can therefore be considered both when the investigated matrix is not responsive enough to the derivatization and to expand the presented method to higher-molecular-weight aldehydes.

2.5. Real Samples Analysis

The method developed was applied to different matrices to prove its ruggedness and feasibility. Samples of green apple, plum, tomato, shampoo, and face wash were prepared as described in the Material and Methods section and analyzed via the cooling MSPME approach (Figure 4). Table 6 reports the FA content for each product, the average CVs% on untreated samples, and the CV% range for the constructed curves, for both SPME and SPME Arrow. FA values were calculated from PFB-FA-oxime data obtained by interpolation of both cooling MSPME and cooling MSPME Arrow curves. Generally, the results found for the matrices examined are in accordance with the previous literature data [19,83,84], confirming the method's suitability for determining FA content in different commercial goods. Figures S6 and S7 of the Supplementary Materials report the curves constructed in matrices investigated using the standard addition method for SPME and SPME Arrow, respectively; the linearity observed in standards slightly worsened, most likely due to the complexity of matrices.

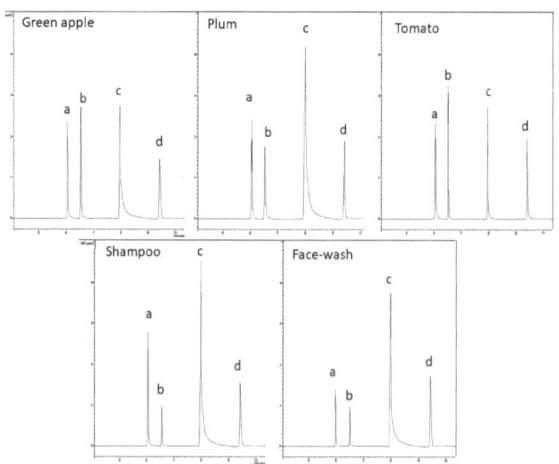

Figure 4. Chromatograms of real samples (green apple, plum, tomato, shampoo, and face wash) analysis: ISs (a and d), PFB-FA-oxime (b), and PFBHA (c) peaks.

Table 6. Mean FA content, R^2, average CV% and CV% range for each matrix investigated.

Matrix	Mean FA Content (mg/L)		R^2		Average CV%		CV% Range	
	SPME	SPME Arrow	SPME	SPME Arrow	SPME	SPME Arrow	SPME	SPME Arrow
Green apple	11.8	11.2	0.9873	0.9990	7.5	6.6	6.8–9.8	5.3–8.8
Plum	9.16	11.4	0.9939	0.9969	6.2	5.9	5.3–8.5	5.2–9.2
Tomato	14.5	20.8	0.9841	0.9930	8.1	5.3	6.2–9.6	4.2–7.3
Shampoo	3.51	6.9	0.9984	0.9882	7.0	4.2	5.9–8.2	3.9–6.7
Face-wash	4.92	8.0	0.9901	0.9987	7.6	4.5	6.5–8.4	3.8–6.9

3. Materials and Methods

3.1. Chemical and Reagents

O-(2,3,4,5,6-pentafluorobenzyl) hydroxylamine hydrochloride (PFBHA·HCl) (CAS 57981-02-9), n-hexane (CAS 110-54-3), 1-bromo-4-fluorobenzene (CAS n. 460-00-4) and p-fluorobenzaldehyde (CAS n. 459-57-4) were purchased from Sigma-Aldrich (Saint Louis, MO, US). Formaldehyde O-(pentafluorobenzyl)oxime (PFB-FA-oxime) (CAS 86356-73-2) was purchased from GiottoBiotech (Sesto Fiorentino, Italy). Milli-Q water 18 MΩ cm (mQ), further purified to eliminate FA using PURE UV3—4-Stage UV Water Purification System (Pure n Natural Systems Inc., Steamwood, IL, USA), was obtained from Millipore (Darmstadt, Germany). Helium (99.999%) as GC carrier gas was obtained from Air Liquid (Paris, France). For automation of the SPME on-fiber PFBHA derivatization, HeadSpace screw-top 20 mL glass vials (HSV) (Part No: 5188-2753) and Hdsp cap 18 mm magnetic PTFE/Sil (Part No. 5188-2759) were purchased from Agilent Technologies (Santa Clara, CA, USA).

We purchased 23-gauge 100 μm PDMS FFA-SPME fibers (9.40 mm^2 phase area, 600 μL phase volume) and 1.5 mm PDMS 250 μm FFA-SPME Arrow fibers (62.8 mm^2 phase area, 11.8 μL phase volume) from Chromline (Prato, Italy).

3.2. Samples

Fruits samples were purchased in a local market and were green apples (Renette variety), plums (Prunus Domestica Black Amber variety), and tomatoes (Ciliegino and Pachino varieties). Cosmetic products included one shampoo sample and one all-purpose face wash sample.

3.3. PFBHA On-Sample Derivatization Routine and Online SPME Sampling

The FA working solution was prepared at 80 mg L^{-1} in water by diluting a 4% (m/m) stock solution.

All steps of the procedure described in the following were fully automated.

On-sample derivatization was performed at 60 °C for 30 min on five calibration levels, dispensing 0, 125, 250, 500, and 1000 μL of the working solution, respectively, into 2 mL vials, containing variable amounts of water, a 20 mg mL^{-1} PFBHA water solution (100 μL) and 50 μL of 20 mg mL^{-1} of p-Fluorobenzaldehyde in ethanol.

p-Fluorobenzaldehyde was used as IS according to its conformity for SPME-GC analysis of carbonyl compounds derivatized with PFBHA [49,67]. FA final concentrations obtained were 0, 5, 10, 20, and 40 mg L^{-1}. The reacted mixtures (2 mL) were then transferred into a 20 mL HSV, containing 1 g of NaCl and 50 μL of 100 mg L^{-1} 1-bromo-4-fluorobenzene in water solution (IS of process, ISP, 50 μL).

1-Bromo-4-fluorobenzene was used as IS in agreement with Güneş et al. [61], given its retention time in proximity to the PFB-FA-oxime, as a quality check for the subsequent analytical steps

Samples were prepared in a 2 mL HSV by diluting each product into proper quantities of water to a final volume of 1.8 mL. Fruit samples peeled and cut into small pieces, were blended in a 1:10 ratio with water. The juice was filtered through a pleated paper filter.

Shampoo and face wash samples were diluted in water (0.8 g in 1 mL of water) and mixed in an ultrasonic bath for 15 min at room temperature, as indicated by Feher et al. [83].

The subsequent on-sample derivatization was performed by adding 20 mg mL^{-1} PFBHA water solution (100 μL) and 20 mg mL^{-1} ISD (50 μL), at 60 °C for 30 min under stirring; the reacted mixtures were transferred in 20 mL HSV containing NaCl and ISP in the proportion indicated above. Four additional vials were prepared in full automation using the same procedure for each commercial product tested, adding 125, 250, 500, and 1000 μL of the FA working solution, respectively, to the initial mixture.

For HS sampling, the fiber was exposed for 30 min to the HS of samples/standards, previous equilibrium under stirring at 60 °C for 20 min, and desorbed in the injector at 250 °C for 2 min.

As for the DI mode, derivatization was performed at 60 °C for 30 min on five calibration levels, prepared following the same proportion used for HS curves, using a "one-pot" approach, in 20 mL HSV. The fiber was dipped for 20 min in the solution, previous equilibrium for 5 min under stirring, and desorbed in the injector at 250 °C for 2 min.

Calibrators and authentic samples were prepared independently and analyzed fivefold in random order.

Regarding the cooling MSPME approach, the same sets of reference standards and samples were prepared following the HS procedure. Consecutive repetitions were performed on each HSV operating in MSPME mode for both standards and samples. The parameters were fiber exposure temperature set at 10 °C, reacted mixture temperature set at 60 °C, 15 min time of exposure, 6 extraction cycles for SPME, and 3 extraction cycles for SPME Arrow.

3.4. Online Robotic System

Automation of the analytical procedure was achieved using a CTC PAL3 System xyz Autosampler (CTC Analytics AG Industrie Strasse 20 CH-4222, Zwingen, Switzerland) with a 1200 mm bar. The apparatus was equipped with a Multi Fiber eXchange (MFX) system (Chromline, Prato, Italy), Liquid Syringe Tool (CTC Analytics AG, Zwingen, Switzerland), SPME dual layer extraction (SDLE) equipped with Chronos software (Chromline, Prato, Italy), an FFA-SPME holder, a tray with 20 mL and 2 mL slots, a fiber conditioning system, two solvent modules (one for 100 mL solvent bottles and the other for 10 mL solvent bottles), and a wash module, to guarantee an automated routine between the exchange of syringe, FFA-SPME, and FFA-SPME Arrow fibers, and shift between traditional SPME and cooling SPME sampling.

3.5. GC-MS Operating Conditions

The chromatographic method proposed by Dugheri et al. was used [60], with a Varian CP3800 GC system coupled to a Varian Saturn 2200 Ion-Trap as the detector (scan mode, 45–300 m/z, EI energy 70 eV).

The column was a DB 35-MS-UI (Agilent J&W), and the 1079 injector port (SCION, Instruments, Amundsenweg, The Netherlands) was provided with a 0.75 mm internal diameter liner. The oven settings were isotherm of 50 °C for 1 min, followed by a linear temperature ramp of 10 °C min^{-1} to 260 °C. Helium was used as the carrier gas, set at a flow rate of 1.2 mL min^{-1}. The absolute quantity of PFB-FA-oxime was calculated on a regression curve obtained via automatic direct injection (1 µL) of hexane solutions in the GC system (0, 5, 10, 20, and 40 mg L^{-1} of PFB-FA-oxime, respectively) to assess the recovery of the method. When operating in cooling SPME mode, the SDLE module, installed on the PAL 6-position agitator, was set at 10 °C using Chronos software.

4. Conclusions

The quantitative determination of FA in an aqueous medium is possible by PFBHA on-sample derivatization and HS- or DI-SPME extraction using a PDMS fiber prior to GC separation and MS detection. The implementation of cooling and multiple SPME approaches to the method leads to satisfactory performances of the methods in terms of sensitivity and recovery. The full automation of sample preparation and analysis makes the procedure well-suited for the routine determination of FA.

Optimal extraction conditions were selected using a 20 mg mL^{-1} PFBHA water solution, followed by 30 min of derivatization time and subsequent SPME or SPME Arrow extraction, with a PDMS coating, for 15 min. Under these conditions, the LOD and LOQ methods were 11 and 36 ng L^{-1} and 8 and 26 ng L^{-1} for SPME and SPME Arrow, respectively. The methods' response functions showed good linearity over the ranges tested, i.e., LOQ of 40 mg L^{-1}, CV% ranging from 5.7 to 10.2 and 4.8 to 9.6, and FA recoveries of 97.4% and 96.3% for SPME and SPME Arrow, respectively.

The method's effectiveness was also demonstrated by testing fruits and cosmetic products. The results obtained confirmed the general suitability of the method proposed here for routine analysis of real-life samples.

Supplementary Materials: The following supporting information can be downloaded at: https://www.mdpi.com/article/10.3390/molecules28145441/s1, Table S1: Design of the cooling-assisted extraction experiment with SPME and SPME Arrow; Equation (S1): Model equation obtained for SPME; Equation (S2): Model equation obtained for SPME Arrow; Table S2: Mean predicted area in the center point of the experimental domain; Figure S1: Calibration curves for SPME (a) and SPME Arrow (b); Figure S2: Normalized quadratic effects plots of the models computed for SPME; Figure S3: Normalized quadratic effects plots of the models computed for SPME Arrow; Figure S4: Calibration curves were constructed using HS and cooling HS for SPME and SPME Arrow; Figure S5: Calibration curves constructed using cooling MSPME and cooling MSPME Arrow; Figure S6: FA-spiked curves constructed using SPME for green apple (a), plum (b), tomato (c), shampoo (d), and face wash (e); Figure S7: FA-spiked curves constructed using SPME Arrow for green apple (a), plum (b), tomato (c), shampoo (d), and face wash (e).

Author Contributions: Conceptualization, S.D. and N.F.; methodology, G.C.; software, J.C.; validation, G.M. and D.S.; formal analysis, N.F.; investigation, S.D.; resources, N.M.; data curation, S.D.; writing—original draft preparation, S.D.; writing—review and editing, N.F., G.C. and J.C.; visualization, V.T.; supervision, R.G.; project administration, G.A. All authors have read and agreed to the published version of the manuscript.

Funding: This research received no external funding.

Data Availability Statement: Not applicable.

Acknowledgments: The study group thanks the Laboratorio per l'Innovazione e per l'applicazione della Robotica nel Monitoraggio degli Ambienti Naturali, di vita e di lavoro (LIROMAN) of PIN—Polo Universitario città di Prato (Prato, Italy) for providing the spaces to carry out the experiments. The publication was made with the contribution of the researcher Jacopo Ceccarelli with a research contract co-funded by the European Union—PON Research and Innovation 2014-2020 in accordance with Article 24, paragraph 3a), of Law No. 240 of December 30, 2010, as amended, and Ministerial Decree No. 1062 of 10 August 2021.

Conflicts of Interest: The authors declare no conflict of interest.

Sample Availability: Not available.

References

1. Formacare. Available online: https://www.formacare.eu/ (accessed on 22 March 2023).
2. Kong, L.; Li, B.; Zhao, L.; Zhang, R.; Wang, C. Density, Viscosity, Surface Tension, Excess Property and Alkyl Chain Length for 1, 4-Butanediol (1)+ 1, 2-Propanediamine (2) Mixtures. *J. Mol. Liq.* **2021**, *326*, 115107. [CrossRef]
3. Dreyfors, J.M.; Jones, S.B.; Sayed, Y. Hexamethylenetetramine: A Review. *Am. Ind. Hyg. Assoc. J.* **1989**, *50*, 579–585. [CrossRef] [PubMed]
4. Ince, M.; Ince, O.K.; Ondrasek, G.; Ince, M.; Ince, O.K.; Ondrasek, G. *Biochemical Toxicology: Heavy Metals and Nanomaterials*; IntechOpen: London, UK, 2020; ISBN 978-1-78984-697-3.
5. Dugheri, S.; Massi, D.; Mucci, N.; Berti, N.; Cappelli, G.; Arcangeli, G. Formalin Safety in Anatomic Pathology Workflow and Integrated Air Monitoring Systems for the Formaldehyde Occupational Exposure Assessment. *Int. J. Occup. Med. Environ. Health* **2021**, *34*, 319–338. [CrossRef] [PubMed]
6. Special Rapporteur on Toxics and Human Rights. Available online: https://www.ohchr.org/en/special-procedures/sr-toxics-and-human-rights (accessed on 24 March 2023).
7. Nielsen, G.D.; Larsen, S.T.; Wolkoff, P. Re-Evaluation of the WHO (2010) Formaldehyde Indoor Air Quality Guideline for Cancer Risk Assessment. *Arch. Toxicol.* **2017**, *91*, 35–61. [CrossRef]
8. Yu, L.; Wang, B.; Cheng, M.; Yang, M.; Gan, S.; Fan, L.; Wang, D.; Chen, W. Association between Indoor Formaldehyde Exposure and Asthma: A Systematic Review and Meta-Analysis of Observational Studies. *Indoor Air* **2020**, *30*, 682–690. [CrossRef] [PubMed]
9. Xu, W.; Zhang, W.; Zhang, X.; Dong, T.; Zeng, H.; Fan, Q. Association between Formaldehyde Exposure and Miscarriage in Chinese Women. *Medicine* **2017**, *96*, e7146. [CrossRef]
10. A Review of Human Carcinogens. Available online: https://monographs.iarc.who.int/wp-content/uploads/2018/06/mono100F.pdf (accessed on 15 April 2023).

11. Formaldehyde Can Cause Rare Cancers, New EPA Analysis Finds (3). Available online: https://news.bloomberglaw.com/environment-and-energy/formaldehyde-can-cause-rare-cancers-new-epa-analysis-finds (accessed on 24 March 2023).
12. *12460-3:2020(En)*; Wood-Based Panels—Determination of Formaldehyde Release—Part 3: Gas Analysis Method. ISO: Geneva, Switzerland, 2020.
13. Branch, L.S. Consolidated Federal Laws of Canada, Formaldehyde Emissions from Composite Wood Products Regulations. Available online: https://laws-lois.justice.gc.ca/eng/regulations/SOR-2021-148/FullText.html (accessed on 24 March 2023).
14. European Commission. Directorate General for Health and Food Safety. Scientific Advice on the Threshold for the Warning 'Contains Formaldehyde' in Annex V, Preamble Point 2 for Formaldehyde-Releasing Substances. Available online: https://data.europa.eu/doi/10.2875/269855 (accessed on 24 March 2023).
15. De Groot, A.C.; White, I.R.; Flyvholm, M.-A.; Lensen, G.; Coenraads, P.-J. Formaldehyde-Releasers in Cosmetics: Relationship to Formaldehyde Contact Allergy. *Contact Dermat.* **2010**, *62*, 2–17. [CrossRef]
16. Circular No. 21/2017/TT-BCT Formaldehyde, Aromatic Amines in Textile Products. Available online: https://english.luatvietnam.vn/circular-no-21-2017-tt-bct-on-promulgating-national-technical-regulations-on-content-of-117773-doc1.html (accessed on 24 March 2023).
17. Piccinini, P.; Senaldi, C.; Summa, C. European Survey on the Release of Formaldehyde from Textiles. Available online: https://publications.jrc.ec.europa.eu/repository/handle/JRC36150 (accessed on 24 March 2023).
18. *14184-1:2011*; Textiles—Determination of Formaldehyde—Part 1: Free and Hydrolised Formaldehyde (Water Extraction Methods). ISO: Geneva, Switzerland, 2011.
19. Nowshad, F.; Islam, M.N.; Khan, M.S. Concentration and Formation Behavior of Naturally Occurring Formaldehyde in Foods. *Agric. Food Secur.* **2018**, *7*, 17. [CrossRef]
20. European Food Safety Authority. Endogenous Formaldehyde Turnover in Humans Compared with Exogenous Contribution from Food Sources. *EFSA J.* **2014**, *12*, 3550. [CrossRef]
21. Health, C. for D. and R. Code of Federal Regulations—Title 21—Food and Drugs. Available online: https://www.fda.gov/medical-devices/medical-device-databases/code-federal-regulations-title-21-food-and-drugs (accessed on 24 March 2023).
22. Perna, R.B.; Bordini, E.J.; Deinzer-Lifrak, M. A Case of Claimed Persistent Neuropsychological Sequelae of Chronic Formaldehyde Exposure: Clinical, Psychometric, and Functional Findings. *Arch. Clin. Neuropsychol. Off. J. Natl. Acad. Neuropsychol.* **2001**, *16*, 33–44. [CrossRef]
23. Chemicals of High Concern to Children—Washington State Department of Ecology. Available online: https://ecology.wa.gov/Regulations-Permits/Reporting-requirements/Childrens-Safe-Products-Act-Reporting/Chemicals-of-high-concern-to-children (accessed on 29 March 2023).
24. Understanding Formaldehyde in Children's and Consumer Products. Available online: https://www.health.state.mn.us/communities/environment/childenvhealth/docs/edu_formaldehyde.pdf (accessed on 24 March 2023).
25. Mo, C. Formaldehyde Regulations in the United States: An Overview. Available online: https://www.compliancegate.com/formaldehyde-regulations-united-states/ (accessed on 24 March 2023).
26. Gao, P. The Exposome in the Era of One Health. *Environ. Sci. Technol.* **2021**, *55*, 2790–2799. [CrossRef] [PubMed]
27. Dugheri, S.; Bonari, A.; Gentili, M.; Cappelli, G.; Pompilio, I.; Bossi, C.; Arcangeli, G.; Campagna, M.; Mucci, N. High-Throughput Analysis of Selected Urinary Hydroxy Polycyclic Aromatic Hydrocarbons by an Innovative Automated Solid-Phase Microextraction. *Molecules* **2018**, *23*, 1869. [CrossRef]
28. Lin, E.Z.; Esenther, S.; Mascelloni, M.; Irfan, F.; Godri Pollitt, K.J. The Fresh Air Wristband: A Wearable Air Pollutant Sampler. *Environ. Sci. Technol. Lett.* **2020**, *7*, 308–314. [CrossRef]
29. Dugheri, S.; Mucci, N.; Cappelli, G.; Trevisani, L.; Bonari, A.; Bucaletti, E.; Squillaci, D.; Arcangeli, G. Advanced Solid-Phase Microextraction Techniques and Related Automation: A Review of Commercially Available Technologies. *J. Anal. Methods Chem.* **2022**, *2022*, e8690569. [CrossRef]
30. Poole, C.; Mester, Z.; Miró, M.; Pedersen-Bjergaard, S.; Pawliszyn, J. Glossary of Terms Used in Extraction (IUPAC Recommendations 2016). *Pure Appl. Chem.* **2016**, *88*, 517–558. [CrossRef]
31. Arthur, C.L.; Pawliszyn, J. Solid Phase Microextraction with Thermal Desorption Using Fused Silica Optical Fibers. *Anal. Chem.* **1990**, *62*, 2145–2148. [CrossRef]
32. Herrington, J.S.; Gómez-Ríos, G.A.; Myers, C.; Stidsen, G.; Bell, D.S. Hunting Molecules in Complex Matrices with SPME Arrows: A Review. *Separations* **2020**, *7*, 12. [CrossRef]
33. SPME Arrow Tip = Less Septa Piercing Force. Available online: https://www.restek.com/it/chromablography/chromablography/spme-arrow-tip(-)(-)less-septa-piercing-force/ (accessed on 24 March 2023).
34. Dennenlöhr, J.; Thörner, S.; Maxminer, J.; Rettberg, N. Analysis of Selected Staling Aldehydes in Wort and Beer by GC-EI-MS/MS Using HS-SPME with On-Fiber Derivatization. *J. Am. Soc. Brew. Chem.* **2020**, *78*, 284–298. [CrossRef]
35. Reyes-Garcés, N.; Gionfriddo, E.; Gómez-Ríos, G.A.; Alam, M.N.; Boyacı, E.; Bojko, B.; Singh, V.; Grandy, J.; Pawliszyn, J. Advances in Solid Phase Microextraction and Perspective on Future Directions. *Anal. Chem.* **2018**, *90*, 302–360. [CrossRef]
36. Crucello, J.; Sampaio, N.M.F.M.; Junior, I.M.; Carvalho, R.M.; Gionfriddo, E.; Marriott, P.J.; Hantao, L.W. Automated Method Using Direct-Immersion Solid-Phase Microextraction and on-Fiber Derivatization Coupled with Comprehensive Two-Dimensional Gas Chromatography High-Resolution Mass Spectrometry for Profiling Naphthenic Acids in Produced Water. *J. Chromatogr. A* **2023**, *1692*, 463844. [CrossRef]

37. Martos, P.A.; Pawliszyn, J. Sampling and Determination of Formaldehyde Using Solid-Phase Microextraction with On-Fiber Derivatization. *Anal. Chem.* **1998**, *70*, 2311–2320. [CrossRef]
38. Cancilla, D.A.; Hee, S.S.Q. O-(2,3,4,5,6-Pentafluorophenyl)Methylhydroxylamine Hydrochloride: A Versatile Reagent for the Determination of Carbonyl-Containing Compounds. *J. Chromatogr. A* **1992**, *627*, 1–16. [CrossRef] [PubMed]
39. Marini, F.; Bellugi, I.; Gambi, D.; Pacenti, M.; Dugheri, S.; Focardi, L.; Tulli, G. Compound A, Formaldehyde and Methanol Concentrations during Low-Flow Sevoflurane Anaesthesia: Comparison of Three Carbon Dioxide Absorbers. *Acta Anaesthesiol. Scand.* **2007**, *51*, 625–632. [CrossRef] [PubMed]
40. Dugheri, S.; Cappelli, G.; Fanfani, N.; Ceccarelli, J.; Trevisani, L.; Sarti, M.; Squillaci, D.; Bucaletti, E.; Gori, R.; Mucci, N.; et al. Formaldehyde Analysis by On-Fiber Derivatization: A Study of the Kinetic Models of Adsorption for Divinylbenzene. *Quím. Nova* **2023**, in press. [CrossRef]
41. Dugheri, S.; Massi, D.; Mucci, N.; Marrubini, G.; Cappelli, G.; Speltini, A.; Bonferoni, M.C.; Arcangeli, G. Exposure to Airborne Formaldehyde: Sampling and Analytical Methods—A Review. *Trends Environ. Anal. Chem.* **2021**, *29*, e00116. [CrossRef]
42. Dugheri, S.; Bonari, A.; Pompilio, I.; Colpo, M.; Mucci, N.; Montalti, M.; Arcangeli, G. Development of an Innovative Gas Chromatography-Mass Spectrometry Method for Assessment of Formaldehyde in the Workplace Atmosphere. *Acta Chromatogr.* **2017**, *29*, 511–514. [CrossRef]
43. Vaz, E.B.; Santos, M.F.C.; de Jesus, E.G.; Vieira, K.M.; Osório, V.M.; Menini, L. Development of Methodology for Detection of Formaldehyde-DNPH in Milk Manager by Central Composite Rotational Design and GC/MS. *Res. Soc. Dev.* **2022**, *11*, e16411931575. [CrossRef]
44. Shin, H.-S.; Lim, H.-H. Simple Determination of Formaldehyde in Fermented Foods by HS-SPME-GC/MS. *Int. J. Food Sci. Technol.* **2012**, *47*, 350–356. [CrossRef]
45. Kim, H.-J.; Shin, H.-S. Simple and Automatic Determination of Aldehydes and Acetone in Water by Headspace Solid-Phase Microextraction and Gas Chromatography-Mass Spectrometry. *J. Sep. Sci.* **2011**, *34*, 693–699. [CrossRef]
46. Bao, M.; Joza, P.J.; Masters, A.; Rickert, W.S. Analysis of Selected Carbonyl Compounds in Tobacco Samples by Using Pentafluorobenzylhydroxylamine Derivatization and Gas Chromatography-Mass Spectrometry. *Contrib. Tob. Nicotine Res.* **2014**, *26*, 86–97. [CrossRef]
47. Wardencki, W.; Orlita, J.; Namieśnik, J. Comparison of Extraction Techniques for Gas Chromatographic Determination of Volatile Carbonyl Compounds in Alcohols. *Fresenius J. Anal. Chem.* **2001**, *369*, 661–665. [CrossRef] [PubMed]
48. Hudson, E.D.; Ariya, P.A.; Gélinas, Y.; Hudson, E.D.; Ariya, P.A.; Gélinas, Y. A Method for the Simultaneous Quantification of 23 C1–C9 Trace Aldehydes and Ketones in Seawater. *Environ. Chem.* **2011**, *8*, 441–449. [CrossRef]
49. Bao, M.; Pantani, F.; Griffini, O.; Burrini, D.; Santianni, D.; Barbieri, K. Determination of Carbonyl Compounds in Water by Derivatization–Solid-Phase Microextraction and Gas Chromatographic Analysis. *J. Chromatogr. A* **1998**, *809*, 75–87. [CrossRef] [PubMed]
50. Tessini, C.; Müller, N.; Mardones, C.; Meier, D.; Berg, A.; von Baer, D. Chromatographic Approaches for Determination of Low-Molecular Mass Aldehydes in Bio-Oil. *J. Chromatogr. A* **2012**, *1219*, 154–160. [CrossRef]
51. Chericoni, S.; Battistini, I.; Dugheri, S.; Pacenti, M.; Giusiani, M. Novel Method for Simultaneous Aqueous in Situ Derivatization of THC and THC-COOH in Human Urine Samples: Validation and Application to Real Samples. *J. Anal. Toxicol.* **2011**, *35*, 193–198. [CrossRef]
52. Pacenti, M.; Dugheri, S.; Traldi, P.; Degli Esposti, F.; Perchiazzi, N.; Franchi, E.; Calamante, M.; Kikic, I.; Alessi, P.; Bonacchi, A.; et al. New Automated and High-Throughput Quantitative Analysis of Urinary Ketones by Multifiber Exchange-Solid Phase Microextraction Coupled to Fast Gas Chromatography/Negative Chemical-Electron Ionization/Mass Spectrometry. *J. Autom. Methods Manag. Chem.* **2010**, *2010*, 972926. [CrossRef]
53. Ghiasvand, A.R.; Hajipour, S.; Heidari, N. Cooling-Assisted Microextraction: Comparison of Techniques and Applications. *TrAC Trends Anal. Chem.* **2016**, *77*, 54–65. [CrossRef]
54. Ghiasvand, A.; Yazdankhah, F.; Paull, B. Heating-, Cooling- and Vacuum-Assisted Solid-Phase Microextraction (HCV-SPME) for Efficient Sampling of Environmental Pollutants in Complex Matrices. *Chromatographia* **2020**, *83*, 531–540. [CrossRef]
55. Koster, E.H.M.; de Jong, G.J. Multiple Solid-Phase Microextraction. *J. Chromatogr. A* **2000**, *878*, 27–33. [CrossRef]
56. Tena, M.T.; Carrillo, J.D. Multiple Solid-Phase Microextraction: Theory and Applications. *TrAC Trends Anal. Chem.* **2007**, *26*, 206–214. [CrossRef]
57. Schmarr, H.-G.; Sang, W.; Ganß, S.; Fischer, U.; Köpp, B.; Schulz, C.; Potouridis, T. Analysis of Aldehydes via Headspace SPME with ON-Fiber Derivatization to Their O-(2,3,4,5,6-Pentafluorobenzyl)oxime Derivatives and Comprehensive 2D-GC-MS. *J. Sep. Sci.* **2008**, *31*, 3458–3465. [CrossRef] [PubMed]
58. Hudson, E.D.; Okuda, K.; Ariya, P.A. Determination of Acetone in Seawater Using Derivatization Solid-Phase Microextraction. *Anal. Bioanal. Chem.* **2007**, *388*, 1275–1282. [CrossRef] [PubMed]
59. Shin, E.-M.; Senthurchelvan, R.; Munoz, J.; Basak, S.; Rajeshwar, K.; Benglas-Smith, G.; Howell, B.C. Photolytic and Photocatalytic Destruction of Formaldehyde in Aqueous Media. *J. Electrochem. Soc.* **1996**, *143*, 1562. [CrossRef]
60. Dugheri, S.; Cappelli, G.; Ceccarelli, J.; Fanfani, N.; Trevisani, L.; Squillaci, D.; Bucaletti, E.; Gori, R.; Mucci, N.; Arcangeli, G. Innovative Gas Chromatographic Determination of Formaldehyde by Miniaturized Extraction and On-Fiber Derivatization, via SPME and SPME Arrow. *Quím. Nova* **2022**, *45*, 1236–1244. [CrossRef]

61. Güneş, K.; Can, Z.; Arda, A. Determination of Formaldehyde in Textile Dye and Auxiliary Chemicals with Headspace Gas Chromatography-Flame Ionization Detector. *Turk. J. Chem.* **2022**, *46*, 575–581. [CrossRef]
62. Tuduri, L.; Desauziers, V.; Fanlo, J.L. Potential of Solid-Phase Microextraction Fibers for the Analysis of Volatile Organic Compounds in Air. *J. Chromatogr. Sci.* **2001**, *39*, 521–529. [CrossRef]
63. Louch, D.; Motlagh, S.; Pawliszyn, J. Dynamics of Organic Compound Extraction from Water Using Liquid-Coated Fused Silica Fibers. *Anal. Chem.* **1992**, *64*, 1187–1199. [CrossRef]
64. Selecting the Appropriate SPME Fiber Coating—Effect of Analyte Molecular Weight and Polarity. Available online: https://www.sigmaaldrich.com/IT/en/technical-documents/technical-article/analytical-chemistry/solid-phase-microextraction/selecting-the-appropriate (accessed on 29 March 2023).
65. Pawliszyn, J. Theory of Solid-Phase Microextraction. *J. Chromatogr. Sci.* **2000**, *38*, 270–278. [CrossRef]
66. Pacenti, M.; Dugheri, S.; Villanelli, F.; Bartolucci, G.; Calamai, L.; Boccalon, P.; Arcangeli, G.; Vecchione, F.; Alessi, P.; Kikic, I.; et al. Determination of Organic Acids in Urine by Solid-Phase Microextraction and Gas Chromatography–Ion Trap Tandem Mass Spectrometry Previous 'in Sample' Derivatization with Trimethyloxonium Tetrafluoroborate. *Biomed. Chromatogr.* **2008**, *22*, 1155–1163. [CrossRef]
67. Moreira, N.; Araújo, A.M.; Rogerson, F.; Vasconcelos, I.; Freitas, V.D.; Pinho, P.G. de Development and Optimization of a HS-SPME-GC-MS Methodology to Quantify Volatile Carbonyl Compounds in Port Wines. *Food Chem.* **2019**, *270*, 518–526. [CrossRef]
68. Xu, S.; Li, H.; Wu, H.; Xiao, L.; Dong, P.; Feng, S.; Fan, J. A Facile Cooling-Assisted Solid-Phase Microextraction Device for Solvent-Free Sampling of Polycyclic Aromatic Hydrocarbons from Soil Based on Matrix Solid-Phase Dispersion Technique. *Anal. Chim. Acta* **2020**, *1115*, 7–15. [CrossRef] [PubMed]
69. Ghiasvand, A.R.; Hosseinzadeh, S.; Pawliszyn, J. New Cold-Fiber Headspace Solid-Phase Microextraction Device for Quantitative Extraction of Polycyclic Aromatic Hydrocarbons in Sediment. *J. Chromatogr. A* **2006**, *1124*, 35–42. [CrossRef] [PubMed]
70. Chai, X.; Jia, J.; Sun, T.; Wang, Y.; Liao, L. Application of a Novel Cold Activated Carbon Fiber-Solid Phase Microextraction for Analysis of Organochlorine Pesticides in Soil. *J. Environ. Sci. Health Part B* **2007**, *42*, 629–634. [CrossRef] [PubMed]
71. Carasek, E.; Pawliszyn, J. Screening of Tropical Fruit Volatile Compounds Using Solid-Phase Microextraction (SPME) Fibers and Internally Cooled SPME Fiber. *J. Agric. Food Chem.* **2006**, *54*, 8688–8696. [CrossRef]
72. Ghiasvand, A.R.; Setkova, L.; Pawliszyn, J. Determination of Flavour Profile in Iranian Fragrant Rice Samples Using Cold-Fibre SPME–GC–TOF–MS. *Flavour Fragr. J.* **2007**, *22*, 377–391. [CrossRef]
73. Haddadi, S.H.; Niri, V.H.; Pawliszyn, J. Study of Desorption Kinetics of Polycyclic Aromatic Hydrocarbons (PAHs) from Solid Matrices Using Internally Cooled Coated Fiber. *Anal. Chim. Acta* **2009**, *652*, 224–230. [CrossRef]
74. Banitaba, M.H.; Hosseiny Davarani, S.S.; Kazemi Movahed, S. Comparison of Direct, Headspace and Headspace Cold Fiber Modes in Solid Phase Microextraction of Polycyclic Aromatic Hydrocarbons by a New Coating Based on Poly(3,4-Ethylenedioxythiophene)/Graphene Oxide Composite. *J. Chromatogr. A* **2014**, *1325*, 23–30. [CrossRef]
75. Pati, S.; Tufariello, M.; Crupi, P.; Coletta, A.; Grieco, F.; Losito, I. Quantification of Volatile Compounds in Wines by HS-SPME-GC/MS: Critical Issues and Use of Multivariate Statistics in Method Optimization. *Processes* **2021**, *9*, 662. [CrossRef]
76. Hakkarainen, M. Developments in Multiple Headspace Extraction. *J. Biochem. Biophys. Methods* **2007**, *70*, 229–233. [CrossRef]
77. Wercinski, S.A. *Solid Phase Microextraction: A Practical Guide*; CRC Press: Boca Raton, FL, USA, 1999; ISBN 978-1-4398-3238-7.
78. Wang, C.-H.; Su, H.; Chou, J.-H.; Lin, J.-Y.; Huang, M.-Z.; Lee, C.-W.; Shiea, J. Multiple Solid Phase Microextraction Combined with Ambient Mass Spectrometry for Rapid and Sensitive Detection of Trace Chemical Compounds in Aqueous Solution. *Anal. Chim. Acta* **2020**, *1107*, 101–106. [CrossRef]
79. Pérez-Olivero, S.J.; Pérez-Pont, M.L.; Conde, J.E.; Pérez-Trujillo, J.P. Determination of Lactones in Wines by Headspace Solid-Phase Microextraction and Gas Chromatography Coupled with Mass Spectrometry. *J. Anal. Methods Chem.* **2014**, *2014*, e863019. [CrossRef] [PubMed]
80. Ezquerro, Ó.; Pons, B.; Tena, M.T. Multiple Headspace Solid-Phase Microextraction for the Quantitative Determination of Volatile Organic Compounds in Multilayer Packagings. *J. Chromatogr. A* **2003**, *999*, 155–164. [CrossRef] [PubMed]
81. Cancilla, D.A.; Chou, C.-C.; Barthel, R.; Hee, S.S.Q. Characterization of the 0-(2,3,4,5,6-Pentafluorobenzyl)- Hydroxylamine Hydrochloride (PFBOA) Derivatives of Some Aliphatic Mono- and Dialdehydes and Quantitative Water Analysis of These Aldehydes. *J. AOAC Int.* **1992**, *75*, 842–853. [CrossRef]
82. Ojala, M.; Kotiaho, T.; Siirilä, J.; Sihvonen, M.-L. Analysis of Aldehydes and Ketones from Beer as O-(2,3,4,5,6-Pentafluorobenzyl) hydroxylamine Derivatives. *Talanta* **1994**, *41*, 1297–1309. [CrossRef]
83. Feher, I.; Schmutzer, G.; Voica, C.; Moldovan, Z. Determination of Formaldehyde in Romanian Cosmetic Products Using Coupled GC/MS System after SPME Extraction. *AIP Conf. Proc.* **2013**, *1565*, 294–297. [CrossRef]
84. Tang, X.; Bai, Y.; Duong, A.; Smith, M.T.; Li, L.; Zhang, L. Formaldehyde in China: Production, Consumption, Exposure Levels, and Health Effects. *Environ. Int.* **2009**, *35*, 1210–1224. [CrossRef]

Disclaimer/Publisher's Note: The statements, opinions and data contained in all publications are solely those of the individual author(s) and contributor(s) and not of MDPI and/or the editor(s). MDPI and/or the editor(s) disclaim responsibility for any injury to people or property resulting from any ideas, methods, instructions or products referred to in the content.

Article

Magnetic Persimmon Leaf Composite: Preparation and Application in Magnetic Solid-Phase Extraction of Pesticides in Water Samples

Yuyue Zang, Na Hang, Jiale Sui, Senlin Duan, Wanning Zhao, Jing Tao and Songqing Li *

Beijing Key Laboratory for Forest Pest Control, College of Forestry, Beijing Forestry University, No. 35 Qinghua East Road, Beijing 100083, China; zangyuyue@126.com (Y.Z.); 15044764521@163.com (N.H.); suijiale1019@163.com (J.S.); dslm0807@bjfu.edu.cn (S.D.); furongzhuoqiuyu@bjfu.edu.cn (W.Z.); taojing1029@hotmail.com (J.T.)
* Correspondence: songqingli@bjfu.edu.cn

Abstract: In recent years, the utilization of biomass materials for the removal and detection of water pollutants has garnered considerable attention. This study introduces, for the first time, the preparation of Fe_3O_4/persimmon leaf magnetic biomass composites. The magnetic composites were employed in a magnetic solid-phase extraction method, coupled with gas chromatography-electron capture detection (GC-ECD), for the analysis of four pesticides (trifluralin, triadimefon, permethrin, and fenvalerate) in environmental water samples. The innovative magnetic persimmon leaf composites were synthesized by in situ generation of Fe_3O_4 nanoparticles through coprecipitation and loaded onto persimmon leaves. These composites exhibit superparamagnetism with a saturation magnetization of 12.8 emu g^{-1}, facilitating rapid phase separation using a magnetic field and reducing the extraction time to 10 min. Desorption can be achieved within 30 s by aspirating 20 times, eliminating the need for time-consuming and labor-intensive experimental steps like filtration and centrifugation. The specific surface area of the magnetic composite adsorbent increased from 1.3279 m^2 g^{-1} for the original persimmon leaf to 5.4688 m^2 g^{-1}. The abundant hydroxyl and carboxyl groups on the composites provide ample adsorption sites, resulting in adsorption capacities ranging from 55.056 mg g^{-1} to 73.095 mg g^{-1} for the studied pesticides. The composites exhibited extraction recoveries ranging from 80% to 90% for the studied pesticides. Compared to certain previously reported MSPE methods, this approach achieves equivalent or higher extraction recoveries in a shorter operation time, demonstrating enhanced efficiency and convenience. Good linearity of the target analytes was obtained within the range of 0.75–1500 µg L^{-1}, with a determination of coefficient (R^2) greater than 0.999. These findings contribute to the use of magnetic persimmon leaf biomass materials as effective and environmentally friendly adsorbents for pollutant determination in water samples.

Keywords: biomass adsorbent; Fe_3O_4/persimmon leaf magnetic composite; magnetic solid-phase extraction; pesticides; water

Citation: Zang, Y.; Hang, N.; Sui, J.; Duan, S.; Zhao, W.; Tao, J.; Li, S. Magnetic Persimmon Leaf Composite: Preparation and Application in Magnetic Solid-Phase Extraction of Pesticides in Water Samples. *Molecules* **2024**, *29*, 45. https://doi.org/10.3390/molecules29010045

Academic Editor: Hiroyuki Kataoka

Received: 14 November 2023
Revised: 15 December 2023
Accepted: 19 December 2023
Published: 20 December 2023

Copyright: © 2023 by the authors. Licensee MDPI, Basel, Switzerland. This article is an open access article distributed under the terms and conditions of the Creative Commons Attribution (CC BY) license (https://creativecommons.org/licenses/by/4.0/).

1. Introduction

Chemical control is one of the primary measures for managing plant diseases and pests. Nitroaniline herbicides, triazole fungicides, and pyrethroid insecticides are commonly employed pesticides in forestry, demonstrating significant efficacy in controlling various prevalent pests and diseases in the forest ecosystem [1–3]. After pesticides are sprayed, they inevitably enter the environment, leading to environmental pollution and the extermination of beneficial organisms. In various environmental media, water bodies are particularly susceptible to pesticide residue pollution. Pesticide residues have strong mobility and enrichment capabilities in flowing bodies of water, including streams, lakes, ponds, and groundwater [4,5]. Pesticides such as imidacloprid, chlorpyrifos, and cypermethrin are not only highly toxic to aquatic organisms like fish and shrimp but also pose a significant threat to pollinators like bees and silkworms [6]. This can result in long-term adverse effects on the environment [7]. Moreover, pesticides entering the environment can undergo bioaccumulation through the food chain, eventually posing health risks to humans. The World Health Organization (WHO) has established maximum acceptable concentrations (MAC) for chemical pollutants in drinking water. For instance, the MAC for triadimefon is set at 20 µg L^{-1}, and the MAC for permethrin is derived to be 300 µg L^{-1} [8]. Therefore, conducting monitoring and detection of pesticides in water bodies is of paramount importance for green conservation and prevention.

Sample pretreatment plays a critical role in the entire process of pesticide residue analysis [9,10]. Solid-phase extraction (SPE) is one of the most widely used sample pretreatment methods and is capable of separating target analytes from samples [11,12]. Nonetheless, it comes with inherent drawbacks, including labor-intensive and time-consuming procedures, limitations in sample flow rates, and the risk of adsorbent pores getting clogged [13,14]. Therefore, new methods are continuously being developed to rapidly separate analytes from samples. Solid-phase microextraction (SPME) is a solvent-free sample preparation technique that combines sampling, separation, and enrichment into a single step [15,16]. This method was introduced in 1990 to overcome the major drawbacks of conventional SPE, including complexity in automation and instrument design [17]. This method eliminates the need for organic solvents and significantly shortens the analysis time [18]. SPME has gained wide attention due to its advantages, such as high sensitivity, solvent-free methods, and simplicity of operation. Many successful separations and enrichments of target analytes have been performed using this method [19–22].

Magnetic solid-phase extraction (MSPE) is a novel technique composed of magnetic inorganic materials and non-magnetic adsorbents. It is based on traditional SPE and SPME technology [23]. Since its introduction in 1999 by Safarikova [24], it has garnered significant interest and has become one of the main branches of sample preparation methods in recent years [25,26]. The use of magnetic adsorbents enables rapid sample separation under an external magnetic field, streamlining and expediting the operational steps [27–29]. MSPE retains the advantages of SPE/SPME and offers several benefits: (1) it has excellent magnetic adsorption capacity, resulting in a short extraction time and significant time savings; (2) selected magnetic adsorbents are highly selective; (3) it enhances the enrichment effect on target compounds, enabling automation of the entire process. These advantages indicate the significant potential of MSPE technology for trace-level analysis of various compounds. Additionally, the adsorption capacity of the magnetic material mainly depends on the nature of the adsorbent material, and the type of adsorbent directly influences the extraction process. Magnetic materials are widely applied in sample preparation due to their excellent extraction performance [30–33]. For instance, Fe_3O_4@C-NFs composites were synthesized through a one-pot synthesis method, achieving efficient separation of tetracycline (TC) in less than 5 min with the assistance of ultra-high performance liquid chromatography (UHPLC) [34]. Another illustration involves the utilization of magnetic molecularly imprinted polymers (MMIP) for the absorption of bisphenol A (BPA) from wastewater samples. The excellent selectivity of MMIP for BPA renders it effective as an MSPE adsorbent in conjunction with spectroscopic instruments [35]. Furthermore, in another study, magnetic cyclodextrins cross-linked with tetrafluoroterephthalonitrile (Fe_3O_4@TFN-CDPs) were developed for detecting pesticides in medicinal plants. The study revealed limits of detection (LOD) ranging from 0.011 to 0.106 µg kg^{-1} for the target pesticides [36]. Hence, a key focus in MSPE research is developing highly efficient magnetic adsorbents with robust properties.

Biomass can be broadly defined as any organic material derived from renewable biological sources [37]. Waste biomass possesses irregular surfaces, varying pore sizes, and active functional groups and is abundant in nature [38]. Converting waste, such as biomass, into materials suitable for environmental applications is a green and eco-friendly waste management approach. The use of adsorbent materials developed from abundant plant residues has the potential to rationalize forest resources [39]. In recent years, there has been increasing attention towards waste biomass, mainly composed of cellulose, hemicellulose, lignin, pectin, or tannin. These materials are rich in functional groups like -COOH, -OH, -C=O, -OCH$_3$, -NH$_2$, -CONH$_2$, etc. [40]. Using waste biomass as adsorbents has the advantages of wide availability, short growth cycles, and minimal secondary pollution to the environment. They also offer abundant functional groups and adsorption sites, making them potential biomass adsorbents [41]. Preparing functional materials from plant samples aligns with the principles of green chemistry, organic agriculture, and sustainable agriculture and has gained increasing attention [42–44].

Persimmon (*Diospyros kaki*) belongs to the family *Ebenaceae*. It is widely distributed in East Asian countries such as China, Japan, and the Republic of Korea, with a cultivation history of several hundred years [45,46]. Persimmon fruits are typically harvested from September to November, leaving a significant amount of persimmon leaves after the harvest [47]. Persimmon leaves are a natural material rich in hydroxyl groups, making them highly suitable for adsorption. Being a byproduct of persimmon trees, they represent a natural adsorbent [48]. From a material cost and environmental perspective, it seems promising to develop persimmon leaves as an adsorbent for various applications. Previous research by Yu and Choi [49] involved the preparation of hybrid beads by mixing chitosan with persimmon leaves. These hybrid beads were structurally capable of adsorbing heavy metals easily due to their carboxyl and carbonyl functional groups. This method was used for the removal of Pb (II) and Cd (II) from aqueous solutions.

The utilization of persimmon leaves as a magnetic biomass adsorbent, coupled with MSPE technology, holds significant promise for the effective adsorption of pesticides in environmental water. Firstly, the abundance of hydroxyl groups in persimmon leaves provides numerous binding sites, enhancing their compatibility with Fe$_3$O$_4$. On the other hand, the inclusion of Fe$_3$O$_4$ nanoparticles elevates the specific surface area and adsorption sites of the magnetic composites. Additionally, the hydroxyl and carboxyl groups in persimmon leaves facilitate intermolecular hydrogen bonds with analytes containing carbonyl and ester groups, while the phenyl structure in persimmon leaves readily forms π-π interactions with pesticides featuring phenyl or heterocyclic groups, thereby bolstering the adsorption process. Consequently, magnetic persimmon leaf composites were synthesized in this study. Waste agricultural and forestry persimmon leaves were utilized as an adsorbent in MSPE, developing a fast and convenient MSPE method. This method was applied in combination with gas chromatography-electron capture detection (GC-ECD) for the detection of four pesticides in environmental water samples. The influence factors on the extraction performance were optimized, and the optimized experimental conditions were determined. The method was evaluated using parameters such as the linear range, LOD, limit of quantitation (LOQ), and other analytical performance parameters. In addition, the reusability of the magnetic persimmon leaf composites was investigated.

2. Results and Discussion

2.1. Characterization of Fe$_3$O$_4$/Persimmon Leaf Magnetic Composite

2.1.1. Fourier Transform Infrared Spectroscopy (FTIR) Analysis

FTIR was employed in the range of 4000–400 cm^{-1} to investigate the structural characteristics of functional groups on the material surfaces. Figure 1A displays the infrared spectra of both unmodified persimmon leaf particles and magnetic persimmon leaf composites. For unmodified persimmon leaf particles, the peak at 3430.23 cm^{-1} corresponds to the phenolic hydroxyl O-H stretching vibration, and the peak at 2925.54 cm^{-1} corresponds to the C-H stretching vibration, representing functional groups within the persimmon tannin's skeletal structure. These observations align with previous studies, affirming the abundance of phenolic hydroxyl groups in persimmon leaf tannins [50,51]. The peak at 1624.49 cm^{-1} represents the C=C stretching vibration, while the peak at 1727.22 cm^{-1} represents the C=O stretching vibration in carbonyl. Upon closer spectral examination, characteristic peaks of O-H and C=O stretching vibrations in the magnetic persimmon leaf composites underwent a blue shift to 3430.23 cm^{-1} and 1731.25 cm^{-1}, respectively.

This shift is possibly attributed to the formation and loading of Fe$_3$O$_4$ on the persimmon leaf surface, disrupting the intermolecular hydrogen bonding originally present in the persimmon leaves. This disruption led to a relatively weak blue shift of the hydroxyl and carbonyl peaks. Additionally, a characteristic peak of Fe-O is evident at 577.43 cm^{-1}, confirming the successful loading of Fe$_3$O$_4$ onto the persimmon leaf surface.

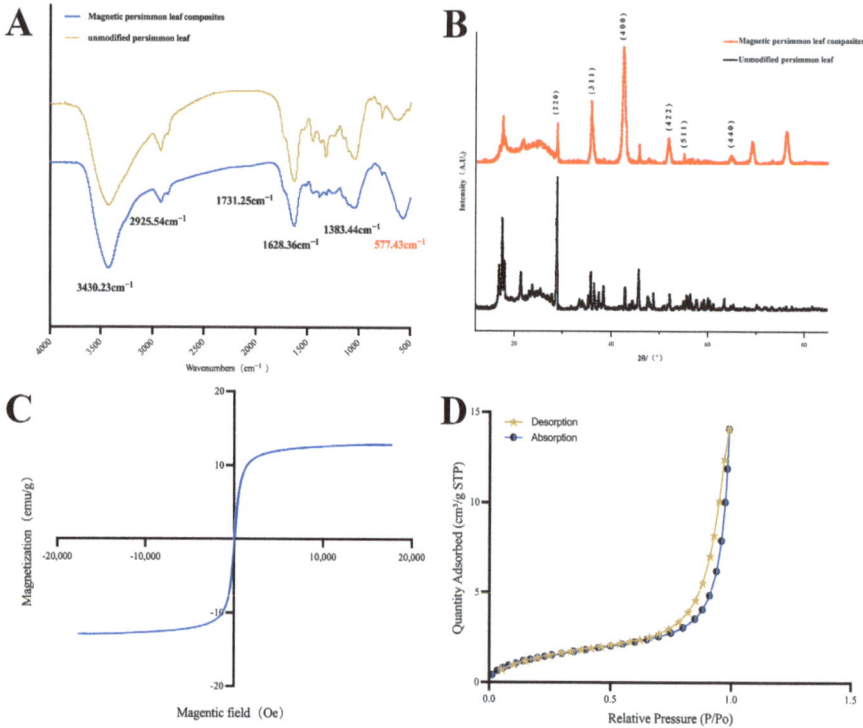

Figure 1. (**A**) Fourier transform infrared spectrum of magnetic persimmon leaf composites and unmodified persimmon leaf; (**B**) X-ray diffraction plot of magnetic persimmon leaf composites and unmodified persimmon leaf; (**C**) Vibrating sample magnetometer of magnetic persimmon leaf composites; (**D**) N$_2$ adsorption-desorption isotherms of magnetic persimmon leaf composites.

2.1.2. X-ray Diffraction (XRD) Analysis

XRD analysis was conducted to further examine the structure of the prepared materials. The XRD patterns of the synthesized magnetic composites are illustrated in Figure 1B, showcasing diffraction peaks at 2θ angles of 29.9°, 35.8°, 42.38°, 52.1°, 56°, and 63.8°, corresponding to the (220), (311), (400), (422), (511), and (440) facets, respectively. These values align with the crystal planes of Fe$_3$O$_4$, as reported in previous studies [52]. This substantiates the successful preparation of the magnetic composites.

2.1.3. Vibrating Sample Magnetometer (VSM) Analysis

VSM was used to test the magnetic properties of the magnetic composites. Figure 1C presents the hysteresis loop of the magnetic composites. As observed, the magnetic composites exhibit a hysteresis loop characteristic of superparamagnetism, and the magnetic saturation value reaches 12.8 emu·g^{-1}. The experimental results indicate that magnetic composites can be rapidly separated from the solution under the influence of an external magnetic field, meeting the requirements for fast MSPE.

2.1.4. Specific Surface Area and Pore Structure (BET) Analysis

The adsorption–desorption of N_2 in the magnetic composites was tested to determine the specific surface area. As shown in Figure 1D the specific surface area of the magnetic composites increases with increasing pressure, accompanied by an increase in pore volume. The change in pore volume exhibits a slow increase when the relative pressure is below 0.9, followed by a rapid increase from 0.9 to 1. The range of 0.6–0.9 represents a hysteresis loop of pore volume change. The N_2 adsorption–desorption isotherm of the magnetic composites can be classified as a Type IV isotherm. The presence of hysteresis indicates the presence of mesopores within the materials. Compared to the original persimmon leaves with a specific surface area of 1.3279 $m^2\ g^{-1}$, the magnetic composite adsorbent has a specific surface area of 5.4688 $m^2\ g^{-1}$, showing a significant increase that provides more adsorption sites.

2.1.5. Scanning Electron Microscopy (SEM) Analysis

SEM was employed to examine the microstructure of the prepared materials. As illustrated in Figure 2a,b, untreated persimmon leaves display a layered structure with irregular surface features and discernible pores. These irregularities enhance the probability of contact between the material and analyte in the solution, facilitating adsorption. Figure 2c,d highlight numerous cracks and folds on the surface of the magnetic composites, providing additional adsorption sites and larger pores for analytes. Consequently, the increased specific surface area proves advantageous for effective adsorption.

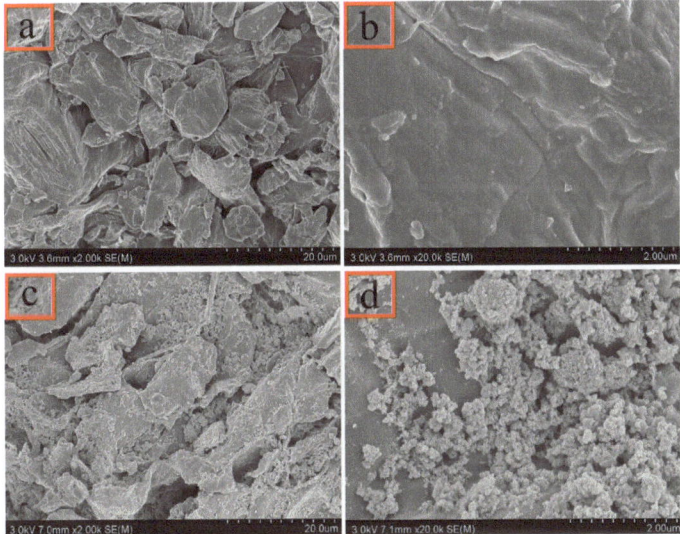

Figure 2. Scanning electron microscopy: (**a,b**) unmodified persimmon leaf particles; (**c,d**) magnetic persimmon leaf composites.

2.2. Adsorbent Performance

To elucidate the adsorption performance of the prepared magnetic composites, this study conducted a comparative analysis of the extraction performance of 100-mesh unmodified persimmon leaf particles, magnetic persimmon leaf composites, and Fe_3O_4 nanoparticles synthesized via coprecipitation. Employing an equal quantity of these three adsorbents under consistent extraction conditions, the extraction recoveries for the target analytes were compared, as illustrated in Figure 3A. The results indicate that the magnetic persimmon leaf composites exhibited superior extraction performance compared to the other two materials. This superiority can be attributed to several factors: firstly, the presence of

Fe$_3$O$_4$ nanoparticles increased the specific surface area and adsorption sites of the magnetic composites; secondly, the hydroxyl and carboxyl groups in persimmon leaves facilitated intermolecular hydrogen bonds with the target analytes, and the phenyl structure in persimmon leaves formed π-π interactions with the target analytes, thereby enhancing adsorption. Additionally, the experiments demonstrated that the magnetic persimmon leaf composites simplify the MSPE process. It offers the advantage of utilizing an external magnetic field for separation, resulting in improved extraction recoveries.

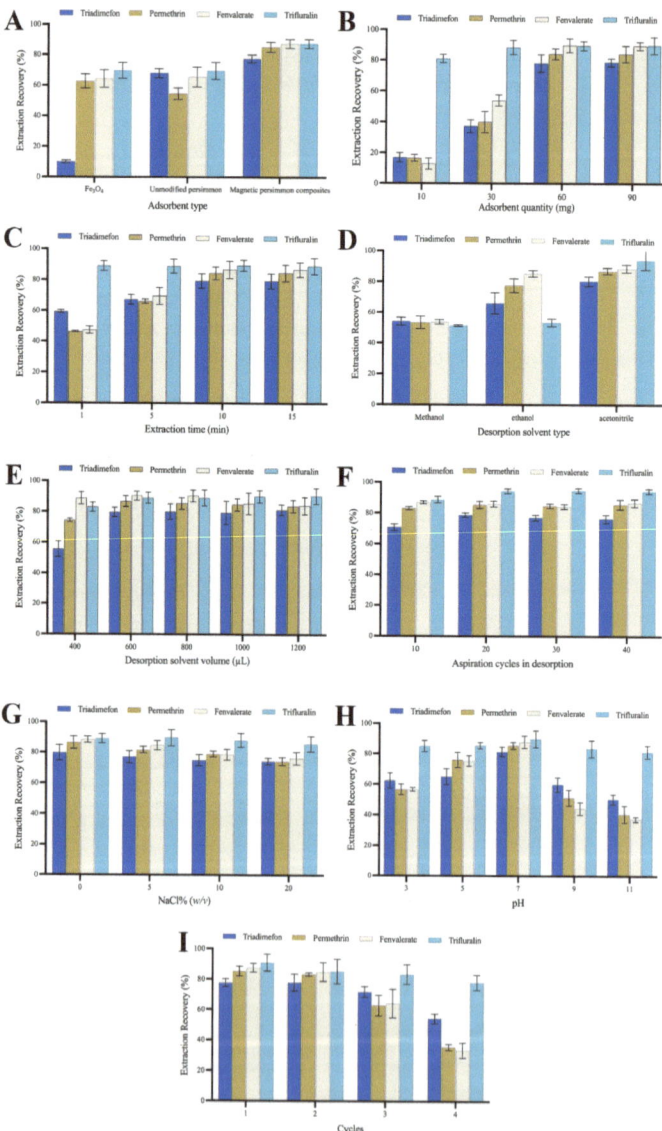

Figure 3. (**A**) Effect of the type of adsorbent; (**B**) Effect of the adsorbent amount; (**C**) Effect of the extraction time; (**D**) Effect of the desorption solvent type; (**E**) Effect of the desorption solvent volume; (**F**) Effect of aspiration cycles in desorption; (**G**) Effect of the ionic strength; (**H**) Effect of the pH; (**I**) Extraction reutilization of magnetic persimmon leaf composites.

2.3. Study of Effective Parameters on Experimental

2.3.1. Adsorbent Amount

The quantity of adsorbent plays a pivotal role in achieving optimized adsorption performance. The adsorbent amount was investigated in the range of 10–90 mg to determine the optimized usage. Adsorbent amounts of 10, 30, 60, and 90 mg were tested. As depicted in Figure 3B, the recoveries of the four analytes exhibit a significant increase within the 10–60 mg range of adsorbent amounts. As the adsorbent amount increases, the recoveries gradually improve. This is likely because a higher amount of adsorbent provides more efficient adsorption sites for the target analytes, enhancing adsorption performance. Beyond 60 mg of adsorbent, the abundance of adsorption sites becomes excessive. To mitigate unnecessary usage, 60 mg was chosen as the optimized amount for magnetic persimmon leaf adsorbents in subsequent experiments.

2.3.2. Extraction Time

MSPE relies on adsorption equilibrium. Appropriate extraction time ensures sufficient contact between the magnetic adsorbent and the analytes. Extraction time can affect the extraction performance of analytes before adsorption equilibrium is reached. To achieve higher extraction performance, extraction times of 1, 5, 10, and 15 min were tested. As shown in Figure 3C, with increasing extraction time from 1 to 10 min, the recoveries gradually improved. At an adsorption time of 10 min, the extraction recovery reached its maximum. Based on these results, an extraction time of 10 min was set.

2.3.3. Desorption Solvent Type and Volume

Effective desorption solvent is a crucial component in the sample pretreatment process, and the choice of desorption solvent is key to improving extraction performance [53]. In this experiment, the influence of three organic solvents, namely methanol, ethanol, and acetonitrile, on desorption performance was compared. Figure 3D illustrates that acetonitrile provides the highest recovery compared to other desorption solvents. Consequently, acetonitrile was selected as the desorption solvent for subsequent experiments. The impact of acetonitrile volume (400 µL, 600 µL, 800 µL, 1000 µL, and 1200 µL) on the extraction performance of analytes was analyzed through optimization. As depicted in Figure 3E, the experimental results indicate that as the volume of acetonitrile increases from 400 µL to 600 µL, the recovery of all four pesticides gradually rises. When the desorption volume is set at 600 µL, the recoveries remain constant, achieving optimal desorption performance. Moreover, with a further increase in desorption volume, the recoveries of all pesticides do not significantly increase, and for some analytes, there is a slight decrease. Excess solvent not only leads to solvent wastage but also affects the enrichment effect. Therefore, 600 µL of acetonitrile was selected as the optimized desorption solvent in the subsequent experiments.

2.3.4. Aspiration Cycles in Desorption

Ensuring effective desorption is crucial for achieving higher extraction recoveries of analytes during the operation. Appropriate aspiration cycles allow analytes to be adequately desorbed from the adsorbent. However, excessive aspiration cycles not only fail to improve desorption efficiency but may also lead to the loss of desorption solvent, thereby reducing extraction recoveries. We systematically optimized the number of aspiration cycles—10, 20, 30, and 40 cycles—each lasting approximately 1.5 s. This corresponds to different desorption times of 15 s, 30 s, 45 s, and 60 s. The results are illustrated in Figure 3F. The experimental findings revealed that employing 20 aspiration cycles achieved the desired extraction recovery level. Further increasing the number of aspirations did not result in a significant change in extraction recoveries. Consequently, we selected 20 aspirations, equivalent to a desorption time of 30 s, for subsequent experiments.

2.3.5. Ionic Strength

The solubility of analytes in aqueous solutions may decrease with an increase in ionic strength, potentially leading to increased extraction performance. However, high ionic strength may reduce the diffusion rate of target analytes, resulting in decreased extraction performance. A slight decrease in the recoveries of the four analytes was observed as NaCl concentrations ranged from 0 to 20% (w/v). This might be attributed to the increase in sample solution viscosity with added salt concentration, reducing the rate at which analytes are adsorbed onto the adsorbent. Therefore, based on the results in Figure 3G, NaCl was not added in subsequent experiments.

2.3.6. Sample Solution pH

The pH of the sample solution can influence both the existing form of analytes and the surface charge of the adsorbent. Therefore, the pH value of the sample solution was optimized to enhance extraction and recovery performance. The influence of pH values between 3.0 and 11.0 on the extraction performance of analytes was evaluated (see experimental results in Figure 3H). Within the pH range of 3.0–7.0, an increase in the sample solution's pH led to an increase in recoveries. When the sample solution's pH reached 7.0, the recoveries were maximized. Subsequently, the recoveries decreased as the sample solution's pH continued to rise. A lower pH leads to a positively charged surface of Fe_3O_4, causing repulsion with the target analytes, and impeding effective adsorption. Extremely low pH may also compromise the stability of Fe_3O_4, thereby reducing its adsorption performance. As the pH increases and creates an alkaline environment, certain pyrethroid pesticides tend to decompose, leading to a significant reduction in their recoveries. For the other two analytes, trifluralin molecules feature two nitro groups and one trifluoromethyl group on the phenyl ring, all of which are strong electron-withdrawing groups. This configuration leads to easier dissociation of the two hydrogen atoms on the phenyl ring compared to when there are no strong electron-withdrawing groups. This results in the partial formation of trifluralin bases, exhibiting weak negative charges. Triadimenfon molecules undergo a similar process with the hydrogen on the triazole group dissociating under alkaline conditions, partially forming triadimenfon bases with weak negative charges. Meanwhile, the protons from the phenolic hydroxyl and carboxyl groups in persimmon leaves may dissociate, forming partial negative ions. This not only generates electrostatic repulsion but also diminishes π-π interactions between the phenyl structure in persimmon leaves and phenyl in pesticides. Additionally, alkaline conditions induce a negative charge on the Fe_3O_4 surface. The increased repulsive force between the adsorbent and weakly negatively charged analytes is unfavorable for adsorption. Consequently, as the pH increases, the adsorption performance of trifluralin and triadimenfon diminishes. Therefore, a pH of 7.0 was chosen as the optimum pH for the sample solution.

2.3.7. Reusability

The number of reuse cycles is an important criterion for evaluating the stability and economic feasibility of magnetic persimmon leaf adsorbents. Choosing an adsorbent with good reusability can effectively reduce experimental costs. After each adsorption-desorption cycle, the magnetic persimmon leaf adsorbent was washed three times with ethanol and dried to ensure no residual analytes before the next MSPE process. As observed in Figure 3I, after two adsorption-desorption cycles, the adsorption performance did not significantly change. However, during the third and fourth cycles, a noticeable decrease in extraction recoveries was observed, possibly due to the loss of adsorption groups on the adsorbent's surface. This suggests that the magnetic persimmon leaf adsorbent prepared in this study has good reusability for up to two cycles, demonstrating its environmentally friendly and cost-effective characteristics.

2.4. Adsorption Isotherms

The interaction between magnetic persimmon leaf composites and four pesticides was investigated through adsorption isotherms. Adsorption experiments covered a range of initial pesticide concentrations from 10 to 300 mg L^{-1}. As the initial concentration increased, the adsorption capacity of the magnetic persimmon leaf composites exhibited a rapid and equilibrating rise. This phenomenon is attributed to the elevated initial target concentration, which provides an enhanced driving force to overcome mass transfer resistance between solvent and solute surfaces [54]. The Langmuir model, suitable for simulating monolayer adsorption on surfaces with a limited number of similar sites, and the Freundlich model, designed for adsorption on amorphous surfaces assuming non-homogeneity, were employed [55,56]. The results, presented in Figure 4 and Table 1, reveal a well-fitting Langmuir isotherm with R^2 values up to 0.97, indicating homogeneous adsorption of pesticides. The maximum adsorption capacities, calculated using the Langmuir isotherm model, were 73.75 mg g^{-1} for trifluralin, 58.07 mg g^{-1} for triadimefon, 65.35 mg g^{-1} for permethrin, and 63.82 mg g^{-1} for fenvalerate, closely aligning with actual data. Additionally, R$_L$ values between 0 and 1, as per experimental results, suggest that the synthesized magnetic persimmon leaf adsorbent effectively facilitated the adsorption of the target analytes [57].

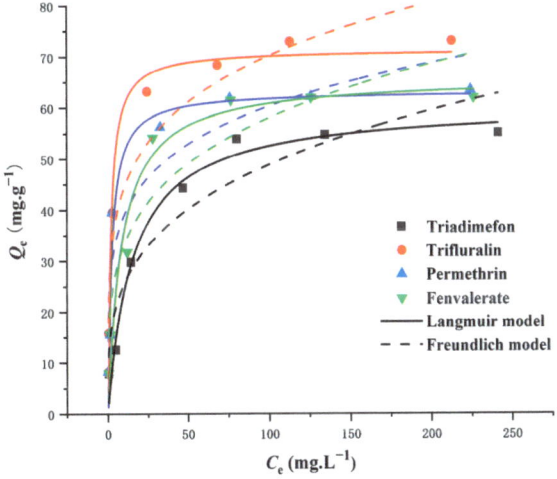

Figure 4. Adsorption isotherms of Langmuir and Freundlich for pesticide adsorption onto magnetic composites.

Table 1. Adsorption isotherm parameters by using Langmuir and Freundlich models.

	Langmuir Model				Freundlich Model		
	Q$_m$ (mg g^{-1})	K$_L$ (L mg^{-1})	R$_L$	R^2	K$_F$ (mg g^{-1} L$^{1/n}$ mg$^{-1/n}$)	1/n	R^2
Trifluralin	73.75	0.4209	0.008–0.192	0.9780	21.00	0.279	0.9074
Triadimefon	58.07	0.0903	0.036–0.525	0.9742	9.42	0.362	0.8880
Permethrin	65.35	0.1346	0.024–0.427	0.9453	20.21	0.223	0.8643
Fenvalerate	63.82	0.1926	0.017–0.342	0.9480	13.46	0.329	0.8763

2.5. Method Validation

Under the optimized conditions, the linearity range, determination of coefficient, LOD, and LOQ of the method were evaluated. Table 2 lists the determination of coefficient and linear equations for the four analytes within a linear range of 0.75–1500 µg L^{-1}. All determinations of coefficients were greater than 0.999. For the LOD, when the signal-to-noise ratio (S/N) of a chromatographic peak for a studied analyte reaches 3, the concentration corresponding to the peak area is defined as the LOD for that analyte. The LODs range from 0.25 to 1.1 µg L^{-1}, and the LOQs range from 0.75 to 3.4 µg L^{-1} based on an S/N of 10. The LODs and LOQs were experimentally tested. In this experiment, the repeatability of the developed method for pesticide detection in environmental water samples was assessed through intra-day and inter-day precision. The intra-day RSD values range from 2.7% to 4.3%, and for inter-day precision, they range from 3.3% to 4.5%. This suggests that the method is feasible, reliable, and stable.

Table 2. Performance characteristics of magnetic persimmon leaf composites-based MSPE method combined with (GC-ECD).

Analyte	Linear Range (µg L^{-1})	Linear Equation	R^2	LOD (µg L^{-1})	LOQ (µg L^{-1})	RSD (%) Intra-Day (n = 3)	RSD (%) Inter-Day (n = 6)	Extraction Recovery (%)
Trifluralin	0.75–100	y = 341.33 + 216.54	0.9994	0.25	0.75	3.1	4.3	90
Triadimefon	1.15–100	y = 414.38 + 192.73	0.999	0.38	1.15	4.3	4.2	80
Permethrin	3.4–1500	y = 79.979 + 223.08	0.9998	1.1	3.4	3.3	3.3	85
Fenvalerate	2.2–1500	y = 80.694 + 880.11	0.9996	0.73	2.2	2.7	4.5	87

2.6. Application of MSPE Based on Fe$_3$O$_4$/Persimmon Leaf Magnetic Composite to Real Water Samples

To further assess the applicability of the developed materials, magnetic composites combined with MSPE and GC-ECD were used to determine four pesticides in real water samples. The developed method was evaluated by sample recovery experiments on real samples at three different concentrations (20, 200, and 300 µg L^{-1}). The results are presented in Table 3. Figure 5 shows the chromatograms of blank and spiked samples. The pesticides were not detected in lake water samples from Olympic Forest Park, Chaoyang Park, and Campus Linzhixin, and the relative recoveries of the spiked samples were between 80% and 95%, with RSD values ranging from 1.4% to 8.0%. This demonstrates the repeatability and reliability of the method for the determination of analytes in environmental water samples.

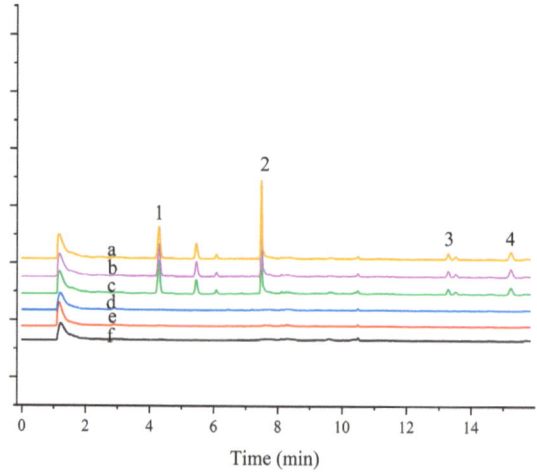

Figure 5. Application of magnetic persimmon leaf composite in environmental water sample analysis: (a) Pretreated Olympic Forest Park water sample spiked at 200 µg L^{-1}; (b) Pretreated Beijing Sun Park water sample spiked at 200 µg L^{-1}; (c) Pretreated Campus water sample spiked at 200 µg L^{-1}; (d) Olympic Forest Park water sample not spiked; (e) Beijing Sun Park water sample water sample not spiked; (f) Campus water sample not spiked. Peaks: (1) Trifluralin (2) Triadimefon (3) Permethrin (4) Fenvalerate.

Table 3. Relative recoveries of the studied pesticides in spiked water samples.

Analyte	Spiked Level ($\mu g\ L^{-1}$)	Olympic Forest Park		Chaoyang Park		BFU Campus	
		RR (%)	RSD (%)	RR (%)	RSD (%)	RR (%)	RSD (%)
Trifluralin	20	82	4.1	82	1.4	84	2.3
	200	82	8.0	81	4.0	81	5.2
	300	91	3.1	89	2.8	86	3.7
Triadimefon	20	83	4.7	83	4.6	86	6.4
	200	80	2.0	80	3.2	81	2.3
	300	89	3.5	88	1.5	92	5.1
Permethrin	20	89	4.5	82	7.1	83	4.6
	200	93	5.3	93	3.4	95	3.1
	300	89	5.1	85	2.5	88	2.5
Fenvalerate	20	88	4.4	82	2.6	91	6.7
	200	93	2.8	93	2.8	94	2.2
	300	86	3.3	91	5.8	86	4.1

2.7. Comparison of the Developed Method with Previously Reported Methods

The analytical performance of the established method was compared with other methods reported in the literature (Table 4). The results indicate that the proposed method offers a more satisfactory linear range, higher recoveries, and lower detection limits. MSPE avoids the need for centrifugation and filtration, making the extraction process simple. By using Fe_3O_4/persimmon leaf composite as the magnetic adsorbent, target analytes can be easily and rapidly separated from the sample solution, and the extraction time is only 10 min. The advantages of the magnetic persimmon leaf adsorbent used in this experiment include its simple synthesis, ease of operation, low toxicity, and cost-effectiveness.

Table 4. Comparison between magnetic persimmon leaf composite-based magnetic solid-phase extraction and other analytical methods.

Method	Detection	Extraction Solvents	Sample	LOD ($\mu g\ L^{-1}$)	Total Sample Preparation (min)	Extraction Recovery (%)	References
SPE [a]	GC-MS	Carbon nanotubes	Olive oil	1.5–3.0	45	79–105	[58]
DSPE–DLLME [b]	GC–FID	L-cysteine and sorbitol	Fruit juice	0.49–0.98	7	68-92	[59]
MSPE [c]	GC-ECD	Carbon nanofibers	Water	1.44–5.15	16	70.0–120.6	[60]
MSPE	HLPC	Magnetic corn stalk biochar	Water and zucchini	0.03 (ng/g); 0.2–0.5 (ng/g)	20	86–113	[61]
MSPE	GC-ECD	Magnetic cork composites	Water	0.3–2.02	15	46–84	[44]
MSPE	GC-ECD	Magnetic persimmon leaf composites	Water	0.25–1.1	10	80–94	This work

[a] Solid phase extraction (SPE); [b] Dispersive solid phase extraction-Dispersive liquid-liquid microextraction (DSPE-DLLME); [c] Magnetic solid-phase extraction (MSPE).

3. Materials and Methods

3.1. Reagents and Materials

Pesticide standards, including triadimefon, permethrin, fenvalerate, and trifluralin, and chemicals such as acetonitrile, polyethylene glycol, glutaraldehyde, ethanol, $Fe_2(SO_4)_3$, and $FeSO_4 \cdot 7H_2O$, were purchased from Aladdin Bio-Chem Technology Co., Ltd. (Shanghai, China). Sodium hydroxide (NaOH) was obtained from Modern Oriental (Beijing, China) Technology Development Co., Ltd. (Beijing, China), while methanol and NaCl were sourced from Macklin Biochemical Co., Ltd. (Shanghai, China).

Data Availability Statement: The data that support the findings of this study are available from the corresponding author upon reasonable request.

Conflicts of Interest: The authors declare no conflict of interest.

References

1. Trimnell, D.; Shasha, B.S.; Doane, W.M. Release of trifluralin from starch xanthide encapsulated formulations. *J. Agric. Food Chem.* **1981**, *29*, 637–640. [CrossRef]
2. Zeglen, S.; Pronos, J.; Merler, H. Silvicultural management of white pines in western North America. *For. Pathol.* **2010**, *40*, 347–368. [CrossRef]
3. Nowak, J.T.; McCravy, W.K.; Fettig, J.C.; Berisford, C.W. Susceptibility of adult hymenopteran parasitoids of the Nantucket pine tip moth (Lepidoptera: Tortricidae) to broad-spectrum and biorational insecticides in a laboratory study. *J. Econ. Entomol.* **2001**, *94*, 1122–1127. [CrossRef]
4. Celeiro, M.; Facorro, R.; Dagnac, T.; Llompart, M. Simultaneous determination of trace levels of multiclass fungicides in natural waters by solid—Phase microextraction—Gas chromatography-tandem mass spectrometry. *Anal. Chim. Acta* **2018**, *1020*, 51–61. [CrossRef] [PubMed]
5. Celeiro, M.; Vazquez, L.; Sergazina, M.; Docampo, S.; Dagnac, T.; Vilar, V.J.P.; Llompart, M. Turning cork by-products into smart and green materials for solid-phase extraction—Gas chromatography tandem mass spectrometry analysis of fungicides in water. *J. Chromatogr. A* **2020**, *1628*, 461437. [CrossRef]
6. Xie, W.; Zhao, J.; Zhu, X.; Chen, S.; Yang, X. Pyrethroid bioaccumulation in wild fish linked to geographic distribution and feeding habit. *J. Hazard. Mater.* **2022**, *430*, 128470. [CrossRef]
7. Cui, J.; Tian, S.; Gu, Y.; Wu, X.; Wang, L.; Wang, J.; Chen, X.; Meng, Z. Toxicity effects of pesticides based on zebrafish (Danio rerio) models: Advances and perspectives. *Chemosphere* **2023**, *340*, 139825. [CrossRef]
8. World Health Organization. *Guidelines for Drinking-Water Quality*, 4th ed.; World Health Organization: Geneva, Switzerland, 2011.
9. Song, X.; Li, F.; Yan, T.; Tian, F.; Ren, L.; Jiang, C.; Wang, Q.; Zhang, S. Research progress in the sample pretreatment techniques and advanced quick detection methods of pesticide residues. *Process Saf. Environ. Prot.* **2022**, *165*, 610–622. [CrossRef]
10. Lin, S.; Zhao, Z.; Lv, Y.; Shen, S.; Liang, S. Recent advances in porous organic frameworks for sample pretreatment of pesticide and veterinary drug residues: A review. *Analyst* **2021**, *146*, 7394–7417. [CrossRef]
11. Ötles, S.; Kartal, C. Solid-phase extraction (SPE): Principles and applications in food samples. *Acta Sci. Pol. Technol. Aliment.* **2016**, *15*, 5–15. [CrossRef]
12. Zhang, L.; Liu, S.; Cui, X.; Pan, C.; Zhang, A.; Chen, F. A review of sample preparation methods for the pesticide residue analysis in foods. *Open Chem.* **2012**, *10*, 900–925. [CrossRef]
13. Wierucka, M.; Biziuk, M. Application of magnetic nanoparticles for magnetic solid-phase extraction in preparing biological, environmental and food samples. *TrAC Trends Anal. Chem.* **2014**, *59*, 50–58. [CrossRef]
14. Wu, A.; Zhao, X.; Wang, J.; Tang, Z.; Zhao, T.; Niu, L.; Yu, W.; Yang, C.; Fang, M.; Lv, H.; et al. Application of solid-phase extraction based on magnetic nanoparticle adsorbents for the analysis of selected persistent organic pollutants in environmental water: A review of recent advances. *Crit. Rev. Environ. Sci. Technol.* **2021**, *51*, 44–112. [CrossRef]
15. Jalili, V.; Barkhordari, A.; Ghiasvand, A. A comprehensive look at solid-phase microextraction technique: A review of reviews. *Microchem. J.* **2020**, *152*, 104319. [CrossRef]
16. Kataoka, H. Current developments and future trends in solid-phase microextraction techniques for pharmaceutical and biomedical analyses. *Anal. Sci.* **2011**, *27*, 893–905. [CrossRef] [PubMed]
17. Arthur, C.L.; Pawliszyn, J. Solid phase microextraction with thermal desorption using fused silica optical fibers. *Anal. Chem.* **1990**, *62*, 2145–2148. [CrossRef]
18. Fedotov, P.S.; Malofeeva, G.I.; Savonina, E.Y.; Spivakov, B.Y. Solid-phase extraction of organic substances: Unconventional methods and approaches. *J. Anal. Chem.* **2019**, *74*, 205–212. [CrossRef]
19. Kataoka, H.; Nakayama, D. Online in-tube solid-phase microextraction coupled with liquid chromatography–tandem mass spectrometry for automated analysis of four sulfated steroid metabolites in saliva samples. *Molecules* **2022**, *27*, 3225. [CrossRef]
20. Papageorgiou, M.; Lambropoulou, D.; Morrison, C.; Namieşnik, J.; Płotka-Wasylka, J. Direct solid phase microextraction combined with gas chromatography–mass spectrometry for the determination of biogenic amines in wine. *Talanta* **2018**, *183*, 276–282. [CrossRef]
21. Ishizaki, A.; Ozawa, K.; Kataoka, H. Simultaneous analysis of carcinogenic N-nitrosamine impurities in metformin tablets using on-line in-tube solid-phase microextraction coupled with liquid chromatography-tandem mass spectrometry. *J. Chromatogr. A* **2023**, *1710*, 464416. [CrossRef]
22. Da Silva Sousa, J.; Do Nascimento, H.O.; De Oliveira Gomes, H.; Do Nascimento, R.F. Pesticide residues in groundwater and surface water: Recent advances in solid-phase extraction and solid-phase microextraction sample preparation methods for multiclass analysis by gas chromatography-mass spectrometry. *Microchem. J.* **2021**, *168*, 106359. [CrossRef]
23. Jia, Y.; Wang, Y.; Yan, M.; Wang, Q.; Xu, H.; Wang, X.; Zhou, H.; Hao, Y.; Wang, M. Fabrication of iron oxide@MOF-808 as a sorbent for magnetic solid phase extraction of benzoylurea insecticides in tea beverages and juice samples. *J. Chromatogr. A* **2020**, *1615*, 460766. [CrossRef] [PubMed]

24. Safarik, I.; Safarikova, M. Use of magnetic techniques for the isolation of cells. *J. Chromatogr. B Biomed. Sci. Appl.* **1999**, *722*, 33–53. [CrossRef] [PubMed]
25. Spietelun, A.; Marcinkowski, L.; De La Guardia, M.; Namiesnik, J. Recent developments and future trends in solid phase microextraction techniques towards green analytical chemistry. *J. Chromatogr. A.* **2013**, *1321*, 1–13. [CrossRef] [PubMed]
26. Yu, M.; Wang, L.; Hu, L.; Li, Y.; Luo, D.; Mei, S. Recent applications of magnetic composites as extraction adsorbents for determination of environmental pollutants. *TrAC Trends Anal. Chem.* **2019**, *119*, 115611. [CrossRef]
27. Li, M.; Liu, W.; Meng, X.; Li, S.; Wang, Q.; Guo, Y.; Wu, Y.; Hao, L.; Yang, X.; Wang, Z.; et al. Facile synthesis of magnetic hypercrosslinked polymer for the magnetic solid-phase extraction of benzoylurea insecticides from honey and apple juice samples. *Food Chem.* **2022**, *395*, 133596. [CrossRef] [PubMed]
28. Mohamed, A.H.; Yahaya, N.; Mohamad, S.; Kamaruzaman, S.; Osman, H.; Nishiyama, N.; Hirota, Y. Synthesis of oil palm empty fruit bunch-based magnetic-carboxymethyl cellulose nanofiber composite for magnetic solid-phase extraction of organophosphorus pesticides in environmental water samples. *Microchem. J.* **2022**, *183*, 108045. [CrossRef]
29. Zhao, Y.; Du, D.; Li, Q.; Chen, W.; Li, Q.; Zhang, Q.; Liang, N. Dummy-surface molecularly imprinted polymers based on magnetic graphene oxide for selective extraction and quantification of pyrethroids pesticides in fruit juices. *Microchem. J.* **2020**, *159*, 105411. [CrossRef]
30. Dong, J.; Feng, Z.A.; Kang, S.S.; An, M.; Wu, G.D. Magnetic solid-phase extraction based on magnetic amino modified multiwalled carbon nanotubes for the fast determination of seven pesticide residues in water samples. *Anal. Methods* **2020**, *12*, 2747–2756. [CrossRef]
31. Wu, G.; Zhang, C.; Liu, C.; Li, X.; Cai, Y.; Wang, M.; Chu, D.; Liu, L.; Meng, T.; Chen, Z. Magnetic tubular nickel@silica-graphene nanocomposites with high preconcentration capacity for organothiophosphate pesticide removal in environmental water: Fabrication, magnetic solid-phase extraction, and trace detection. *J. Hazard. Mater.* **2023**, *457*, 131788. [CrossRef]
32. Wang, C.; Liu, L.; Zhang, Z.; Wu, Q.; Wang, Z. Magnetic biomass activated carbon-based solid-phase extraction coupled with high performance liquid chromatography for the determination of phenylurea herbicides in bottled rose juice and water samples. *Food Anal. Methods.* **2016**, *9*, 80–87. [CrossRef]
33. Xie, C.; Wei, S.; Chen, D.; Lan, W.; Yan, Z.; Wang, Z. Preparation of magnetic ion imprinted polymer with waste beer yeast as functional monomer for Cd (II) adsorption and detection. *RSC Adv.* **2019**, *9*, 23474–23483. [CrossRef] [PubMed]
34. Li, P.; Bai, J.; He, P.; Zeng, J. One pot synthesis of nanofiber-coated magnetic composites as magnetic dispersive solid-phase extraction adsorbents for rapid determination of tetracyclines in aquatic food products. *Molecules* **2023**, *28*, 7421. [CrossRef] [PubMed]
35. Karrat, A.; Amine, A. Solid-phase extraction combined with a spectrophotometric method for determination of Bisphenol-A in water samples using magnetic molecularly imprinted polymer. *Microchem. J.* **2021**, *168*, 106496. [CrossRef]
36. Senosy, I.A.; Lu, Z.; Zhou, D.; Abdelrahman, T.M.; Chen, M.; Zhuang, L.; Liu, X.; Cao, Y.; Li, J.; Yang, Z. Construction of a magnetic solid-phase extraction method for the analysis of azole pesticides residue in medicinal plants. *Food Chem.* **2022**, *386*, 132743. [CrossRef] [PubMed]
37. Ahorsu, R.; Medina, F.; Constantí, M. Significance and challenges of biomass as a suitable feedstock for bioenergy and biochemical production: A review. *Energies* **2018**, *11*, 3366. [CrossRef]
38. Fernández López, J.A.; Doval Miñarro, M.D.; Angosto, J.M.; Fernández-Lledó, J.; Obón, J.M. Adsorptive and surface characterization of mediterranean agrifood processing wastes: Prospection for pesticide removal. *Agronomy* **2021**, *11*, 561. [CrossRef]
39. Vieira, Y.; Dos Santos, J.M.N.; Georgin, J.; Oliveira, M.L.S.; Pinto, D.; Dotto, G.L. An overview of forest residues as promising low-cost adsorbents. *Gondwana Res.* **2022**, *110*, 393–420. [CrossRef]
40. Schwantes, D.; Goncalves, A.C.; Fuentealba, D.; Carneiro, M.F.H.; Tarley, C.R.T.; Prete, M.C. Removal of chlorpyrifos from water using biosorbents derived from cassava peel, crambe meal, and pinus bark. *Chem. Eng. Res. Des.* **2022**, *188*, 142–165. [CrossRef]
41. Borukhova, S. Biomass for sustainable applications: Pollution remediation and energy. *Green Process. Synth.* **2014**, *3*, 305–306. [CrossRef]
42. Behbahan, A.K.; Mahdavi, V.; Roustaei, Z.; Bagheri, H. Preparation and evaluation of various banana-based biochars together with ultra-high performance liquid chromatography-tandem mass spectrometry for determination of diverse pesticides in fruiting vegetables. *Food Chem.* **2021**, *360*, 130085. [CrossRef] [PubMed]
43. Nazir, N.A.M.; Raoov, M.; Mohamad, S. Spent tea leaves as an adsorbent for micro-solid-phase extraction of polycyclic aromatic hydrocarbons (PAHs) from water and food samples prior to GC-FID analysis. *Microchem. J.* **2020**, *159*, 105581. [CrossRef]
44. Hang, N.; Yang, Y.; Zang, Y.; Zhao, W.; Tao, J.; Li, S. Magnetic cork composites as biosorbents in dispersive solid-phase extraction of pesticides in water samples. *Anal. Methods* **2023**, *15*, 3510–3521. [CrossRef] [PubMed]
45. Hossain, A.; Moon, H.K.; Kim, J.K. Antioxidant properties of Korean major persimmon (Diospyros Kaki) leaves. *Food Sci. Biotechnol.* **2018**, *27*, 177–184. [CrossRef] [PubMed]
46. Song, Y.R.; Han, A.R.; Lim, T.G.; Kang, J.H.; Hong, H.D. Discrimination of structural and immunological features of polysaccharides from persimmon leaves at different maturity stages. *Molecules* **2019**, *24*, 356. [CrossRef] [PubMed]
47. Chang, Y.L.; Lin, J.T.; Lin, H.L.; Liao, P.L.; Wu, P.J.; Yang, D.J. Phenolic compositions and antioxidant properties of leaves of eight persimmon varieties harvested in different periods. *Food Chem.* **2019**, *289*, 74–83. [CrossRef] [PubMed]
48. Lee, S.Y.; Choi, H.J. Persimmon leaf bio-waste for adsorptive removal of heavy metals from aqueous solution. *J. Environ. Manag.* **2018**, *209*, 382–392. [CrossRef] [PubMed]

49. Yu, S.W.; Choi, H.J. Application of hybrid bead, persimmon leaf and chitosan for the treatment of aqueous solution contaminated with toxic heavy metal ions. *Water Sci. Technol.* **2018**, *78*, 837–847. [CrossRef]
50. Bacelo, H.A.M.; Santos, S.C.R.; Botelho, C.M.S. Tannin-based biosorbents for environmental applications—A review. *Chem. Eng. J.* **2016**, *303*, 575–587. [CrossRef]
51. Pangeni, B.; Paudyal, H.; Inoue, K.; Ohto, K.; Kawakita, H.; Alam, S. Preparation of natural cation exchanger from persimmon waste and its application for the removal of cesium from water. *Chem. Eng. J.* **2014**, *242*, 109–116. [CrossRef]
52. Mohammadkhani, F.; Montazer, M.; Latifi, M. Microwave absorption characterization and wettability of magnetic nano iron oxide/recycled PET nanofibers web. *J. Text. Inst.* **2019**, *110*, 989–999. [CrossRef]
53. Yang, Y.; Liu, W.; Hang, N.; Zhao, W.; Lu, P.; Li, S. On-site sample pretreatment: Natural deep eutectic solvent-based multiple air-assisted liquid–liquid microextraction. *J. Chromatogr. A* **2022**, *1675*, 463136. [CrossRef] [PubMed]
54. Yang, Q.; Wu, P.; Liu, J.; Rehman, S.; Ahmed, Z.; Ruan, B.; Zhu, N. Batch interaction of emerging tetracycline contaminant with novel phosphoric acid activated corn straw porous carbon: Adsorption rate and nature of mechanism. *Environ. Res.* **2020**, *181*, 108899. [CrossRef] [PubMed]
55. Afshin, S.; Rashtbari, Y.; Vosough, M.; Dargahi, A.; Fazlzadeh, M.; Behzad, A.; Yousefi, M. Application of box–behnken design for optimizing parameters of hexavalent chromium removal from aqueous solutions using Fe3O4 loaded on activated carbon prepared from alga: Kinetics and equilibrium study. *J. Water Process Eng.* **2021**, *42*, 102113. [CrossRef]
56. Niri, M.V.; Mahvi, A.H.; Alimohammadi, M.; Shirmardi, M.; Golastanifar, H.; Mohammadi, M.J.; Naeimabadi, A.; Khishdost, M. Removal of natural organic matter (NOM) from an aqueous solution by NaCl and surfactant-modified clinoptilolite. *J. Water Health* **2015**, *13*, 394–405. [CrossRef] [PubMed]
57. Wei, M.; Marrakchi, F.; Yuan, C.; Cheng, X.; Jiang, D.; Zafar, F.F.; Fu, Y.; Wang, S. Adsorption modeling, thermodynamics, and DFT simulation of tetracycline onto mesoporous and high-surface-area NaOH-activated macroalgae carbon. *J. Hazard. Mater.* **2022**, *425*, 127887. [CrossRef]
58. López-Feria, S.; Cárdenas, S.; Valcárcel, M. One step carbon nanotubes-based solid-phase extraction for the gas chromatographic–mass spectrometric multiclass pesticide control in virgin olive oils. *J. Chromatogr. A* **2009**, *1216*, 7346–7350. [CrossRef]
59. Farajzadeh, M.A.; Mohebbi, A.; Izadyar, M.; Mogaddam, M.R.A.; Pezhhanfar, S. Facile preparation of nitrogen–doped amorphous carbon nanocomposite as an efficient sorbent in dispersive solid phase extraction. *Int. J. Environ. Anal. Chem.* **2023**, *103*, 1020–1038. [CrossRef]
60. Singh, M.; Pandey, A.; Singh, S.; Singh, S.P. Iron nanoparticles decorated hierarchical carbon fiber forest for the magnetic solid-phase extraction of multi-pesticide residues from water samples. *Chemosphere* **2021**, *282*, 131058. [CrossRef]
61. Wang, Y.; Ma, R.; Xiao, R.; Wu, Q.; Wang, C.; Wang, Z. Preparation of a magnetic porous carbon with hierarchical structures from waste biomass for the extraction of some carbamates. *J. Sep. Sci.* **2017**, *40*, 2451–2458. [CrossRef]

Disclaimer/Publisher's Note: The statements, opinions and data contained in all publications are solely those of the individual author(s) and contributor(s) and not of MDPI and/or the editor(s). MDPI and/or the editor(s) disclaim responsibility for any injury to people or property resulting from any ideas, methods, instructions or products referred to in the content.

Article

Preparation of COPs Mixed Matrix Membrane for Sensitive Determination of Six Sulfonamides in Human Urine

Ying Liu [1], Yong Zhang [1], Jing Wang [2], Kexin Wang [1], Shuming Gao [1], Ruiqi Cui [1], Fubin Liu [1] and Guihua Gao [1,*]

[1] School of Pharmacy, Jining Medical University, Rizhao 276826, China; ying1737883492@126.com (Y.L.); zhang_2211@163.com (Y.Z.); tawkx589@163.com (K.W.); 17860202743@163.com (S.G.); 13697839528@163.com (R.C.); liufblvy@163.com (F.L.)

[2] School of Pharmacy, Shandong University of Traditional Chinese Medicine, Jinan 250355, China; wangjing20210903@163.com

* Correspondence: guihua526@163.com; Fax: +86-633-2983690

Citation: Liu, Y.; Zhang, Y.; Wang, J.; Wang, K.; Gao, S.; Cui, R.; Liu, F.; Gao, G. Preparation of COPs Mixed Matrix Membrane for Sensitive Determination of Six Sulfonamides in Human Urine. *Molecules* 2023, 28, 7336. https://doi.org/10.3390/molecules28217336

Academic Editor: Rosa Herráez Hernández

Received: 17 October 2023
Revised: 26 October 2023
Accepted: 27 October 2023
Published: 30 October 2023

Copyright: © 2023 by the authors. Licensee MDPI, Basel, Switzerland. This article is an open access article distributed under the terms and conditions of the Creative Commons Attribution (CC BY) license (https:// creativecommons.org/licenses/by/ 4.0/).

Abstract: In this study, TpDMB-COPs, a specific class of covalent organic polymers (COPs), was synthesized using Schiff-base chemistry and incorporated into a polyvinylidene fluoride (PVDF) polymer for the first time to prepare COPs mixed matrix membranes (TpDMB-COPs-MMM). A membrane solid-phase extraction (ME) method based on the TpDMB-COPs-MMM was developed to extract trace levels of six sulfonamides from human urine identified by high-performance liquid chromatography (HPLC). The key factors affecting the extraction efficiency were investigated. Under the optimum conditions, the proposed method demonstrated an excellent linear relationship in the range of 3.5–25 ng/mL ($r^2 \geq 0.9991$), with the low limits of detection (LOD) between 1.25 ng/mL and 2.50 ng/mL and the limit of quantification (LOQ) between 3.50 ng/mL and 7.00 ng/mL. Intra-day and inter-day accuracies were below 5.0%. The method's accuracy was assessed by recovery experiments using human urine spiked at three levels (7–14 ng/mL, 10–15 ng/mL, and 16–20 ng/mL). The recoveries ranged from 87.4 to 112.2% with relative standard deviations (RSD) $\leq$ 8.7%, confirming the applicability of the proposed method. The developed ME method based on TpDMB-COPs-MMM offered advantages, including simple operation, superior extraction affinity, excellent recycling performance, and easy removal and separation from the solution. The prepared TpDMB-COPs-MMM was demonstrated to be a promising adsorbent for ME in the pre-concentration of trace organic compounds from complex matrices, expanding the application of COPs and providing references for other porous materials in sample pre-treatment.

Keywords: covalent organic polymer; mixed matrix membranes; TpDMB-COPs; membrane solid-phase extraction; HPLC; sulfonamides

1. Introduction

Sulfonamides (SAs) have been widely used to prevent bacterial infectious diseases in food-borne animals due to their efficacy, stability, and affordability [1]. However, excessive and prolonged use of SAs has led to residues in animal products, which can be ingested by humans through the food chain [2,3]. This may result in various toxic side effects, such as allergic reactions, drug resistance, and teratogenicity, as well as carcinogenic and mutagenic effects. Approximately ninety percent of SAs and their metabolites are excreted in human and animal urine after application due to incomplete absorption and metabolism [4,5]. As urine is the primary excretory pathway for SAs, its content can indirectly indicate the degree of human contamination SAs [6,7]. Numerous countries and organizations have already restricted or prohibited the use of SAs. Therefore, there is a pressing demand for an efficient, rapid, and accurate method to analyze SAs in human urine.

Modern analytical technologies, including high-performance liquid chromatography (HPLC) [8,9], fluorescence detection [10], capillary electrophoresis with ultraviolet detection (CE-UV) [11,12], and liquid chromatograph-tandem mass spectrometer (LC–MS/MS) [13–16], have been used to determine SAs drugs in recent decades. Among them, HPLC is the most commonly employed method for the determination of SAs, due to its sensitivity, specificity, and ability to identify unknown substances. However, direct measurement of low SAs concentrations in complex human urine is challenging [17]. Sample pre-treatment is a crucial step in the analytical process due to the low concentrations of SAs and complex matrix in human urine. Various methods have been utilized to pre-treat SAs residues in different matrices, such as liquid-liquid extraction [18,19], solid-phase extraction [20], and solid-phase microextraction [21]. Classical extraction methods, such as solid-phase extraction (SPE) and dispersive solid-phase extraction (dSPE), are commonly preferred, while certain drawbacks, including tedious and time-consuming procedures, high operational costs, and significant solvent consumption, cannot be ignored. Therefore, a suitable and efficient sample pre-treatment is necessary. Mixed-matrix membrane solid-phase extraction (ME) [22] has garnered significant attention as an interesting method in recent years. It involves incorporating the adsorbent into the polymer matrix and allowing the mixed matrix membrane (MMM) to tumble freely through the sample under vortex force to complete the extraction of the target analytes. This approach avoids several issues, such as centrifugation for solid-liquid separation and material loss, while offering the advantages of good reproducibility, simple operation, and easy acquisition. The key component of ME is the adsorbent, and it exhibits broad applicability in the separation and purification of samples due to the variety of available adsorbents. Adsorbents with excellent performance can ensure effective enrichment and purification of samples. Various studies have explored adsorbents, such as carbon nanotubes [23], activated carbon [24], metallic organic frameworks [25], covalent organic polymers (COPs) [26], and others.

COPs, an emerging class of porous material, have gained significant attention owing to their mild synthesis conditions, excellent thermal stability, porosity, high specific surface area, and moderate to high chemical stability [27]. In general, COPs are prepared with monomers of light elements. They are constructed using covalent bonds, such as C=C, C-C, and C=N, and can be categorized as crystalline covalent organic frameworks (COFs) and amorphous COPs [28]. Although crystalline COFs are challenging to synthesize due to harsh reaction conditions, amorphous COPs can be easily prepared through a simple synthetic process. They have been widely applied in catalysis [29], gas absorption [30], sample pretreatment [31], and drug delivery [32]. The existence of carbon, oxygen, and nitrogen atoms in COP blocks enhances their enrichment efficiency as the adsorbents, owing to the inter-molecular forces with target compounds [33]. Therefore, COPs hold excellent potential for sample pretreatment. Ma et al. [34] developed a novel covalent magnetic sulfonated organic polymer for solid-phase magnetic extraction of protoberberine alkaloids. Wei et al. [35] synthesized an imine-based porous 3D COP, which proved an effective solid-phase sorbent for detecting amphenicols in water samples. These findings have demonstrated the practical and promising applications of COPs in sample pretreatment [36]. However, the use of COPs as mixed matrix membrane extraction adsorbents has seldom been explored. Previous reports have indicated that amorphous COPs are typically synthesized by polymerizing designed organic monomers, including various aldehydes and amides, through the Schiff-base reaction [37].

In this study, a COP named TpDMB-COPs was synthesized through a Schiff-base reaction using 2,4,6-triformylphloroglucinol (Tp) and 3,3′-dimethylbenzidine (DMB) as organic monomers. Subsequently, a TpDMB-COPs mixed matrix membrane (TpDMB-COPs-MMM) was prepared by embedding the TpDMB-COPs into polyvinylidene fluoride (PVDF). Based on the obtained TpDMB-COPs-MMM, a membrane solid-phase extraction method (ME) for the quantitative extraction of SAs from human urine was developed. The SAs analyzed in this method included sulfamerazine (SMR), sulfamethoxypyridazine (SMP), sulfamethizole (SMX), sulfadimethoxine (SDM), sulfadoxine (SDX), and sulfaquinoxaline (SQ), which were determined by HPLC.

2. Results and Discussion

2.1. Characterization of Tp-DMB-COPs and Tp-DMB-COPs-MMM

The TpDMB-COPs was characterized using TGA, BET, and SEM. The TGA curve shown in Figure S2 indicated that the TpDMB-COPs possessed high thermal stability and could be stable at 412 °C. The N_2 adsorption isotherm of TpDMB-COPs was illustrated in Figure S3. The TpDMB-COPs display a representative type-IV isotherm at relative 0.3–1.0 pressure, thereby suggesting the existence of porous construction and the BET surface area was 142 m^2/g^{-1}. The surface morphology of synthesized TpDMB-COPs was evaluated using the SEM shown in Figure 1B. The high magnification SEM image of TpDMB-COPs revealed a porous network and nanoskeleton structure, presenting the successful synthesis of TpDMB-COPs.

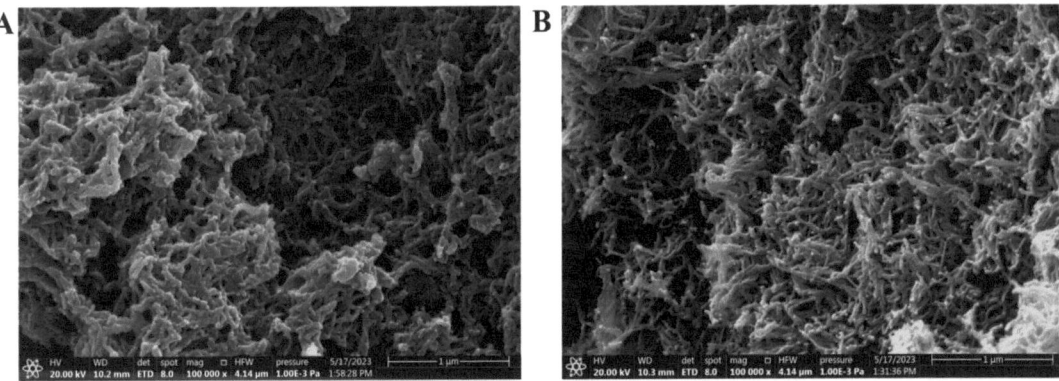

Figure 1. The SEM images of TpDMB-COPs-MMM (**A**) and TpDMB-COPs powder (**B**) (Magnification: 100,000×).

The TpDMB-COPs-MMM was also characterized through SEM. The SEM image was presented in Figure 1A, showing a clear powder morphology on the surface, which demonstrated successful preparation of TpDMB-COPs-MMM while maintaining the powder's morphology and properties. The SEM also indicated that the TpDMB-COPs were well-dispersed on the TpDMB-COPs-MMM surface, leading to an increased roughness of its surface. Furthermore, due to the hydrophobic nature of TpDMB-COPs, when it was added to PVDF to prepare a mixed matrix membrane, might cause an increase in the hydrophobicity of TpDMB-COPs-MMM. This can be confirmed by the contact angle measurement.

The contact angle measurements were conducted at room temperature using the sessile drop method. The results indicated an increase in the TpDMB-COPs-MMM contact angle (Figure 2B) from 60.5 ± 1.5° to 91.9 ± 1.1° compared with the pristine bare membrane (Figure 2A). This increase can be attributed to the increased surface roughness and hydrophobicity after treatment with TpDMB-COPs [20]. The surface contact angle of TpDMB-COPs-MMM was increased due to the hydrophobic nature of the carbon-carbon double bond skeleton of TpDMB-COPs. These results confirmed the successful synthesis of TpDMB-COPs-MMM [38].

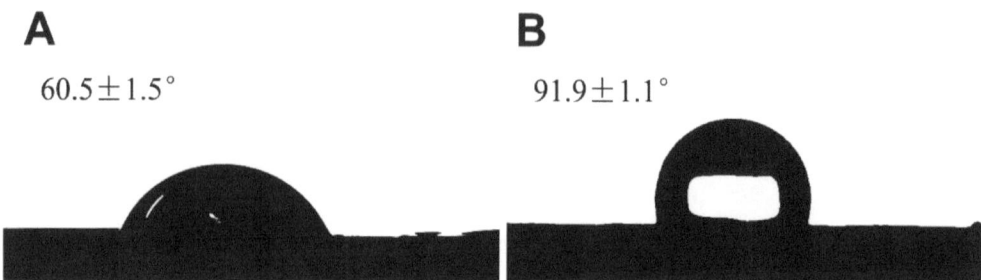

Figure 2. The static water contact angles of (**A**) pristine bare membrane and (**B**) TpDMB-COPs-MMM.

2.2. Optimization of ME Procedures

To enhance extraction efficiency, we investigated several key factors, including pH, salt concentration, extraction time, type of eluent, eluent volume, and elution time, using the recoveries of the six SAs varieties as an evaluation indicator.

2.2.1. Effect of the Sample pH

The pH of the extraction solution plays a crucial role in shaping the molecular or ionic states of SAs, in addition to affecting their interaction with the absorber (TpDMB-COPs-MMM). Thus, optimizing the pH was the first parameter studied. The effect of pH on extraction efficiency was investigated across various pH values ranging from 3.0 to 12.0. Figure 3A indicated that recoveries increased as the pH increased from 3.0 to 6.0, with the highest recoveries achieved at pH 6.0, and for most SAs, the recoveries decreased with further increases in pH from 6.0 to 10.0, though increased again at pH 10.0 for SDM and SQ. Nevertheless, for SDM and SQ, the recoveries at pH 10.0 remained lower than those at pH 6.0. And a significant decrease in the recoveries was observed when pH exceeded 10.0. This can be attributed to the amphoteric nature of SAs, where different pH levels dictated the prevalence of ionized and molecular forms [39–41]. They might exist in an ionic state at pH values below 6.0 or above 10.0, and be molecular states at 6.0 and 10.0 which are more likely to be adsorbed due to the hydrophobic force. Consequently, a sample pH of 6.0 was selected for subsequent analyses.

2.2.2. Effect of the Salt Concentration

The impact of salt concentration on extraction efficiency is primarily observed in the salt effect. In order to investigate the influence of ionic strength on SAs extraction, different concentrations of NaCl were examined, with added NaCl ranging from 0% to 2% (w/v) (Figure 3B). For most SAs, the highest recoveries were achieved when 1.5% NaCl (w/v) was added to the sample. The extraction performance did not improve with further additional NaCl, probably due to increased sample viscosity. While the recoveries of SMP and SDM were the highest when adding 1.0% NaCl (w/v); however, the recoveries were not significantly different from those when 1.5% NaCl (w/v) was added. Therefore, subsequent experiments were conducted by adding 1.5% (w/v) NaCl to the sample solution.

2.2.3. Effect of the Extraction Time

Extraction time refers to the duration required for the analytes to be adsorbed from the sample solution onto the sorbent. As the adsorption process followed an extraction procedure based on equilibrium, determining the optimal extraction time could guarantee efficient equilibration of analytes between the adsorbent and the aqueous phase. Figure 3C demonstrated that the extraction equilibrium was attained within roughly 30 min under continuous vortexing. Notably, the recoveries of most targets decreased with prolonged adsorption time. Hence, a 30 min extraction time was considered to be the optimal.

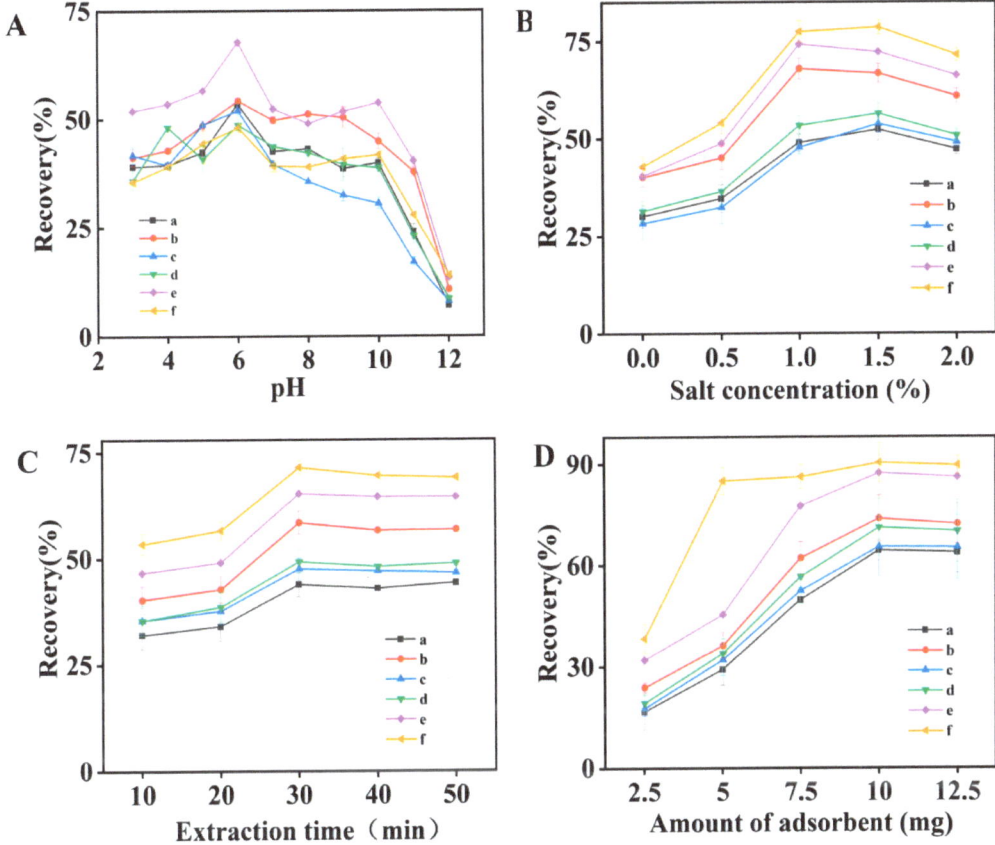

Figure 3. The investigation of sample pH (**A**); salt concentration (**B**); amount of adsorbent (**C**), and extraction time (**D**) (a: SMR, b: SMP, c: SMX, d: SDX, e: SDM, and f: SQ).

2.2.4. Effect of Adsorbent Amount

The effect of the adsorbent amount on the recovery of six spiked SAs was investigated within the range of 2.5 to 12.5 mg. As displayed in Figure 3D, the highest recoveries was obtained when the amount of adsorbent was 10.0 mg, and the recoveries did not increase significantly when the amount of adsorbent exceeded 10.0 mg. It was concluded that the 10.0 mg of adsorbent was considered sufficient for subsequent experiments. The results also suggested that the developed sorbent exhibited high adsorption efficiency and could achieve extraction efficacy with a small amount of adsorbent, such that 10.0 mg of adsorbent was established as the optimal resin amount for subsequent experiments.

2.2.5. Effect of Eluent Type and Volume

Since the adsorption of SAs onto TpDMB-COPs-MMM relied on π-π interactions, hydrogen bond forces, and hydrophobic interactions, it was necessary to consider desorption solvents with different polarities. Four eluents were examined, namely methanol, formic acid-methanol (1:1000, v/v), acetonitrile, and formic acid-acetonitrile (1:1000, v/v). Figure 4A illustrated that formic acid-methanol (1:1000, v/v) resulted in higher recoveries, making it the optimal eluent. Furthermore, the elution volume within the 0.5–5.0 mL was analyzed. The results (Figure 4B) revealed that the elution volume gradually increased with the volume of formic acid-methanol (1:1000, v/v) until reaching 1.0 mL. Further increases in volume beyond 1.0 mL did not lead to higher recoveries of the six SAs. Hence, a volume of 1.0 mL of formic acid-methanol (1:1000, v/v) was selected as the appropriate eluent.

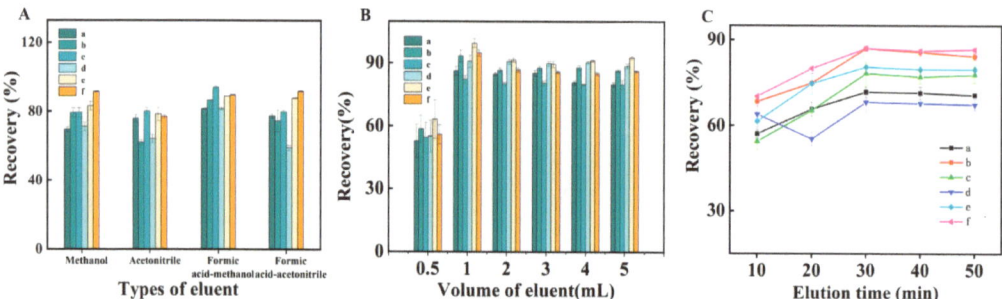

Figure 4. The investigation of types of eluent (**A**), volume of eluent (**B**), and elution time (**C**) (a: SMR, b: SMP, c: SMX, d: SDX, e: SDM, and f: SQ).

2.2.6. Effect of Elution Time

An appropriate elution time is crucial for ensuring the complete elution of an analyte and to maximizing extraction efficiency. In order to determine the optimal elution time, the effects of elution time between 10 and 50 min were studied. As depicted in Figure 4C, the recoveries increased rapidly between 10 and 30 min. However, when the elution time was further prolonged, the recoveries decreased slightly. This could be attributed to the equilibrium-based nature of the elution process, where over time, the targets might reattach to the adsorbent, resulting in a slight decrease in the elution efficiency. Thus, the elution time was fixed at 30 min.

2.2.7. Reusability of TpDMB-COPs-MMM

Under optimal extraction conditions, the reusability of TpDMB-COPs-MMM as an adsorbent was assessed through five adsorption and desorption cycles. TpDMB-COPs-MMM was used as an adsorbent for a blank sample spiked with six SAs at concentration of 10 ng/mL. After completing the whole adsorption and desorption process, the TpDMB-COPs-MMM was taken out and washed with water and methanol. Then, it was put into spiked sample solution again for adsorption and desorption. As illustrated in Figure S4, the recoveries were slightly decreased after the fifth cycle. This indicated that the prepared TpDMB-COPs-MMM could be reused for at least five cycles.

2.2.8. Comparison of Adsorption Capacity of TpDMB-COPs-MMM and TpDMB-COPs Powder

Under the optimal extraction conditions, the adsorption capacity of TpDMB-COPs-MMM was compared with TpDMB-COPs powder of the same mass. Figure 5 demonstrates that there is no significant difference in adsorption capacity between the TpDMB-COPs powder and TpDMB-COPs-MMM for six SAs. This finding provided evidence that the extraction performance of membrane was comparable to that of the powder. It indicated that converting the powder into a membrane did not change its adsorption ability and may even offer added convenience by eliminating the need for high-speed centrifugation for solid-liquid separation, and reducing adsorbent loss. Therefore, TpDMB-COPs-MMM was selected for detecting SAs in human urine.

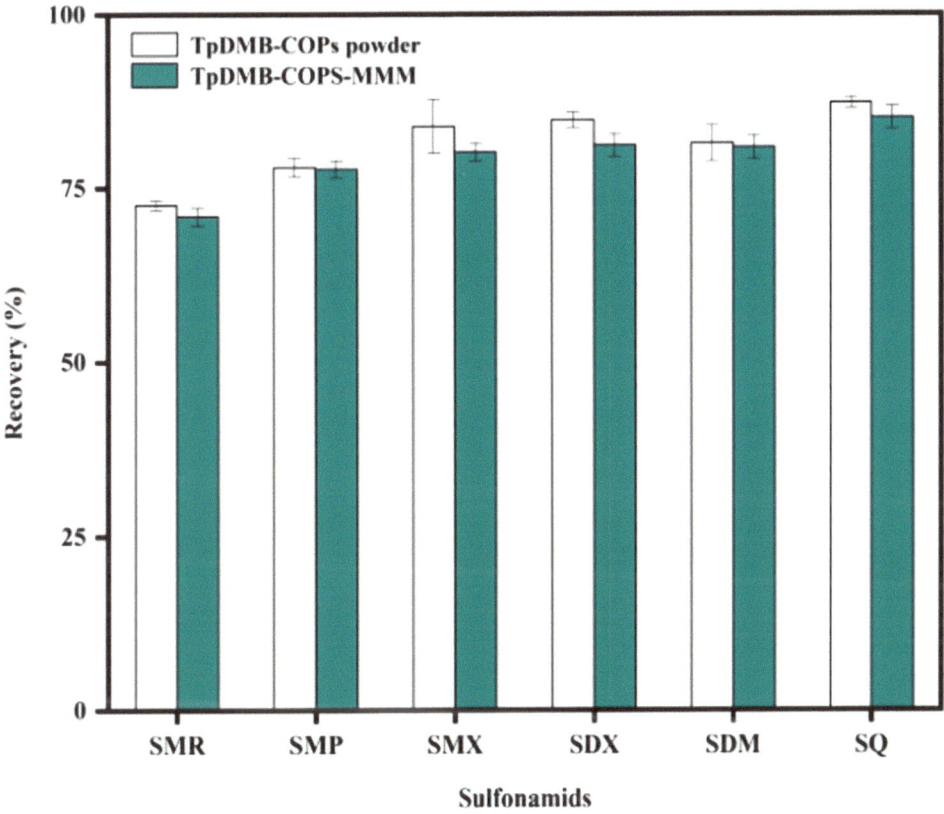

Figure 5. The comparison of adsorption performance of TpDMB-COPs-MMM material and TpDMB-COPs powder.

2.3. Method Validation and Application to Real Samples

The developed ME-HPLC methodology was evaluated for linearity, the limits of detection (LOD), repeatability, the limits of quantification (LOQ), precision, and accuracy under optimized conditions. With pure water added standard mixed solutions of the six compounds at different concentrations were prepared as mock water samples at five concentrations levels within a specified range. These samples were then extracted under the optimized ME conditions, and analyzed by HPLC. The obtained working curves for the ME-HPLC extraction method yielded correlation coefficients (r^2) ranging from 0.9991 to 0.9998. The detection and quantification limits for the six SAs ranged from 1.25 to 2.5 ng/mL and 3.5 to 7.0 ng/mL, respectively, with signal-to-noise ratios of 3 and 10. The results are presented in Table 1.

Table 1. Linearity and range and results of LOD, LOQ.

Analytes	LOD (ng/mL)	LOQ (ng/mL)	Range (ng/mL)	r^2
SMR	1.25	3.50	3.5–20	0.9995
SMP	1.88	5.00	5–25	0.9998
SMX	2.50	7.00	7–25	0.9991
SDX	2.50	7.00	7–25	0.9997
SDM	2.50	7.00	7–25	0.9995
SQ	1.25	3.50	3.5–20	0.9993

The precision of the method was evaluated based on intra-day and inter-day experiments, determined by assessing the relative standard deviation (RSD). Six replicates of spiked samples were applied to analyze the RSD on the same day, and three replicates of spiked samples were analyzed on three consecutive days. Detailed precision data can be found in Table S2. Accuracy experiments were conducted by spiking three concentrations of six SAs standard samples into blank human urine samples. The recoveries ranged from 87.4% to 112.2%, with RSDs less than 8.7%, demonstrating the reliability and applicability of the method. The results are presented in Table S3.

In order to assess the practicality of this approach, urine samples from volunteers were analyzed. The urine samples' supernatant underwent pretreatment and were tested using the ME-HPLC method and the chromatograms were shown in Figure S4. Table 2 presents the results, indicating that SQ and SMR were detected in sample No. 2, with SQ at a concentration of 5.59 ng/mL and SMR below the LOQ. Sample No. 3 presented SMR at 17.65 ng/mL. In sample No. 4, SMR was detected at 14.39 ng/mL, and SQ was below the LOQ. Sample No. 6 indicated SMR at a concentration of 9.86 ng/mL. Sample No. 7 exhibited SDM at 19.82 ng/mL, while SMR and SMP were below the LOQ. SDM was detected in samples No. 7 but below the LOQ. No sulfonamides were detected in samples No. 1, No. 5, No. 8, and No. 10.

Table 2. Determination of the target analytes in human urine.

Number	Concentration (ng/mL)					
	SMR	SMP	SMX	SDX	SDM	SQ
1	—	—	—	—	—	—
2	<LOQ	—	—	—	—	5.59
3	17.65	—	—	—	—	—
4	14.39	—	—	—	—	<LOQ
5	—	—	—	—	—	—
6	9.86	—	—	—	—	—
7	<LOQ	<LOQ	—	—	19.82	—
8	—	—	—	—	—	—
9	—	—	—	—	<LOQ	—
10	—	—	—	—	—	—

—: not detected.

2.4. Comparison of the Developed Method with Previously Reported Methods

In order to highlight the application of this method in determining SAs in urine samples, a comparison with other techniques for SAs extraction was presented in Table 3. The developed method exhibited comparable or superior analytical performance in adsorbent amount and LOD compared with reported methods. Moreover, the developed TpDMB-COPs-MMM adsorbent used in this method was eco-friendly, low toxicity, and reusable. The comparison of analytical performance suggested that the membrane solid-phase extraction process with the TpDMB-COPs-MMM adsorbent was appropriate for the extraction and quantitative analysis of SAs in complex matrices.

Table 3. Comparison of the developed method with previously reported methods for determination of sulfonamides.

Extraction and Analytical Method	Analytes	Sample Matrix	Amount of Adsorbent	LOD	References
d-SPE/HPLC-DAD	SMR, SMP, SMX, SDM	Chicken, Lamb, Beef	15 mg	2.4~4.9 ng/g	[42]
MLC-DAD	SMR	Medicated feeds	—	32.7 mg/kg	[43]
LPME/HPLC	SMR, SMX	Urine	—	33~57 ng/mL	[44]
HPLC-MS/MS	SDX	Herbal products	—	0.002 µg/mL	[45]

Table 3. Cont.

Extraction and Analytical Method	Analytes	Sample Matrix	Amount of Adsorbent	LOD	References
MSPE/HPLC-DAD	SDM	Milk	50 mg	2.5 µg/kg	[46]
MSPE/HPLC–DAD	SMX, SDM, SQ	Milk	10 mg	12~14 µg/L	[47]
d-SPE/HPLC-FLD	SMX, SDM, SDX	Chicken, Muscle, Egg	300 mg	4.1~5.5 µg/kg	[48]
VUA-DLLMEDES/HPLC-PDA	SDM	Fish	0.6 mL	5.5 µg/kg	[49]
SPME/HPLC	SMP	Water	—	2 ng/mL	[50]
ME/HPLC-DAD	SMR, SMP, SMX, SDX, SDM, SQ	Urine	10 mg	1.25~2.50 ng/mL	This work

"—" stands for the relevant content not mentioned in the article; d-SPE: dispersive solid phase extraction; MLC: micellar liquid chromatography; LPME: liquid phase microextraction; MSPE: magnetic solid-phase extraction; VUA-DLLMEDES: vortexultrasonic assisted dispersive liquid-liquid microextraction in combination with hydrophobic deep eutectic solvents; SPME: solid phase microextraction; DAD: diode-array detector; LC-MS/MS: liquid chromatograph-tandem mass spectrometer; HPLC-FLD: high performance liquid chromatography-fluorescence detection; PDA: photo-diode array.

3. Materials and Methods

3.1. Chemicals and Reagents

Six sulfonamide standards, including SMR, SMP, SMX, SDX, SDM, and SQ with purities of >98% were purchased from Aladdin Bio-Chem Technology Co., Ltd. (Shanghai, China), and their structures were illustrated in Table S1. Sodium chloride (NaCl), ethanol (EtOH), and acetone (ACE) were procured from Tianjin Kermel Chemical Reagent Co., Ltd. (Tianjin, China). PVDF and 2,4,6-triformylphloroglucinol (Tp) were purchased from Aladdin Bio-Chem Technology Co., Ltd. (Shanghai, China). p-Toluenesulfonic acid (PTSA) and 3,3′-dimethylbenzidine (DMB) were purchased from Xilong Scientific Co., Ltd. (Shantou, China). Dimethylformamide (DMF) was purchased from Beijing Tongguang Fine Chemical Company. (Beijing, China). All materials and reagents were of at least analytical grade and used as received. Methanol and acetonitrile with chromatographically pure were purchased from Shanghai Xingke Bio-Chem Technology Co., Ltd. (Shanghai, China). Ultra-pure water was obtained from a Thermo LabTower EDI 15 ultra-pure water system (Thermo Scientific, Stockholm, Sweden).

3.2. Instrumentation

The adsorbent's morphological analysis under a scanning electron microscope (SEM) was performed using an Aprco S (Thermo Fisher Scientific, Waltham, MA, USA). Thermogravimetric analysis (TGA) was conducted by using TGA/DSC 1 analyzer (Mettler Toledo, Greifensee, Switzerland). The Brunauer-Emmett-Teller (BET) specific surface area was analyzed by using an ASAP 2010 apparatus (Micromeritics, Norcross, GA, USA). Analysis via chromatography was performed using an Agilent 1260 HPLC system with diode array detection (Agilent Technologies Co., Ltd., Santa Clara, CA, USA). The pH values were determined with a PHS-3C pH meter (Shanghai Leici Scientific Instrument Co., Ltd., Shanghai, China). For TpDMB-COPs-MMM preparation, an electronic hot plate (Shanghai Electronic Technology Co., Ltd., Shanghai, China), an ultrasonic cleaner (Shanghai Shangyi Instrument Equipment Co., Ltd., Shanghai, China), and a Vortex-2500 MT vortex mixer (Lichen Instrument Technology Co., Ltd., Shaoxing, China) were employed. Stock solutions were prepared in methanol and stored at −20 °C, while working solutions were prepared daily by diluting the stock solutions with ultrapure water.

3.3. Synthesis of TpDMB-COPs

In this study, the TpDMB-COPs were synthesized following a previous report by heating reflux [51]. Specifically, 0.021 g of Tp (0.1 mmol), 0.032 g of DMB (0.15 mmol), and 0.077 g of PTSA (0.45 mmol) were mixed with 150 mL of DMF and stirred for 12 h at 165 °C under N_2 atmosphere. The resulting orange-red powder was collected after solvent exchange with DMF and EtOH, and subsequent overnight drying under vacuum.

3.4. Preparation of TpDMB-COPs-MMM

The TpDMB-COP powder (15 mg) was dispersed in ACE (2 mL) by sonication in a bath for 5 min, resulting in a TpDMB-COPs suspension. This suspension was subsequently mixed with a PVDF solution (0.6 g, 7.5 wt% in DMF) to achieve a final TpDMB-COPs to PVDF ratio of 1:3 (w/w). The mixture was agitated for 15 min at 65 °C on an electronic hot plate to remove ACE, obtaining a suspension of TpDMB-COPs and PVDF in DMF. The DMF suspension was evenly coated onto a glass substrate (1.8 cm × 1.8 cm), which was then heated to 65 °C on the plate to remove solvent. The coated films were delaminated from the glass substrate using a slide and stored. The preparation procedure was shown in Figure S1.

3.5. Sample Collection

Urine samples were collected from 10 volunteers. All human urine samples were filtered through a 0.22 μm membrane to remove suspended particles and then stored in brown glass vials at 4 °C in a refrigerator before analysis. Analysis of the samples was performed within 3 days.

3.6. ME Procedure

After activation with methanol and water in sequence, TpDMB-COPs-MMM was loaded into a 15 mL centrifuge tube containing 10 mL of sample solution with pre-adjusted pH and salinity. The mixture was vortexed for 30 min. Subsequently, TpDMB-COPs-MMM was easily removed with tweezers, dried with filter paper, and transferred to a 2 mL centrifuge tube. Precisely 1.0 mL of formic acid-methanol (1:1000, v/v) was used for eluting the target compounds from TpDMB-COPs-MMM by vigorous vortexing for 30 min. Finally, the TpDMB-COPs-MMM was removed, the eluent was filtered through a 0.22 μm filter membrane, and an aliquot of 10 μL was injected into the HPLC-DAD system for analysis.

3.7. HPLC Conditions

Chromatographic separation was performed using an Agilent 1260 HPLC system equipped with a DAD detector. An ODS-100V column (150 mm × 4.6 mm, 3 μm) was employed at 30 °C, with an injection volume of 10 μL. The detection wavelength was set at 268 nm. The mobile phase consisted of an aqueous solution of 0.1% formic acid (solvent A) and methanol (solvent B) at a flow rate of 1.0 mL/min. Prior to use, the mobile phase was filtered through a 0.22 μm filter. The system employed the following linear gradient: 0 min (72% A, 28% B), 8.0 min (72% A, 28% B), 12.0 min (62% A, 38% B), 17.0 min (52% A, 48% B), 19.0 min (72% A, 28% B), and 30.0 min (72% A, 28% B). All six compounds were successfully separated within 30 min.

4. Conclusions

In this study, a novel mixed matrix membrane named TpDMB-COPs-MMM was prepared by combining covalent organic framework polymers (TpDMB-COPs) with PVDF for the first time. TpDMB-COPs-MMM proved to be an effective sorbent for extracting SAs from human urine, followed by the HPLC analysis. The developed ME technology offered the advantage of simple operation and TpDMB-COPs-MMM can be easily separated from water without additional equipment, such as a centrifuge or vacuum pump. Notably, TpDMB-COPs-MMM possessed several advantages. It was cost-effective, easy to prepare, and can be reused at least 5 times without losing extraction efficiency, thus reducing total analysis time and cost. The method has a low detection limit, excellent precision, is accurate and reproducible, which makes it suitable for the detection of trace amounts of contamination in human urine samples. These properties make it a promising candidate for further development in the context of food safety and public health applications. Furthermore, the preparation of COPs-mixed matrix membrane also provided a potential avenue for the utilization of other COPs. This process served as a valuable reference for extending the application of COPs in sample preparation.

Supplementary Materials: The following supporting information can be downloaded at: https://www.mdpi.com/article/10.3390/molecules28217336/s1. Figure S1. The preparation procedure of TpDMB-COPs-MMM. Figure S2. The TGA curve of TpDMB-COPs powder. Figure S3. The N_2 adsorption-desorption isotherms of TpDMB-COPs powder. Figure S4. Reusability of TpDMB-COPs-MMM. Figure S5. The chromatograms of spiked and real samples. Table S1. The chemical structure of six sulfonamides. Table S2. The results of precision. Table S3. The results of recoveries experiment.

Author Contributions: Conceptualization, Y.L. and Y.Z.; methodology, K.W. and R.C.; validation, K.W., S.G. and F.L.; investigation, Y.Z. and Y.L.; data curation, Y.L. and J.W.; writing—original draft preparation, Y.L.; writing—review and editing, G.G.; supervision, G.G.; project administration, G.G.; funding acquisition, G.G. All authors have read and agreed to the published version of the manuscript.

Funding: The work was financially supported by the National Natural Science Foundation of China (82003706).

Institutional Review Board Statement: Not applicable.

Informed Consent Statement: Not applicable.

Data Availability Statement: The data presented in this study are available on request from the corresponding author.

Acknowledgments: The authors would like to thank all the reviewers who participated in the review during the preparation of this manuscript.

Conflicts of Interest: The authors declare no conflict of interest.

References

1. Farooq, M.U.; Su, P.; Yang, Y. Applications of a Novel Sample Preparation Method for the Determination of Sulfonamides in Edible Meat by CZE. *Chromatographia* **2009**, *69*, 1107–1111. [CrossRef]
2. Stoev, G.; Michailova, A. Quantitative determination of sulfonamide residues in foods of animal origin by high-performance liquid chromatography with fluorescence detection. *J. Chromatogr. A* **2000**, *871*, 37–42. [CrossRef] [PubMed]
3. Xie, X.; Huang, S.; Zheng, J.; Ouyang, G. Trends in sensitive detection and rapid removal of sulfonamides: A review. *J. Sep. Sci.* **2020**, *43*, 1634–1652. [CrossRef]
4. Wei, Z.; Cheng, X.; Li, J.; Wang, G.; Mao, J.; Zhao, J.; Lou, X. Ultrasensitive evanescent wave optical fiber aptasensor for online, continuous, type-specific detection of sulfonamides in environmental water. *Anal. Chim. Acta* **2022**, *1233*, 340505. [CrossRef]
5. Li Juan, Y.; Ci Dan, Z.X.; Liao, Q.G.; Da Wen, Z.; Lin Guang, L. Pipette-tip solid-phase extraction using cetyltrimethylammonium bromide enhanced molybdenum disulfide nanosheets as an efficient adsorbent for the extraction of sulfonamides in environmental water samples. *J. Sep. Sci.* **2020**, *43*, 905–911. [CrossRef]
6. Feng, Y.; Hu, X.; Zhao, F.; Zeng, B. Fe_3O_4/reduced graphene oxide-carbon nanotubes composite for the magnetic solid-phase extraction and HPLC determination of sulfonamides in milk. *J. Sep. Sci.* **2019**, *42*, 1058–1066. [CrossRef]
7. Tian, H.; Li, R.; Guo, F.; Chen, X. An Efficient Method for the Preparation of Sulfonamides from Sodium Sulfinates and Amines. *ChemistryOpen* **2022**, *11*, e202200097. [CrossRef]
8. Song, Y.; Wu, L.; Lu, C.; Li, N.; Hu, M.; Wang, Z. Microwave-assisted liquid-liquid microextraction based on solidification of ionic liquid for the determination of sulfonamides in environmental water samples. *J. Sep. Sci.* **2014**, *37*, 3533–3538. [CrossRef]
9. Georgiadis, D.E.; Tsalbouris, A.; Kabir, A.; Furton, K.G.; Samanidou, V. Novel capsule phase microextraction in combination with high performance liquid chromatography with diode array detection for rapid monitoring of sulfonamide drugs in milk. *J. Sep. Sci.* **2019**, *42*, 1440–1450. [CrossRef]
10. Yang, L.; Shi, Y.; Li, J.; Luan, T. In situ derivatization and hollow-fiber liquid-phase microextraction to determine sulfonamides in water using UHPLC with fluorescence detection. *J. Sep. Sci.* **2018**, *41*, 1651–1662. [CrossRef]
11. Li, T.; Shi, Z.G.; Zheng, M.M.; Feng, Y.Q. Multiresidue determination of sulfonamides in chicken meat by polymer monolith microextraction and capillary zone electrophoresis with field-amplified sample stacking. *J. Chromatogr. A* **2008**, *1205*, 163–170. [CrossRef] [PubMed]
12. Mala, Z.; Gebauer, P.; Bocek, P. New methodology for capillary electrophoresis with ESI-MS detection: Electrophoretic focusing on inverse electromigration dispersion gradient. High-sensitivity analysis of sulfonamides in waters. *Anal. Chim. Acta* **2016**, *935*, 249–257. [CrossRef]
13. Chen, R.; Yang, Y.; Wang, N.; Hao, L.; Li, L.; Guo, X.; Zhang, J.; Hu, Y.; Shen, W. Application of packed porous nanofibers-solid-phase extraction for the detection of sulfonamide residues from environmental water samples by ultra high performance liquid chromatography with mass spectrometry. *J. Sep. Sci.* **2015**, *38*, 749–756. [CrossRef] [PubMed]
14. Kim, S.C.; Carlson, K. Quantification of human and veterinary antibiotics in water and sediment using SPE/LC/MS/MS. *Anal. Bioanal. Chem.* **2007**, *387*, 1301–1315. [CrossRef] [PubMed]

15. Chico, J.; Rubies, A.; Centrich, F.; Companyo, R.; Prat, M.D.; Granados, M. High-throughput multiclass method for antibiotic residue analysis by liquid chromatography-tandem mass spectrometry. *J. Chromatogr. A* **2008**, *1213*, 189–199. [CrossRef]
16. da Silva, M.R.; Mauro Lancas, F. Evaluation of ionic liquids supported on silica as a sorbent for fully automated online solid-phase extraction with LC-MS determination of sulfonamides in bovine milk samples. *J. Sep. Sci.* **2018**, *41*, 2237–2244. [CrossRef]
17. Osinski, Z.; Patyra, E.; Kwiatek, K. HPLC-FLD-Based Method for the Detection of Sulfonamides in Organic Fertilizers Collected from Poland. *Molecules* **2022**, *27*, 2031. [CrossRef]
18. Fikarova, K.; Horstkotte, B.; Machian, D.; Sklenarova, H.; Solich, P. Lab-In-Syringe for automated double-stage sample preparation by coupling salting out liquid-liquid extraction with online solid-phase extraction and liquid chromatographic separation for sulfonamide antibiotics from urine. *Talanta* **2021**, *221*, 121427. [CrossRef]
19. Wang, Y.; Li, M.; Zhu, L.; Wang, Y. On-line preconcentration and determination of sulfadiazine in food samples using surface molecularly imprinted polymer coating by capillary electrophoresis. *J. Chromatogr. A* **2023**, *1696*, 463965. [CrossRef]
20. Niu, Y.; Yang, T.; Ma, S.; Peng, F.; Yi, M.; Wan, M.; Mao, C.; Shen, J. Label-free immunosensor based on hyperbranched polyester for specific detection of alpha-fetoprotein. *Biosens. Bioelectron.* **2017**, *92*, 1–7. [CrossRef]
21. Zhang, M.; Liu, H.; Han, Y.; Bai, L.; Yan, H. On-line enrichment and determination of aristolochic acid in medicinal plants using a MOF-based composite monolith as adsorbent. *J. Chromatogr. B* **2020**, *1159*, 122343. [CrossRef] [PubMed]
22. Zhang, Y.; Wei, K.; Wang, L.; Gao, G. A membrane solid-phase extraction method based on MIL-53-mixed-matrix membrane for the determination of estrogens and parabens: Polyvinylidene difluoride membrane versus polystyrene-block-polybutadiene membrane. *Biomed. Chromatogr.* **2022**, *36*, e5454. [CrossRef] [PubMed]
23. Ding, J.; Gao, Q.; Li, X.S.; Huang, W.; Shi, Z.G.; Feng, Y.Q. Magnetic solid-phase extraction based on magnetic carbon nanotube for the determination of estrogens in milk. *J. Sep. Sci.* **2011**, *34*, 2498–2504. [CrossRef]
24. Moreno-Marenco, A.R.; Giraldo, L.; Moreno-Pirajan, J.C. Parabens Adsorption onto Activated Carbon: Relation with Chemical and Structural Properties. *Molecules* **2019**, *24*, 4313. [CrossRef] [PubMed]
25. Chen, Z.; Yu, C.; Xi, J.; Tang, S.; Bao, T.; Zhang, J. A hybrid material prepared by controlled growth of a covalent organic framework on amino-modified MIL-68 for pipette tip solid-phase extraction of sulfonamides prior to their determination by HPLC. *Mikrochim. Acta* **2019**, *186*, 393. [CrossRef]
26. Guo, J.; Li, T.; Wang, Q.; Zhang, N.; Cheng, Y.; Xiang, Z. Superior oxygen electrocatalysts derived from predesigned covalent organic polymers for zinc-air flow batteries. *Nanoscale* **2018**, *11*, 211–218. [CrossRef]
27. Ye, X.-L.; Huang, Y.-Q.; Tang, X.-Y.; Xu, J.; Peng, C.; Tan, Y.-Z. Two-dimensional extended π-conjugated triphenylene-core covalent organic polymer. *J. Mater. Chem. A* **2019**, *7*, 3066–3071. [CrossRef]
28. Zhou, W.; Zhao, D.; Wu, Q.; Fan, B.; Dan, J.; Han, A.; Ma, L.; Zhang, X.; Li, L. Amorphous CoP nanoparticle composites with nitrogen-doped hollow carbon nanospheres for synergetic anchoring and catalytic conversion of polysulfides in Li-S batteries. *J. Colloid Interface Sci.* **2021**, *603*, 1–10. [CrossRef]
29. Yang, B.; Han, Q.; Han, L.; Leng, Y.; O'Carroll, T.; Yang, X.; Wu, G.; Xiang, Z. Porous Covalent Organic Polymer Coordinated Single Co Site Nanofibers for Efficient Oxygen-Reduction Cathodes in Polymer Electrolyte Fuel Cells. *Adv. Mater.* **2023**, *35*, e2208661. [CrossRef]
30. Huang, Y.; Hao, X.; Ma, S.; Wang, R.; Wang, Y. Covalent organic framework-based porous materials for harmful gas purification. *Chemosphere* **2022**, *291*, 132795. [CrossRef]
31. Wang, Z.; Zhang, Y.; Chang, G.; Li, J.; Yang, X.; Zhang, S.; Zang, X.; Wang, C.; Wang, Z. Triazine-based covalent organic polymer: A promising coating for solid-phase microextraction. *J. Sep. Sci.* **2021**, *44*, 3608–3617. [CrossRef] [PubMed]
32. Kaur, G.; Kumar, D.; Sundarrajan, S.; Ramakrishna, S.; Kumar, P. Recent Trends in the Design, Synthesis and Biomedical Applications of Covalent Organic Frameworks. *Polymers* **2022**, *15*, 139. [CrossRef] [PubMed]
33. Xin, J.; Wang, X.; Li, N.; Liu, L.; Lian, Y.; Wang, M.; Zhao, R.S. Recent applications of covalent organic frameworks and their multifunctional composites for food contaminant analysis. *Food Chem.* **2020**, *330*, 127255. [CrossRef] [PubMed]
34. Ma, S.; He, Z.; Teng, Q.; Wang, R. One-pot synthesis of magnetic sulfonated covalent organic polymer for extraction of protoberberine alkaloids in herbs and human plasma. *J. Sep. Sci.* **2023**, *46*, e2200613. [CrossRef]
35. Wei, J.; Chen, L.; Zhang, R.; Yu, Y.; Ji, W.; Hou, Z.; Chen, Y.; Zhang, Z. An Imine-Based Porous 3D Covalent Organic Polymer as a New Sorbent for the Solid-Phase Extraction of Amphenicols from Water Sample. *Molecules* **2023**, *28*, 3301. [CrossRef] [PubMed]
36. Zhang, C.; Li, G.; Zhang, Z. A hydrazone covalent organic polymer based micro-solid phase extraction for online analysis of trace Sudan dyes in food samples. *J. Chromatogr. A* **2015**, *1419*, 1–9. [CrossRef]
37. Xiang, Z.; Cao, D. Synthesis of luminescent covalent-organic polymers for detecting nitroaromatic explosives and small organic molecules. *Macromol. Rapid Commun.* **2012**, *33*, 1184–1190. [CrossRef]
38. Gralinski, S.R.; Roy, M.; Baldauf, L.M.; Olmstead, M.M.; Balch, A.L. Introduction of a $(Ph_3P)_2Pt$ group into the rim of an open-cage fullerene by breaking a carbon-carbon bond. *Chem. Commun.* **2021**, *57*, 10218–10221. [CrossRef]
39. Wang, Y.; Ma, Q.; Wang, Y.; Li, Z.; Li, Z.; Yuan, Q. New insights into the structure-performance relationships of mesoporous materials in analytical science. *Chem. Soc. Rev.* **2018**, *47*, 8766–8803. [CrossRef]
40. Jian, N.; Zhao, M.; Liang, S.; Cao, J.; Wang, C.; Xu, Q.; Li, J. High-Throughput and High-Efficient Micro-solid Phase Extraction Based on Sulfonated-Polyaniline/Polyacrylonitrile Nanofiber Mats for Determination of Fluoroquinolones in Animal-Origin Foods. *J. Agric. Food Chem.* **2019**, *67*, 6892–6901. [CrossRef]

41. Xiao, Y.; Xiao, R.; Tang, J.; Zhu, Q.; Li, X.; Xiong, Y.; Wu, X. Preparation and adsorption properties of molecularly imprinted polymer via RAFT precipitation polymerization for selective removal of aristolochic acid I. *Talanta* **2017**, *162*, 415–422. [CrossRef] [PubMed]
42. Xia, L.; Dou, Y.; Gao, J.; Gao, Y.; Fan, W.; Li, G.; You, J. Adsorption behavior of a metal organic framework of University in Oslo 67 and its application to the extraction of sulfonamides in meat samples. *J. Chromatogr. A* **2020**, *1619*, 460949. [CrossRef] [PubMed]
43. Patyra, E.; Kwiatek, K. Application of Micellar Mobile Phase for Quantification of Sulfonamides in Medicated Feeds by HPLC-DAD. *Molecules* **2021**, *26*, 3791. [CrossRef] [PubMed]
44. Dowlatshah, S.; Santigosa, E.; Saraji, M.; Payan, M.R. A selective and efficient microfluidic method-based liquid phase microextraction for the determination of sulfonamides in urine samples. *J. Chromatogr. A* **2021**, *1652*, 462344. [CrossRef] [PubMed]
45. Mwankuna, C.J.; Kiros, F.; Mariki, E.E.; Mabiki, F.P.; Malebo, H.M.; Mdegela, R.H.; Styrishave, B. Optimization of HPLC-MS/MS method for determination of antimalarial adulterants in herbal products. *Anal. Sci.* **2023**, *39*, 407–416. [CrossRef] [PubMed]
46. Jullakan, S.; Bunkoed, O. A nanocomposite adsorbent of metallic copper, polypyrrole, halloysite nanotubes and magnetite nanoparticles for the extraction and enrichment of sulfonamides in milk. *J. Chromatogr. B* **2021**, *1180*, 122900. [CrossRef]
47. Ibarra, I.S.; Miranda, J.M.; Rodriguez, J.A.; Nebot, C.; Cepeda, A. Magnetic solid phase extraction followed by high-performance liquid chromatography for the determination of sulphonamides in milk samples. *Food Chem.* **2014**, *157*, 511–517. [CrossRef]
48. Huertas-Perez, J.F.; Arroyo-Manzanares, N.; Havlikova, L.; Gamiz-Gracia, L.; Solich, P.; Garcia-Campana, A.M. Method optimization and validation for the determination of eight sulfonamides in chicken muscle and eggs by modified QuEChERS and liquid chromatography with fluorescence detection. *J. Pharm. Biomed. Anal.* **2016**, *124*, 261–266. [CrossRef]
49. Li, Q.; Ji, K.; Tang, N.; Li, Y.; Gu, X.; Tang, K. Vortex-ultrasonic assisted dispersive liquid-liquid microextraction for seven sulfonamides of fish samples based on hydrophobic deep eutectic solvent and simultaneous detecting with HPLC-PDA. *Microchem. J.* **2023**, *185*, 108269. [CrossRef]
50. Guo, X.; Yin, D.; Peng, J.; Hu, X. Ionic liquid-based single-drop liquid-phase microextraction combined with high-performance liquid chromatography for the determination of sulfonamides in environmental water. *J. Sep. Sci.* **2012**, *35*, 452–458. [CrossRef]
51. Ding, J.; Liu, Y.; Huang, S.; Wang, X.; Yang, J.; Wang, L.; Xue, M.; Zhang, X.; Chen, J. In Situ Construction of a Multifunctional Quasi-Gel Layer for Long-Life Aqueous Zinc Metal Anodes. *ACS Appl. Mater. Interfaces* **2021**, *13*, 29746–29754. [CrossRef] [PubMed]

Disclaimer/Publisher's Note: The statements, opinions and data contained in all publications are solely those of the individual author(s) and contributor(s) and not of MDPI and/or the editor(s). MDPI and/or the editor(s) disclaim responsibility for any injury to people or property resulting from any ideas, methods, instructions or products referred to in the content.

Article

An Automated Micro Solid-Phase Extraction (µSPE) Liquid Chromatography-Mass Spectrometry Method for Cyclophosphamide and Iphosphamide: Biological Monitoring in Antineoplastic Drug (AD) Occupational Exposure

Stefano Dugheri [1,*], Donato Squillaci [1], Valentina Saccomando [1], Giorgio Marrubini [2], Elisabetta Bucaletti [1], Ilaria Rapi [1], Niccolò Fanfani [1], Giovanni Cappelli [1] and Nicola Mucci [1]

1 Department of Experimental and Clinical Medicine, University of Florence, 50134 Florence, Italy; donato.squillaci@unifi.it (D.S.); valentina.saccomando@edu.unifi.it (V.S.); elisabetta.bucaletti@unifi.it (E.B.); ilaria.rapi@edu.unifi.it (I.R.); niccolo.fanfani@unifi.it (N.F.); giovanni.cappelli@unifi.it (G.C.); nicola.mucci@unifi.it (N.M.)
2 Department of Drug Sciences, University of Pavia, Via Taramelli 12, 27100 Pavia, Italy; giorgio.marrubini@unipv.it
* Correspondence: stefano.dugheri@unifi.it

Abstract: Despite the considerable steps taken in the last decade in the context of antineoplastic drug (AD) handling procedures, their mutagenic effect still poses a threat to healthcare personnel actively involved in compounding and administration units. Biological monitoring procedures usually require large volumes of sample and extraction solvents, or do not provide adequate sensitivity. It is here proposed a fast and automated method to evaluate the urinary levels of cyclophosphamide and iphosphamide, composed of a miniaturized solid phase extraction (µSPE) followed by ultrahigh-performance liquid chromatography-tandem mass spectrometry (UHPLC-MS/MS) analysis. The extraction procedure, developed through design of experiments (DoE) on the ePrep One Workstation, required a total time of 9.5 min per sample, with recoveries of 77–79% and a solvent consumption lower than 1.5 mL per 1 mL of urine sample. Thanks to the UHPLC-MS/MS method, the limits of quantification (LOQ) obtained were lower than 10 pg/mL. The analytical procedure was successfully applied to 23 urine samples from compounding wards of four Italian hospitals, which resulted in contaminations between 27 and 182 pg/mL.

Keywords: antineoplastic drugs; biological monitoring; ultrahigh-performance liquid chromatography; tandem mass spectrometry; µSPEed; micro solid phase extraction

1. Introduction

Antineoplastic drugs (ADs) are a heterogeneous and widely used class of compounds with a rapidly growing market [1]. ADs are mainly used as chemotherapy in the treatment of neoplastic diseases, but they also play an important role in haematology and rheumatology and are used to treat non-cancer diseases such as multiple sclerosis, psoriasis, and systemic lupus erythematosus. Their various mechanisms of action usually involve inflicting genetic damage to cancerous cells without specific targeting, which inevitably leads to important side effects at the expense of healthy cells, both of treated patients and healthcare personnel. Despite the considerable steps taken in the last decade in the context of safety regulations relating to their handling [2,3], Ads' mutagenic effect still presents a tangible risk concerning occupational exposure, due to the possibility of dermal absorption from contaminated surfaces, which is the most common route of exposure in the hospital environment. The European Parliament and Council published the third revision of the Carcinogen and Mutagens Directive (CMD) 2004/37/EC [4] recognizing and prioritizing for the first time this important issue for healthcare workers and patients exposed to hazardous medicinal products (HMPs). In 2020, the European Commission conducted a study and consultation to further amend the CMD [5–7], which resulted in the last revision (Directive 2022/431/EU) [8] in March 2022 being adopted by national laws in all EU member states by 5 April 2024. Both Directive 2022/431/EU and the resulting guidelines, "Guidance for the safe management of hazardous medicinal products at work" [9], published in 2023, encourage the development of monitoring methods and biological surveillance for exposed health professionals. While surface contamination sampling remains an important tool to detect incorrect working procedures and increase the awareness of hospital workers [10–14], biological monitoring is still the only way to truly assess occupational exposure and identify possible pathological correlations. Nowadays, ADs can still often be found in the biological fluids of exposed healthcare personnel [15]. Numerous analytical methods can be encountered in the literature for the extraction of ADs in urine samples, mainly based on solid-phase extraction (SPE) or liquid-liquid extraction (LLE), but they usually require large volumes of samples and extraction solvents, or do not provide adequate sensitivity [16–21] due to the progressive decrease in encountered contaminations [22]. Solvents' consumption shown in previously cited studies can vary from 5 to 45 mL and, in many cases, by the use of not environmentally friendly substances such as dichloromethane and n-hexane. Thus, the development of growingly sensitive, green and fast methods of analysis via the means of innovative technologies is the only way to keep up with the decreasing levels of contamination which can be encountered after the latest safety implements, following the "as low as reasonably achievable" (ALARA) principle.

Many miniaturized solid-phase extraction techniques (μSPE) are being developed and applied to sample preparation in numerous fields of application [23,24]. This sample treatment technique aims to improve selectivity and sensitivity through sample clean-up and pre-concentration. It generally involves miniaturized cartridge-type devices that contain packed solid particles of porous chromatographic material. The mechanism of action of SPE consists in the interaction between the solid sorbent phase and the analytes contained inside a liquid sample solution that is percolated through the SPE bed. The analytes bonded to the sorbent are then recovered thanks to the affinity of a small amount of solvent. In this context, the μSPEed cartridges, patented by ePrep Pty Ltd. (Melbourne, Australia) (United States Patent US 2015/0352543 A1), represent a valid innovation. The cartridge is comparable to a short HPLC column: it presents a 3 μm sorbent particle size, offering a higher surface area (instead of the 50 μm diameter particles traditionally used in SPE) and thus a more efficient separation, along with a pressure-driven one-way check valve that allows an ultra-low dead volume connection. The valve design allows the sample to be drawn into a syringe avoiding transit through the sorbent bed and then passed through the stationary phase by simply pulling and pushing the plunger. The cartridges are reusable (depending on the sample matrix and operating procedure) after adequate rinsing and can be coupled with the digiVOL® Digital Syringe Driver or the ePrep ONE workstation, both marketed by ePrep Pty Ltd. (Melbourne, Australia), to automate the procedure.

The aim of this work was to develop a fast and automatable sample preparation method which could be coupled to ultrahigh-performance liquid chromatography-tandem mass spectrometry (UHPLC-MS/MS) analysis and offer competitive results in comparison to the existing methods, whose detection limits generally vary between 5 pg/mL and 30 ng/mL, with lower environmental impact. The selected target molecules, iphosphamide and cyclophosphamide, are two widely prescribed DNA-alkylating agents which are strongly excreted unchanged in urine and thus commonly employed as markers of exposure to ADs. The approach of Design of Experiments (DoE) was chosen to optimize the multiple parameters involved in the micro-extraction setup, allowing a reduction in the number of experiments needed for the optimization of the method and thus the amount of solvents and resources utilized. Moreover, to evaluate the environmental friendliness of the analytical method proposed, the Green Analytical Procedure Index (GAPI) tool was applied [25].

2. Results and Discussion

For the time being, biological monitoring for AD exposure presents many issues, including the need to detect low urinary levels and use different extraction methods for each class of analytes, which entails complex and time-consuming procedures. Furthermore, the complexity of the dermal route of exposure makes the correlation between surface contamination and urinary AD levels extremely difficult.

The drugs in the study, cyclophosphamide and iphosphamide, represent a convenient starting point for new method development, thanks to their feature of being found unchanged in urine samples from exposed personnel and their strong response in electrospray source (ESI) mass spectrometry. As a matter of fact, for most Ads, only the metabolized drug can be found in urine in detectable concentrations, but since these metabolites are often commercially unavailable, a correct quantification is strictly tied to expensive and specific synthetic procedures.

The developed method offers a fast and sensitive alternative to the currently used urine extraction procedures, along with comparable performances. The DoE approach is in different ways a key factor in the optimization of the mobile phase, minimizing the number of experiments and giving an intuitive response, and may be transposed in the near future for the development of extraction methods for similar analytes. The μSPEed cartridges, containing 3 μm particles, moves the solid-phase extraction (SPE) technique close to HPLC systems, both in regard to the cleanness of performances and automatability. They can be reused multiple times and coupled with the ePrep One workstation, which can eventually be directly connected to the HPLC injection port, allowing the method to be easily expanded to larger batch sizes. At the same time, the growing number of sorbent chemistries and different particle sizes might extend the application of this workflow, theoretically, to any kind of analyte in any biological fluid.

2.1. Method Development

Method development was undertaken by applying a Plackett-Burman 12-run design. From the data obtained by applying a Plackett-Burman design, it was clear that the conditions needed to perform the sample extraction, allowing the highest sensitivity for both the analytes, were mainly affected by the volume of the washing solution step (x_2), the composition of the washing solution (x_4), and the sample loading speed on the cartridge (x_5), as can be seen in Figure 1. The duration of the sample treatment, instead, depends mainly on the speed of loading and eluting steps and, thus, on the volumes utilized in these steps.

The multilinear regression model describing the sensitivity of the internal standard (IS) was not significant, and thus, it is not reported and only briefly mentioned here. None of the model's coefficients were statistically significant, meaning that the selected factors (x_1–x_9) have no impact on the IS peak area. Therefore, slight modifications to the extraction procedure do not affect the IS peak area. This observation supports the conclusion that the method is robust as regards the isolation and detection of the IS. So, this model could be neglected and the study can be focused on the one describing the sensitivity of the analytes.

The model describing the sensitivity of cyclophosphamide has been validated (confidence interval of 95%) because the experimental response (65,490 ± 10,678 counts) was not significantly different from the predicted one (59,339 ± 10,678 counts) at the test points. In particular, the cyclophosphamide peak area was greatest when (factors are listed according to the significance of their coefficients):

- the composition of the washing solution is 90:10 of H_2O/MeOH (the highest level for factor x_4);
- the volume of basic water utilized to equilibrate the μSPE cartridge is 0.2 mL (the lowest level for factor x_2);
- the sample loading speed is 25 μL/s (the highest level for factor x_5).

Figure 1. Plots of the models' coefficients' magnitude and sign obtained for each response. (**a**) refers to the model computed for the sensitivity of cyclophosphamide. (**b**) refers to the model describing the sensitivity of iphosphamide. (**c**) refers to the model for the duration of the sample extraction. LEGEND: x_1, volume of methanol utilized to condition the μSPE cartridge. x_2, volume of KOH basified water utilized to equilibrate the μSPE cartridge. x_3, volume of the washing solution. x_4, composition of the washing solution. x_5, sample loading speed. x_6, dispensing speed for the conditioning step. x_7, dispensing speed for the equilibration step. x_8, dispensing speed for the washing step. x_9, dispensing speed for the elution step. Factors x_{10} and x_{11} are fictitious (dummy) factors used to estimate the random error in the experiments. The grey area in the plots shows the magnitude of the random error estimated using the coefficients of the dummy factors. So in (**a**), factors x_1, x_3, x_6, x_7, x_8, and x_9, having coefficients smaller than that of factor x_{10}, are considered insignificant since they have effects smaller than that produced by random error. In (**b**), for the same reason, factors x_1, x_3, x_7, x_8, and x_9, are shown as not significant since they have coefficients smaller than that of dummy factor, x_{10}. In (**c**), factors x_4 and x_7 have coefficients smaller than that of dummy factor x_{11}.

The model describing the sensitivity of iphosphamide is validated (confidence interval of 95%) with an error of prediction of about 12% (experimental response of 65,490 ± 10,678 counts vs. the predicted response of 48,439 ± 10,678 counts). The coefficients' magnitude and sign are perfectly consistent with those of the cyclophosphamide model. In this case, the factors that cause the peak area to increase are (listed in order of significance):

- the composition of the washing solution is 90:10 of $H_2O/MeOH$ (the highest level for factor x_4);

- the volume of water utilized to equilibrate the μSPE cartridge is 0.2 mL (the lowest level for factor x_2);
- the sample loading speed is 25 μL/s (the highest level for factor x_5);
- the dispensing speed4 for the conditioning step is 55 μL/s (the highest level for factor x_6).

The model describing the duration of the sample extraction is not validated (confidence intervals of 95% probability are: for the experimental response, 716 ± 9 s and for the predicted response, 804 ± 9 s) but the reason for this is that the experiments at the test points provided results of high precision (RSD 1.3%). So, in our opinion, although the model is not validated, the screening results provide important information that can be used to minimize the time interval for the extraction. The duration of the extraction is the shortest when:

- the sample loading speed is 25 μL/s (the highest level for factor x_5);
- the volume of washing solution is 0.2 mL (the lowest level for factor x_3);
- the dispensing speed of the washing solution is 25 μL/s (the highest level for factor x_8);
- the volume of methanol utilized to condition the μSPE cartridge is 0.2 mL (the lowest level for factor x_1);
- the volume of water utilized to equilibrate the μSPE cartridge is 0.2 mL (the lowest level for factor x_2);
- the dispensing speed for the conditioning step is 55 μL/s (the highest level for factor x_6);
- the dispensing speed for the eluting step is 25 μL/s (the highest level for factor x_9).

Moreover, the responses for the S/N ratio of cyclophosphamide and iphosphamide were computed, and their models were validated at the confidence interval of 95%. As can be seen in Figure 2, the models are very similar and the factors that affect this ratio the most are x_3 and x_4, which should be set at the highest level in order to maximize the S/N ratio. The other factors have slight differences between the two models, but their contribution to the variation of the S/N ratio is very low.

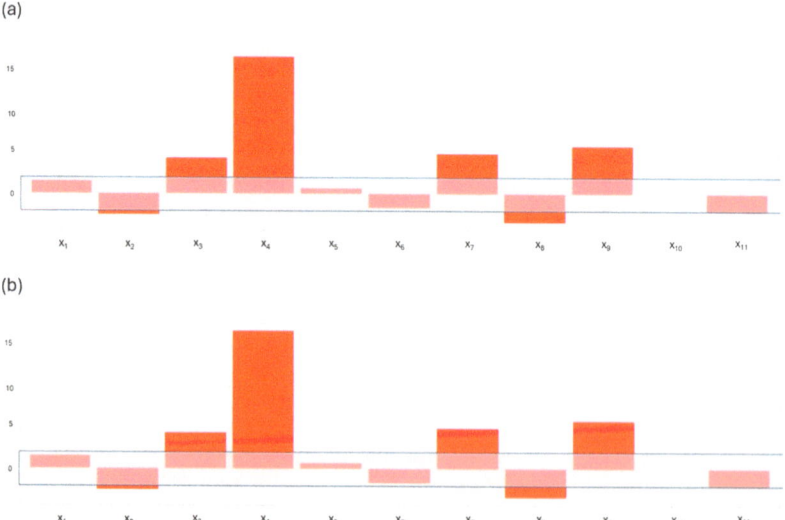

Figure 2. Plots of the coefficients computed for the S/N ratio models of cyclophosphamide (**a**) and iphosphamide (**b**). LEGEND: x_1, volume of methanol utilized to condition the μSPE cartridge. x_2, volume of basified water utilized to equilibrate the μSPE cartridge. x_3, volume of the washing solution. x_4, composition of the washing solution. x_5, sample loading speed. x_6, dispensing speed for

the conditioning step. x_7, dispensing speed for the equilibration step. x_8, dispensing speed for the washing step. x_9, dispensing speed for the elution step. Factors x_{10} and x_{11} are fictitious (dummy) factors used to estimate the random error in the experiments. The grey area shows the magnitude of the random error estimated using the dummy factor x_{10}. Factor x_1, having a coefficient smaller than that of factor x_{10}, is considered as not significant since it has an effect smaller than that produced by random error.

In conclusion, the experimental design provides valid models to describe the sensitivity of the analytes, allowing the setting up of the sample extraction by performing a few experiments.

The experimental conditions selected were: volume of methanol utilized to condition the μSPE cartridge (x_1) = 200 μL; volume of basified water utilized to equilibrate the μSPE cartridge (x_2) = 200 μL; volume of the washing solution (x_3) = 500 μL; composition of the washing solution (x_4) = 90:10 H_2O:MeOH; sample loading speed (x_5) = 25 μL/s; dispensing speed for the conditioning step (x_6) = 55 μL/s; dispensing speed for the equilibration step (x_7) = 55 μL/s; dispensing speed for the washing step (x_8) = 25 μL/s; dispensing speed for the elution step (x_9) = 10 μL/s.

2.2. Sample Preparation and μSPE

The phenyl, polystyrene divinylbenzene (PS/DVB RP), was chosen as sorbent material from among the available ones due to its resistance to the wide range of pH values which need to be tested to maximize the extraction of the analytes. The tests carried out on the pH of extraction showed a decreasing intensity for the signal of the analytes with pH lowering, which might be explained both by the presumed basic nature of the compounds, and thus the higher percentage of the unprotonated form, and by the higher quantity of precipitate that formed during the base addition, which could lead to a higher extraction capacity of the μSPE cartridge. The tests were repeated after the method development phase, and it was confirmed that the best choice for extraction pH of the urine was 11.

2.3. Performance Evaluation

The performance evaluation results, which are summarized in Table 1, show that the required sensitivity was reached, with LOQs of approximately 9 pg/mL. This method, along with the one proposed by Izzo et al. [17], presents the lowest LOQs and solvent volumes that can be encountered in the literature for the analytes, without the need for further concentration steps on the extract. The calculated precision was comprehended between 18.7% and 21.6%, while the accuracy was between 102% and 111%. An example of the obtained experimental chromatograms is reported in Figure 3. The obtained values for matrix effect and recovery are 93% and 77% for iphosphamide and 88% and 79% for cyclophosphamide, respectively.

Table 1. Performance evaluation results, expressed as the limit of detection (LOD), the limit of quantification (LOQ), precision (PR) and accuracy (ACC) of the method, extraction recovery (RE), and matrix effect (ME).

Compound	Slope	Intercept	R^2	LOD (pg/mL)	LOQ (pg/mL)	PR (%)	ACC (%)	RE (%)	ME (%)
Iphosphamide	0.019	0.055	0.996	2.87	8.6	18.7	102	77	93
Cyclophosphamide	0.020	0.094	0.995	3.12	9.4	21.6	111	79	88

As can be seen in Figure 4, the tests on μSPEed showed that, in the developed conditions, a single cartridge can be reused up to five times, with a relative standard deviation lower than 5% for both analytes, before observing a reduction in their performances.

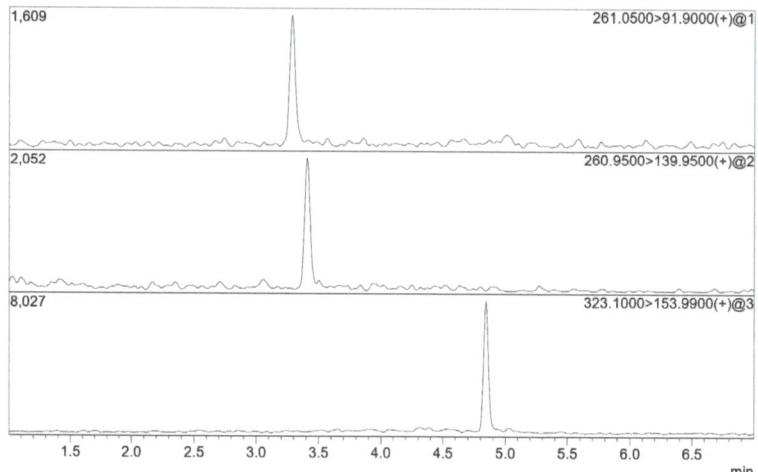

Figure 3. Stacked view of the experimental chromatograms of quantification ions of iphosphamide (**top**), cyclophosphamide (**centre**) and trophosphamide (**bottom**) obtained for the 10 pg/mL calibration level.

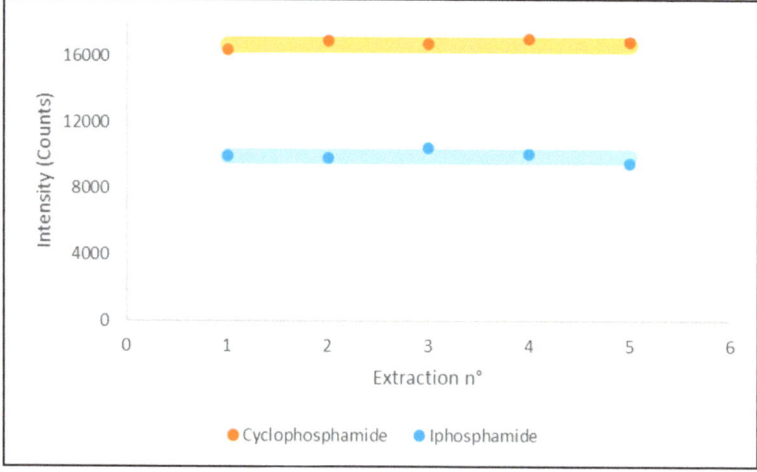

Figure 4. Measured intensities for both analytes for five consecutive extractions on a single μSPEed cartridge.

2.4. On-Field Method Application

Of the 23 samples assayed for the present study, two were positive for iphosphamide (27 and 182 pg/mL), while one was positive for cyclophosphamide (95 pg/mL). Even if the relatively limited number of operators monitored during the present study does not permit extensive considerations, the encountered positivities confirm the need for sensitive and high-throughput biological monitoring surveillance, which will allow further clinical investigation of Ads' occupational hazard and provide the basis for the introduction of safe threshold values.

2.5. Greenness Evaluation

Nowadays, resource sustainability and environmental protection have gained great importance; thus, the evaluation of the green character of an analytical protocol must be taken into account.

Different tools for assessing the greenness of the analytical procedure are available online [26], so an evaluation of the one which best fits the purpose of the authors has been performed. The oldest one, the National Environmental Methods Index (NEMI) [27] was discarded because it works with chemicals which are reported in official lists, such as EPA TRI list, and antiblastic drugs are not included in it. Moreover, this tool does not consider the consumption of chemical and reagent, and the amount of waste generated. Another interesting approach is the Analytical Eco-Scale [28] which is based on penalty points attributable to each step of the analytical process and then subtracted from a base of 100, but the main drawback is the inability to discriminate between the macro- and microscale of method applications. An additional interesting tool to determine the sustainability of a method is the Analytical Method Greenness Score (AMGS) calculator [29], which is not an absolute measure of the method greenness in that it considers only the environmental impact of the instrumental determination of a sample, while the sample pre-treatment is not taken into account.

Last, being the most complete tool to the authors' knowledge, the Green Analytical Procedure Index (GAPI) was utilized [25] to highlight that the microextraction procedure developed in this work in addition to being safer for the operators and faster in the sample preparation, also complies with the green chemistry principles. Figure 5 shows the pictograms of the two methods, and it is quite clear how the procedure involving microextraction has a greener character.

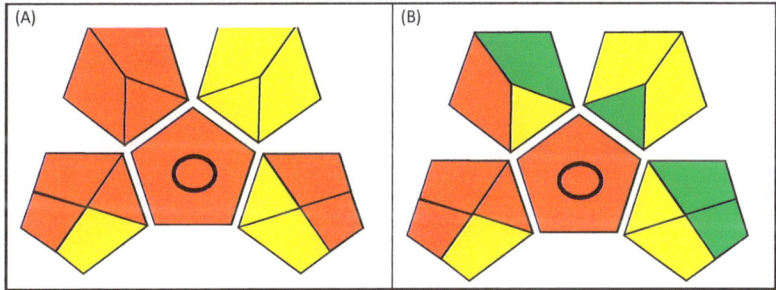

Figure 5. (**A**) reports the GAPI pictogram of the method in use in the authors' laboratory, while (**B**) is one of the proposed microextraction procedures [performed with the software available at: https://mostwiedzy.pl/en/justyna-plotka-wasylka,647762-1/gapi (accessed on 27 September 2023)].

In this approach, each step of the analytical procedure is identified by a pentagram using a colour scale from red to green to indicate high or low environmental impact. As shown in Figure 5, the first pentagram at the bottom left is equal for the two methods because the sample collection and storage are the same for both procedures. While concerning the sample pre-treatment (pentagram at the upper left) and the reagent usage (pentagram at the upper right) we have the main differences, involving the microextraction approach's less amount of sample and reagents, and less sample manipulation by the operator. The pentagram at the bottom right refers to the instrumentation step which differs only in the amount of waste generated and the hazard for the operator.

3. Materials and Methods

3.1. Chemicals

Acetonitrile, water, and methanol ultrahigh-performance liquid chromatography/mass spectrometry (ULC/MS) of purity grade were purchased from Biosolve Chimie SARL (Dieuze, France). Formic acid LC/MS grade was purchased from Carlo Erba reagents (Milan, Italy). Cyclophosphamide monohydrate, iphosphamide, ammonium formate, and potassium hydroxide 1 M solution were gradient-grade HPLC reagents or better, purchased from Merck KgaA (Darmstadt, Germany). Trophosphamide (purity > 95%), used as internal standard (IS), was purchased from CliniSciences (Guidonia Montecelio (RM), Italy).

3.2. Instruments

The LC system consists of a Shimadzu Nexera X2, equipped with a DGU-20A5R degasser unit, two LC-30AD pumps, SIL-30AC autosampler, CBM-20A system controller, and CTO-20AC column oven, coupled through ESI with a Shimadzu LCMS 8050 triple quadrupole (Shimadzu Corp., Kyoto, Japan).

The software, LabSolution® ver. 5.97 (Shimadzu Corp., Kyoto, Japan), was used to perform instrument control and data acquisition.

Sample extraction was performed using an ePrep® ONE workstation, equipped with a 500 µL ePrep eZy-Connect (µSPEed®) Syringe (P.N. 01-09083) and PS/DVB RP, 3 µm/300Å µSPEed® Cartridges (P.N. 01-10151). The software used to operate the workstation was ePrep AXIS Software (ver. 1.24.19).

3.3. Chromatographic Conditions and Mass Spectrometry Parameters

The UHPLC mobile phase consisted of 4mM ammonium formate 0.021% formic acid water solution (A) and an acetonitrile:methanol 90:10 (v/v) mixture with 0.021% formic acid addition (B). The elution was carried out at a constant flow rate of 0.55 mL/min, applying a 7 min linear gradient from 10 to 85% of B.

The chromatographic column was a Cortecs® UPLC T3, 500 × 2.1 mm (Waters Corporation, Milford, MA, USA), packed with material made of core-shell particles of 1.6 µm diameter. The total analysis time was 13.2 min.

The settings of the ESI source, operating in positive ion mode, were the following: interface voltage 4 kV, nebulizing gas flow 3 L/min, heating gas flow 10 L/min, interface temperature 400 °C, desolvation temperature 650 °C, desolvation line temperature 300 °C, heat block temperature 500 °C, and drying gas flow 10 L/min. The tandem mass spectrometry acquisition was set to multiple reaction monitoring (MRM) with a dwell time of 63 msec. The following fragmentations were selected as quantifier and reference transitions, respectively: for cyclophosphamide 260.95 > 139.95 [−22 V], 260.95 > 105.9 [−21 V]; for iphosphamide 261.05 > 91.9 [−23 V], 261.05 > 153.9 [−22 V]; for trophosphamide 323.1 > 153.99 [−24 V], 323.1 > 106.1 [−21 V].

3.4. Standard Solutions and Calibration Levels

Stock solutions of iphosphamide, cyclophosphamide, and trophosphamide were prepared at 1 mg/mL using a mixture of H_2O/MeOH 50:50 (v/v) and were stored at −20 °C. A mix solution containing 1 µg/mL of iphosphamide and cyclophosphamide was prepared in H_2O from the stock solutions, then diluted with water at 5 ng/mL to obtain the STD work solution. A trophosphamide solution was prepared at 1 µg/mL in H_2O from the stock solutions, then diluted with water to 15 ng/mL to obtain the IS work solution.

Starting from stock solutions, a working solution containing 5 ng/mL of iphosphamide and cyclophosphamide in H_2O and a 15 ng/mL of SI working solution in H_2O were prepared.

A six-level calibration curve was prepared by adding 5 µL of IS work solution, the appropriate volume of STD work solution, and a blank urine pool to reach a final volume of 1.5 mL. The analyte concentrations in the calibration solutions were: 0, 5, 10, 20, 35, and 50 pg/mL. An internal quality control solution (CQI) was prepared at 15 pg/mL of analytes and 50 pg/mL of IS.

3.5. Sample Preparation

The samples analyzed for method development were prepared using the following procedure: a 300 mL blank urine pool was collected from workers who were not exposed to antineoplastic drugs; it was thus divided into two portions and spiked with analytes and IS solutions to obtain a 200 mL part containing 100 pg/mL of cyclophosphamide, iphosphamide, IS, and a 100 mL part containing only 100 pg/mL of IS; the two portions were divided in 5 mL aliquots, stored at −80 °C, and analyzed through LC-MS/MS according to the DoE experimental plan.

Due to a lack of uniformity for the analytes' pka values, which can be found in the literature [30–33] or predicted with different software programs [34,35], the pH of extraction was tested before and after the method setup by extracting spiked urine samples at three different pH values, respectively 3, 7 and 11, and monitoring the presence of the analytes in all the steps of the extraction.

A 1.5 mL volume of urine was collected in a 5 mL polypropylene tube and added to 150 µL of a 1M KOH solution and 5 µL of a 15 pg/mL IS solution, then centrifuged at 800× g for 10 min and filtered through a 0.2 µm GHP Acrodisc® syringe filter (Pall Corporation, Long Island, NY, USA) before the extraction procedure. The same procedure was also applied to pooled urine used to prepare calibration levels.

The extraction method was also tested on 23 operators involved in AD handling from four Italian hospitals by retrieving 24 h urine samples. Aliquots from the samples were collected and analyzed through the developed method.

3.6. Chemometric Evaluation

The application of a Plackett-Burman design allowed us to evaluate the effect of 9 factors on sample extraction by performing 12 experiments.

The responses selected to develop the method were four in number: the peak areas of the analytes (cyclophosphamide, iphosphamide, and internal standard) and the duration of the sample extraction. The aim of the experimental design was to maximize the peak areas, both for the analytes and the internal standard, which means maximizing their sensitivity and, at the same time, minimizing the sample extraction duration.

The factors evaluated to study the sample extraction on the ePrep ONE workstation were the volume of methanol utilized to condition the µSPE cartridge (x_1), the volume of basic water utilized to equilibrate the µSPE cartridge (x_2), the volume of the washing solution (x_3), the composition of the washing solution (x_4), the sample loading speed (x_5), the dispensing speed for the conditioning step (x_6), the dispensing speed for the equilibration step (x_7), the dispensing speed for the washing step (x_8), and the dispensing speed for the elution step (x_9). Two dummy factors were added to the experimental plan (Table 2, factors x_{10} and x_{11}) to estimate the random error; additionally, three validation points (experiments 1, 6, and 7) were randomly chosen to assess the concordance between the experimental observations and the model predictions.

The data were collected using Microsoft Excel and processed using Rstudio Version 1.2.1335 ©2009–2019 Rstudio, Inc., as GUI for R version 3.6.1 (5 July 2019) "Action of the Toes", ©2019 (Rstudio PBC, Boston, MA, USA). The R Foundation for Statistical Computing.

Table 2. Experimental plan applied in the study. The experiments are listed in the random order applied during the working session.

#	x_1 (mL)	x_2 (mL)	x_3 (mL)	x_4 (%MeOH, v/v)	x_5 (μL/s)	x_6 (μL/s)	x_7 (μL/s)	x_8 (μL/s)	x_9 (μL/s)	x_{10} (Dummy)	x_{11} (Dummy)
1	0.35	0.35	0.35	5	15	55	55	15	15	0	0
2	0.5	0.2	0.2	0	25	15	55	25	5	1	1
3	0.5	0.2	0.5	10	5	55	55	25	5	−1	−1
4	0.2	0.2	0.5	0	25	55	15	25	25	1	−1
5	0.2	0.5	0.2	10	25	15	55	25	25	−1	−1
6	0.35	0.35	0.35	5	15	55	55	15	15	0	0
7	0.35	0.35	0.35	5	15	55	55	15	15	0	0
8	0.5	0.5	0.2	10	25	55	15	5	5	1	−1
9	0.2	0.2	0.2	0	5	15	15	5	5	−1	−1
10	0.2	0.2	0.2	10	5	55	55	5	25	1	1
11	0.5	0.2	0.5	10	25	15	15	5	25	−1	1
12	0.2	0.5	0.5	0	25	55	55	5	5	−1	1
13	0.2	0.5	0.5	10	5	15	15	25	5	1	1
14	0.5	0.5	0.2	0	5	55	15	25	25	−1	1
15	0.5	0.5	0.5	0	5	15	55	5	25	1	−1

LEGEND: #, experiment number. x_1, volume of methanol utilized to condition the μSPE cartridge. x_2, volume of basic water utilized to equilibrate the μSPE cartridge. x_3, volume of the washing solution. x_4, composition of the washing solution. x_5, sample loading speed. x_6, dispensing speed for the conditioning step. x_7, dispensing speed for the equilibration step. x_8, dispensing speed for the washing step. x_9, dispensing speed for the elution step. x_{10}, x_{11} dummy factors.

3.7. Performance Evaluation

To evaluate the performances of the method, three sets of calibration samples and CQI were freshly prepared and analyzed every day for six days to control the interday precision, whereas for the intraday precision, six sets were prepared and analyzed in a single day. Calibration curves were obtained by plotting the peak area ratios (PAR) between the analyte and IS quantitation ions versus the nominal concentration of the calibration solution. A linear regression analysis was applied to obtain the best-fitting calibration curve. The limits of detection and quantitation (LOD and LOQ) were calculated according to ICH guidelines using the approach based on the standard deviation of the intercept and slope of the regression [36].

The method precision was evaluated through the relative standard deviation (RSD%) of the replicate analysis of CQIs. The accuracy was determined by the recovery ratio percentage (RE%) computed between the determined and theoretical amounts of the replicate analysis of the CQIs.

Three sets of four replicate samples were prepared to evaluate each analyte's matrix effect (ME) and RE%. Set1 was prepared by adding 30 µL of STD work solution to a 1.5 mL blank urine sample to obtain a concentration of 100 pg/mL and then proceeding to extraction; Set2 was prepared by extracting blank pool urine samples and adding 5 µL of ME mix solution to 195 µL of extract; Set3 was prepared by diluting 5 µL of ME mix solution with 0.1% formic acid MeOH:H$_2$O 90:10 solution up to a concentration of 500 pg/mL [37].

ME and RE% figures were calculated for each analyte by comparing the mean results of Set1, Set2, and Set3 according to the formula:

$$\mathrm{ME}(\%) = \frac{\mathrm{Set2}}{\mathrm{Set3}} \times 100 \qquad (1)$$

$$\mathrm{RE}(\%) = \frac{\mathrm{Set1}}{\mathrm{Set2}} \times 100 \qquad (2)$$

3.8. µSPE

The solutions used for µSPE steps were: 250 mL of a KOH 0.01 M, 250 mL of a 1% formic acid MeOH:H$_2$O 90:10, 100 mL of a KOH 0.01 M H$_2$O:MeOH 90:10, and 100 mL of a KOH 0.01 M H$_2$O:MeOH 95:5.

The extraction method consisted of the following steps: 200 µL methanol conditioning (dispense velocity 55 µL/s), 200 µL of KOH 0.01 M equilibration (dispense velocity 55 µL/s), two times 500 µL of basified urine sample (sample loading velocity 25 µL/s) for a total of 1 mL of urine loaded, 500 µL of KOH 0.01 M H$_2$O:MeOH 90:10 wash (dispense velocity 25 µL/s), 200 µL of 1% formic acid MeOH:H$_2$O 90:10 elution (dispense velocity 10 µL/s), and 300 µL of methanol wash (dispense velocity 15 µL/s). All the aspiration flow rates were set at 15 µL/s. After vortex mixing, 5 µL of the eluted solution was injected for the UHPLC-MS/MS analysis. The extraction time for each sample was 9.5 min.

The ePrep ONE workstation was equipped with a 2 mL glass vial sample rack (ePrep Part N. 01-03035), Shimadzu autosampler rack adapter plate (ePrep Part N. 01-03018), 50 mL Reagent Jar Adapter Plate (ePrep Part N. 01-03085), µSPEed Cartridge Rack (ePrep Part N. 01-04160-01), Adapter Plates for common Sample 2 mL Vial Racks (ePrep Part N. 01-03051), 500 µL ePrep Syringe, and µSPEed Connection (ePrep Part N. 01-09083), as shown in Figure 6.

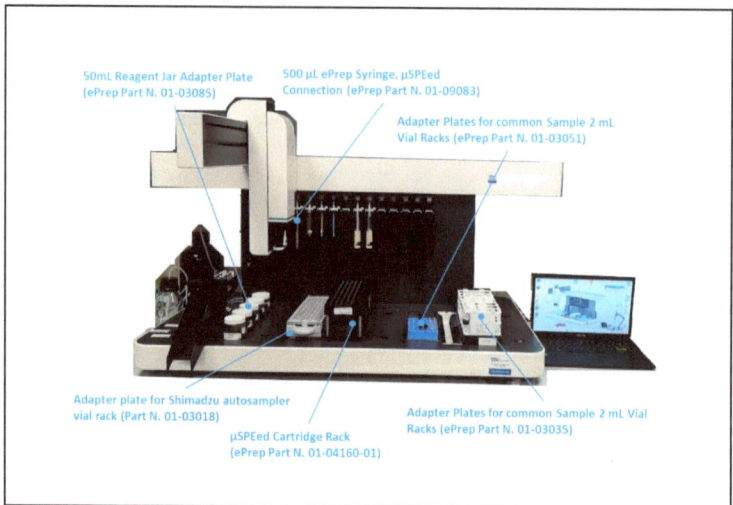

Figure 6. Configuration used for the ePrep ONE workstation.

3.9. Greenness Evaluation

The environmental impact of the analytical method was evaluated by comparing it to the procedure currently in use inside the research laboratory, which involves the use of conventional SPE cartridges [38]. The comparison was performed following the instruction of the Green Analytical Procedure Index (GAPI) tool. All the analytical steps except the sample preparation were held constant to focus the results on the extraction procedure.

4. Conclusions

The healthcare personnel of compounding and administration units are nowadays daily exposed to hazardous medicinal products such as Ads, and the use of these drugs is progressively expanding. Biological monitoring is a strong tool for occupational health, but increasingly sensitive methods are necessary to keep up with the low concentrations which can be encountered in urine samples. Over the years, many methods for the determination of the most common ADs (such as cyclophosphamide and iphosphamide) have been proposed, and yet most of them require great amounts of samples and solvents to reach the sensitivity of pg/mL, along with multiple sample preparation and concentration steps.

We here propose a µSPE automated method that, coupled with UHPLC-MS/MS analysis, can determine urine contaminations of cyclophosphamide and iphosphamide in the order of 10 pg/mL, with an automated extraction time of less than 10 min. The chemometric development strategy, which allows the reduction in the number of experiments needed for the setup to 12, might be used as a scaffold for future applications of ePrep µSPEed cartridges in the expansion of monitored substances.

The developed method shows adequate accuracy and precision in the range of the LOQ. Furthermore, the need for low volumes of solvents and possibility to use a single cartridge up to five times, makes the µSPEed extraction approach in line with the current and future developments of green chemistry.

The analytical procedure was successfully applied to urine samples from compounding wards of four Italian hospitals and could be implemented on a large scale to allow further clinical investigation of Ads' occupational hazard and provide the basis for the introduction of safe threshold values.

Author Contributions: Conceptualization, S.D. and D.S.; methodology, G.M.; software, E.B.; validation, V.S. and I.R.; formal analysis, G.C.; investigation, N.F.; resources, N.M.; data curation, D.S.; writing—original draft preparation, S.D.; writing—review and editing, D.S.; visualization, V.S.; supervision, N.M.; project administration, S.D.; funding acquisition, N.M. All authors have read and agreed to the published version of the manuscript.

Funding: This research received no external funding.

Institutional Review Board Statement: Not applicable.

Informed Consent Statement: Ethical review and approval were waived for this study, due to human biological samples are collected and analyzed according to the protocol within regular and mandatory medical examination for workers exposed to ADs and all subjects were previously informed and agreed to use the results of the analysis for scientific research purposes. Study and protocol were conducted according to the guidelines of the Declaration of Helsinki.

Data Availability Statement: Data are contained within the article.

Conflicts of Interest: The authors declare no conflicts of interest.

References

1. Global Anticancer Drugs Market to Record 6.7% CAGR through 2027. Available online: https://www.biospace.com/article/global-anticancer-drugs-market-to-record-6-7-percent-cagr-through-2027/ (accessed on 7 November 2023).
2. EUR-Lex-32022L0431-EN-EUR-Lex. Available online: https://eur-lex.europa.eu/eli/dir/2022/431 (accessed on 19 April 2023).
3. Amendments to the Carcinogens and Mutagens Directive on Hazardous Drugs and Implications for Change to the Healthcare System in Europe to Ensure Compliance with Its Requirements. Available online: https://www.europeanbiosafetynetwork.eu/wp-content/uploads/2019/03/Amendments-to-CMD3-and-implications.pdf (accessed on 20 March 2023).
4. Directive (EU) 2019/983 of the European Parliament and of the Council of 5 June 2019 Amending Directive 2004/37/EC on the Protection of Workers from the Risks Related to Exposure to Carcinogens or Mutagens at Work. Available online: https://eur-lex.europa.eu/legal-content/EN/TXT/?uri=CELEX:32019L0983 (accessed on 11 June 2023).
5. Protecting Health Workers from Hazardous Products. Available online: https://www.europeanbiosafetynetwork.eu/protecting-health-workers-from-hazardous-products/ (accessed on 23 March 2023).
6. HOSPEEM-EPSU Position in View of the European Commission Study Supporting the Assessment of Different Options Concerning the Protection of Workers from Exposure to Hazardous Medicinal Products, Including Cytotoxic Medicinal Products. Available online: https://www.epsu.org/sites/default/files/article/files/HOSPEEM-EPSU-position-Carcinogens-and-Mutagens-Directive_0.pdf (accessed on 20 March 2023).
7. Study Supporting the Assessment of Different Options Concerning the Protection of Workers from Exposure to Hazardous Medicinal Products, Including Cytotoxic Medicinal Products—Publications Office of the EU. Available online: https://op.europa.eu/en/publication-detail/-/publication/f43015ec-a24f-11eb-b85c-01aa75ed71a1/language-en (accessed on 23 March 2023).
8. Directive (EU) 2022/431 of the European Parliament and of the Council of 9 March 2022 Amending Directive 2004/37/EC on the Protection of Workers from the Risks Related to Exposure to Carcinogens or Mutagens at Work. Available online: https://eur-lex.europa.eu/eli/dir/2022/431/oj (accessed on 19 April 2023).
9. Guidance for the Safe Management of Hazardous Medicinal Products at Work | Safety and Health at Work EU-OSHA. Available online: https://osha.europa.eu/en/publications/guidance-safe-management-hazardous-medicinal-products-work (accessed on 7 November 2023).
10. Kåredal, M.; Jönsson, R.; Wetterling, M.; Björk, B.; Hedmer, M. A Quantitative LC–MS Method to Determine Surface Contamination of Antineoplastic Drugs by Wipe Sampling. *J. Occup. Environ. Hyg.* **2022**, *19*, 50–66. [CrossRef] [PubMed]
11. Sottani, C.; Grignani, E.; Cornacchia, M.; Negri, S.; della Cuna, F.S.R.; Cottica, D.; Bruzzese, D.; Severi, P.; Strocchi, D.; Verna, G.; et al. Occupational Exposure Assessment to Antineoplastic Drugs in Nine Italian Hospital Centers over a 5-Year Survey Program. *Int. J. Environ. Res. Public Health* **2022**, *19*, 8601. [CrossRef] [PubMed]
12. Arnold, S.; Jeronimo, M.; Astrakianakis, G.; Kunz, M.; Petersen, A.; Chambers, C.; Malard Johnson, D.; Zimdars, E.; Davies, H.W. Developing Wipe Sampling Strategy Guidance for Assessing Environmental Contamination of Antineoplastic Drugs. *J. Oncol. Pharm. Pract.* **2023**, *29*, 1816–1824. [CrossRef] [PubMed]
13. Dugheri, S.; Mucci, N.; Bucaletti, E.; Squillaci, D.; Cappelli, G.; Trevisani, L.; Bonari, A.; Cecchi, M.; Mini, E.; Ghiori, A.; et al. Monitoring Surface Contamination for Thirty Antineoplastic Drugs: A New Proposal for Surface Exposure Levels (SELs). *Med. Prac.* **2022**, *73*, 383–396. [CrossRef] [PubMed]
14. Mucci, N.; Dugheri, S.; Farioli, A.; Garzaro, G.; Rapisarda, V.; Campagna, M.; Bonari, A.; Arcangeli, G. Occupational Exposure to Antineoplastic Drugs in Hospital Environments: Potential Risk Associated with Contact with Cyclophosphamide-and Ifosfamide-Contaminated Surfaces. *Med. Prac.* **2020**, *71*, 519–529. [CrossRef] [PubMed]
15. Leso, V.; Sottani, C.; Santocono, C.; Russo, F.; Grignani, E.; Iavicoli, I. Exposure to Antineoplastic Drugs in Occupational Settings: A Systematic Review of Biological Monitoring Data. *Int. J. Environ. Res. Public Health* **2022**, *19*, 3737. [CrossRef] [PubMed]

16. Canal-Raffin, M.; Khennoufa, K.; Martinez, B.; Goujon, Y.; Folch, C.; Ducint, D.; Titier, K.; Brochard, P.; Verdun-Esquer, C.; Molimard, M. Highly Sensitive LC–MS/MS Methods for Urinary Biological Monitoring of Occupational Exposure to Cyclophosphamide, Ifosfamide, and Methotrexate Antineoplastic Drugs and Routine Application. *J. Chromatogr. B* **2016**, *1038*, 109–117. [CrossRef]
17. Izzo, V.; Charlier, B.; Bloise, E.; Pingeon, M.; Romano, M.; Finelli, A.; Vietri, A.; Conti, V.; Manzo, V.; Alfieri, M.; et al. A UHPLC–MS/MS-Based Method for the Simultaneous Monitoring of Eight Antiblastic Drugs in Plasma and Urine of Exposed Healthcare Workers. *J. Pharm. Biomed. Anal.* **2018**, *154*, 245–251. [CrossRef]
18. Poupeau, C.; Tanguay, C.; Plante, C.; Gagné, S.; Caron, N.; Bussières, J.-F. Pilot Study of Biological Monitoring of Four Antineoplastic Drugs among Canadian Healthcare Workers. *J. Oncol. Pharm. Pract.* **2017**, *23*, 323–332. [CrossRef]
19. Fabrizi, G.; Fioretti, M.; Mainero Rocca, L. Dispersive Solid-Phase Extraction Procedure Coupled to UPLC-ESI-MS/MS Analysis for the Simultaneous Determination of Thirteen Cytotoxic Drugs in Human Urine. *Biomed. Chromatogr.* **2016**, *30*, 1297–1308. [CrossRef]
20. Hon, C.-Y.; Teschke, K.; Shen, H.; Demers, P.A.; Venners, S. Antineoplastic Drug Contamination in the Urine of Canadian Healthcare Workers. *Int. Arch. Occup. Environ. Health* **2015**, *88*, 933–941. [CrossRef] [PubMed]
21. Sessink, P.J.; Leclercq, G.M.; Wouters, D.-M.; Halbardier, L.; Hammad, C.; Kassoul, N. Environmental Contamination, Product Contamination and Workers Exposure Using a Robotic System for Antineoplastic Drug Preparation. *J. Oncol. Pharm. Pract.* **2015**, *21*, 118–127. [CrossRef] [PubMed]
22. Koller, M.; Böhlandt, A.; Haberl, C.; Nowak, D.; Schierl, R. Environmental and Biological Monitoring on an Oncology Ward during a Complete Working Week. *Toxicol. Lett.* **2018**, *298*, 158–163. [CrossRef]
23. Dugheri, S.; Marrubini, G.; Mucci, N.; Cappelli, G.; Bonari, A.; Pompilio, I.; Trevisani, L.; Arcangeli, G. A Review of Micro-Solid-Phase Extraction Techniques and Devices Applied in Sample Pretreatment Coupled with Chromatographic Analysis. *Acta Chromatogr.* **2020**, *33*, 99–111. [CrossRef]
24. Dugheri, S.; Mucci, N.; Cappelli, G.; Trevisani, L.; Bonari, A.; Bucaletti, E.; Squillaci, D.; Arcangeli, G. Advanced Solid-Phase Microextraction Techniques and Related Automation: A Review of Commercially Available Technologies. *J. Anal. Methods Chem.* **2022**, *2022*, 8690569. [CrossRef] [PubMed]
25. Płotka-Wasylka, J. A New Tool for the Evaluation of the Analytical Procedure: Green Analytical Procedure Index. *Talanta* **2018**, *181*, 204–209. [CrossRef] [PubMed]
26. Taylor, T. The LCGC Blog: Are We Greenwashing Analytical Chemistry? *Column* **2023**, *19*, 24–28.
27. Keith, L.H.; Gron, L.U.; Young, J.L. Green Analytical Methodologies. *Chem. Rev.* **2007**, *107*, 2695–2708. [CrossRef]
28. Gałuszka, A.; Migaszewski, Z.M.; Konieczka, P.; Namieśnik, J. Analytical Eco-Scale for Assessing the Greenness of Analytical Procedures. *TrAC Trends Anal. Chem.* **2012**, *37*, 61–72. [CrossRef]
29. Hicks, M.B.; Farrell, W.; Aurigemma, C.; Lehmann, L.; Weisel, L.; Nadeau, K.; Lee, H.; Moraff, C.; Wong, M.; Huang, Y.; et al. Making the Move towards Modernized Greener Separations: Introduction of the Analytical Method Greenness Score (AMGS) Calculator. *Green Chem.* **2019**, *21*, 1816–1826. [CrossRef]
30. Sottani, C.; Rinaldi, P.; Leoni, E.; Poggi, G.; Teragni, C.; Delmonte, G.; Minoia, C. Simultaneous Determination of Cyclophosphamide, Ifosfamide, Doxorubicin, Epirubicin and Daunorubicin in Human Urine Using High-Performance Liquid Chromatography/Electrospray Ionization Tandem Mass Spectrometry: Bioanalytical Method Validation. *Rapid Commun. Mass Spectrom.* **2008**, *22*, 2645–2659. [CrossRef]
31. Cyclophosphamide. Available online: https://go.drugbank.com/drugs/DB00531 (accessed on 27 December 2023).
32. Ifosfamide. Available online: https://go.drugbank.com/drugs/DB01181 (accessed on 27 December 2023).
33. Mioduszewska, K.; Dołżonek, J.; Wyrzykowski, D.; Kubik, Ł.; Wiczling, P.; Sikorska, C.; Toński, M.; Kaczyński, Z.; Stepnowski, P.; Białk-Bielińska, A. Overview of Experimental and Computational Methods for the Determination of the PKa Values of 5-Fluorouracil, Cyclophosphamide, Ifosfamide, Imatinib and Methotrexate. *TrAC Trends Anal. Chem.* **2017**, *97*, 283–296. [CrossRef]
34. MolGpKa. Available online: https://xundrug.cn/molgpka (accessed on 27 December 2023).
35. ARChem: Automated Reasoning in Chemistry. Available online: http://www.archemcalc.com/sparc.html (accessed on 27 December 2023).
36. Abraham, J. International Conference on Harmonisation of Technical Requirements for Registration of Pharmaceuticals for Human Use. In *Handbook of Transnational Economic Governance Regimes*; Tietje, C., Brouder, A., Eds.; Brill | Nijhoff: Leiden, The Netherlands, 2010; pp. 1041–1053. ISBN 978-90-04-18156-4.
37. Matuszewski, B.K.; Constanzer, M.L.; Chavez-Eng, C.M. Strategies for the Assessment of Matrix Effect in Quantitative Bioanalytical Methods Based on HPLC–MS/MS. *Anal. Chem.* **2003**, *75*, 3019–3030. [CrossRef] [PubMed]
38. Dugheri, S.; Bonari, A.; Pompilio, I.; Boccalon, P.; Tognoni, D.; Cecchi, M.; Ughi, M.; Mucci, N.; Arcangeli, G. Analytical Strategies for Assessing Occupational Exposure to Antineoplastic Drugs in Healthcare Workplaces. *Med. Pr. Work. Health Saf.* **2018**, *69*, 589–604. [CrossRef] [PubMed]

Disclaimer/Publisher's Note: The statements, opinions and data contained in all publications are solely those of the individual author(s) and contributor(s) and not of MDPI and/or the editor(s). MDPI and/or the editor(s) disclaim responsibility for any injury to people or property resulting from any ideas, methods, instructions or products referred to in the content.

Article

Simultaneous Determination of Tobacco Smoke Exposure and Stress Biomarkers in Saliva Using In-Tube SPME and LC-MS/MS for the Analysis of the Association between Passive Smoking and Stress

Hiroyuki Kataoka *, Saori Miyata and Kentaro Ehara

School of Pharmacy, Shujitsu University, Okayama 703-8516, Japan
* Correspondence: hkataoka@shujitsu.ac.jp

Abstract: Passive smoking from environmental tobacco smoke not only increases the risk of lung cancer and cardiovascular disease but may also be a stressor triggering neuropsychiatric and other disorders. To prevent these diseases, understanding the relationship between passive smoking and stress is vital. In this study, we developed a simple and sensitive method to simultaneously measure nicotine (Nic) and cotinine (Cot) as tobacco smoke exposure biomarkers, and cortisol (CRT), serotonin (5-HT), melatonin (MEL), dopamine (DA), and oxytocin (OXT) as stress-related biomarkers. These were extracted and concentrated from saliva by in-tube solid-phase microextraction (IT-SPME) using a Supel-Q PLOT capillary as the extraction device, then separated and detected within 6 min by liquid chromatography–tandem mass spectrometry (LC–MS/MS) using a Kinetex Biphenyl column (Phenomenex Inc., Torrance, CA, USA). Limits of detection (S/N = 3) for Nic, Cot, CRT, 5-HT, MEL, DA, and OXT were 0.22, 0.12, 0.78, 0.39, 0.45, 1.4, and 3.7 pg mL^{-1}, respectively, with linearity of calibration curves in the range of 0.01–25 ng mL^{-1} using stable isotope-labeled internal standards. Intra- and inter-day reproducibilities were under 7.9% and 14.6% (n = 5) relative standard deviations, and compound recoveries in spiked saliva samples ranged from 82.1 to 106.6%. In thirty nonsmokers, Nic contents positively correlated with CRT contents (R^2 = 0.5264, n = 30), while no significant correlation was found with other biomarkers. The standard deviation of intervals between normal beats as the standard measure of heart rate variability analysis negatively correlated with CRT contents (R^2 = 0.5041, n = 30). After passive smoke exposure, Nic levels transiently increased, Cot and CRT levels rose over time, and 5-HT, DA, and OXT levels decreased. These results indicate tobacco smoke exposure acts as a stressor in nonsmokers.

Keywords: passive smoking; biomarkers; cortisol; in-tube solid-phase microextraction (IT-SPME); liquid chromatography–tandem mass spectrometry (LC–MS/MS)

1. Introduction

Passive smoking from environmental tobacco smoke, including secondhand smoke from cigarettes and exhaled smoke from smokers, and thirdhand smoke from indoor surfaces such as wallpaper, curtains, and clothing, is a serious public health concern [1–6]. It is a major risk factor for lung cancer, cardiovascular disease, and respiratory disease among nonsmokers. Passive smoking brings a 1.3-fold higher risk of developing lung cancer [7], and spouses of smokers have a 2-fold higher risk of developing lung cancer [8]. Additionally, the disgust caused by tobacco smoke exposure may induce stress, which is linked to various neuropsychiatric disorders [9] such as anxiety [10,11], depression [10,11], burnout [12], and stress adjustment disorder [13], as well as lifestyle-related diseases such as cardiovascular disease [14], eating disorders [15], and obesity [16].

Many studies have shown a relationship between smoking and the hypothalamus-pituitary-adrenal cortex system, a critical pathway involved in both stress response and nicotine (Nic) dependence [17]. Nic is known to induce cortisol (CRT) production, a stress biomarker [18]. However, the effects of passive smoking on stress remain unclear due to individual differences in susceptibility. Recently, we conducted a lifestyle questionnaire among nonsmokers and measured Nic and its metabolite cotinine (Cot) in hair, showing accumulation even in those unaware of tobacco smoke exposure [19]. We also found that nonsmokers exposed to tobacco smoke had increased levels of CRT and dehydroepiandrosterone sulfate (DHEAS) in saliva, indicating a stress response [20]. Therefore, it is necessary to measure tobacco smoke exposure biomarkers and stress-related biomarkers simultaneously to accurately assess the effects of passive smoking on stress.

Stress is mediated mainly by the hypothalamus-pituitary-adrenal cortex axis and the hypothalamus-sympathetic-adrenocortical system [21–24], with various endocrine hormones and neurotransmitters serving as biomarkers [24–30]. For example, endocrine steroid hormones such as CRT, testosterone (TES), DHEA, and DHAES [9,11,20,27], α-amylase [31], and chromogranin A [32] are widely used as biomarkers to estimate stress states. Mental stress fluctuates with emotions and can be acute or chronic. Dopamine (DA) causes elation, serotonin (5-HT) causes calmness, and oxytocin (OXT) causes warmth [24–26,30,33,34], making them useful biomarkers for relaxation states. Due to the low in vivo levels of these biomarkers, highly sensitive analytical equipment and reliable sample preparation methods are essential for precise measurement. In addition, while blood biomarker levels can quantify acute stress responses, the act of blood collection itself can be stressful. Urinary biomarkers may not reflect immediate stress due to delayed metabolic processes. In contrast, saliva collection is easy, non-invasive, and stress-free, with biomarkers transferring from blood to saliva in about 30 min. Although biomarker concentrations in saliva are lower than in blood [11,27,30,34], they are highly correlated with serum and plasma levels [27].

Other stress and relaxation assessment methods include psychological tools such as medical interviews and tests [24], but these are subjective. Objective evaluation methods based on physiological indicators such as heart rate, electroencephalography (EEG), fingertip pulse wave, blood pressure, skin temperature, sweating, and skin electrical activity are used [24,35–41]. Advances in sensors, software, and signal processing methods have led to miniaturized, portable devices for continuous monitoring [24,35,36,42–45]. Heart rate variability (HRV) is a noninvasive tool for assessing autonomic nervous system function and diagnosing stress [37–45]. HRV analysis using photoplethysmography [39,40,44,46] in wearable devices such as smartphones is convenient and inexpensive [45,46]. However, digital measurements can be temporary and influenced by various factors. In contrast, biochemical indicators provide objective and quantitative assessments of stress and relaxation through biomarker analysis in biological samples. Combining biochemical and physiological measurements can yield more accurate assessments of the effects of passive smoking.

Table 1 summarizes the characteristics of the main analytical methods previously reported for the determination of tobacco smoke exposure and stress biomarkers in saliva samples. Nic and Cot concentrations in saliva have been determined using immunological assay [47], gas chromatography–mass spectrometry (GC–MS/MS) [48], high-performance liquid chromatography (HPLC) [49], LC–mass spectrometry (LC–MS) [50], and LC–MS/MS [51,52]. Similarly, immunological methods [53–55], GC–MS [56], HPLC [57–59], and LC–MS/MS [60–64] are commonly used to measure stress- and relaxation-related biomarkers. Commercial immunosensor, enzyme immunoassay, and enzyme-linked immunosorbent assay (ELISA) kits can detect biomarkers with high sensitivity, but they have considerable analysis times per sample. While ELISA can process many samples at high throughput, it is relatively expensive. Immunological methods also require specific antibodies and are prone to cross-reactivity, making the simultaneous analysis of multiple biomarkers difficult. HPLC methods have low sensitivity and selectivity.

GC–MS/MS methods offer excellent selectivity but require time consuming derivatization to convert analytes to volatile compounds. Therefore, LC–MS and LC–MS/MS methods, which do not require derivatization and provide excellent sensitivity and selectivity, are suitable for high-throughput analysis of these biomarkers in biological samples. However, most methods need tedious and time-consuming off-line sample preparation, such as liquid–liquid extraction (LLE) and solid-phase extraction (SPE) to remove co-existing substances in biological samples. To date, no method has been reported for the simultaneous measurement of tobacco smoke exposure biomarkers and stress- and relaxation-related biomarkers.

We previously developed an in-tube solid-phase microextraction method (IT-SPME) [65] that uses a fused silica capillary coated on the inner surface with an adsorbent as an extraction device. This method is superior to the conventional off-line methods (Table 1) of introducing compounds to LC after pretreatment with LLE or SPE, as it is simple, labor-saving, and environmentally friendly because compounds can be efficiently extracted and concentrated without the use of organic solvents in a fully automated manner and introduced to LC on-line. This method has been applied to the highly sensitive and selective automated analysis of several compounds related to smoking [19,66], stress, and relaxation [20,67–71] by on-line coupling with LC–MS/MS. However, these methods do not simultaneously measure tobacco smoke exposure and acute stress status.

This study aims to develop a non-invasive and sensitive analytical method for the simultaneous analysis of tobacco smoke exposure biomarkers (Nic and Cot) and stress- and relaxation-related biomarkers (CRT, DA, 5-HT, MEL, and OXT) using IT-SPME LC–MS/MS methods. Additionally, the relationship between tobacco smoke exposure and stress/relaxation responses was evaluated by analyzing these salivary biomarkers in nonsmokers and by assessing HRV and autonomic balance based on fingertip pulse wave measurements.

Table 1. Main analytical methods previously reported for the determination of tobacco smoke exposure and stress biomarkers in saliva samples.

Biomarker [1]	Analytical Method [2]	Linearity Range (ng mL^{-1})	LOD (pg mL^{-1})	LOQ (pg mL^{-1})	Precision RSD (%)	Recovery (%)	Remarks	Ref.
Nic, Cot, OH-Cot	SPE GC-MS/MS	0.5–1000	500	500	1.56–9.62	89–92	Oral fluid 0.2 mL, derivatization	[48]
Nic, Cot	HPLC-UV						Saliva 5 mL	[49]
Nic, Cot, alkaloids	IT-SPME LC-MS	0.5–20	15–40	-	0.53–4.73	83–98	Saliva 0.1–0.2 mL	[50]
Nic, Cot, alkaloids	SPE LC-MS/MS	1–100	250–1000	1000	≤10	80–119	Oral fluid 0.5 mL	[51]
CRT	Immuno FET sensor	0.01–15	5		≤10	104	Saliva 0.05 mL	[55]
CRT, CRN	SPE HPLC-UV	2.0–40	36–72		2.7–7.0	88–99	Saliva 0.05 mL	[57]
CRT, CRN	IL-DDDME LC-UV/Vis	5–500	110–160	370–540	2.8–5.5	83–116	Saliva 1–1.5 mL	[58]
CRT, CRN, corticosterone	MEPS-HPLC-DAD	5–100	1500	5000	2.6–4.9	82–86	Saliva 0.4 mL	[59]
CRT, CRN, MEL	TF LC-MS/MS	0.2–10 0.001–0.1 (MEL)	-	200–1900 1.4 (MEL)	≤5	95–106	Saliva 0.05 mL	[60]
CRT, TES, DHEA	IT-SPME LC-MS	0.002–100	0.3–8.9	10–290	1.0–4.9	94–106	Saliva 0.1 mL	[67]
CRT, DHEA-S	LC-MS/MS	1.0–25.0	20–30	30–60	4.6–17.9	95–110	Saliva 0.4 mL	[61]
CRT, CRN	SPE UPLC-MS/MS	0.005–10	5	10–50	2–4	95–103	Saliva 0.2 mL	[62]
CRT, other steroids	SPE UPLC-MS/MS	0.005–5.0	-	50 (CRT)	7.2 (CRT)	-	Saliva 0.2 mL	[63]
CRT, TES, DHEA, DHEA-S	IT-SPME LC-MS	0.01–20	0.40–8.5	36–768	0.9–6.1	95–106	Saliva 0.05 mL	[20]
CRT, CRN, MEL	LLE UPLC-MS/MS	0.2–10 0.001–0.1 (MEL)	36–54 0.7 (MEL)	181–360 2.3 (MEL)	7–14	86–99	Saliva 0.25 mL	[64]
Nic, Cot, CRT, DA, 5-HT, MEL, OXT	IT-SPME LC-MS	0.01–25	0.12–3.7	4–124	1.3–7.9	82–107	Saliva 0.05 mL	This study

[1] CRN: cortisone; TES: testosterone; DHEA: dehydroepiandrosterone; DHEA-S: DHEA sulfate; [2] IL-DDDME: ionic liquid dispersive liquid–liquid microextraction; Immuno FET: immuno field-effect transistor; TF: turbulent flow; UPLC: ultra-performance LC.

2. Results and Discussion

2.1. Optimization of IT-SPME and Desorption of Biomarkers

The IT-SPME system (Figure 1) is programmed with autosampler software to manage each step from extraction, concentration, and desorption of the analytes in a capillary column, to introduction into an LC separation column. This enables on-line sequential analysis of multiple samples without using organic solvents [65]. The capillary column is conditioned by washing it with methanol and water before extracting compounds from the sample solution, which prevents carryover effects from previous sample analyses. An air gap is necessary during conditioning to prevent mixing of the mobile phase and sample solution, and to facilitate desorption of analytes from the capillary coating by the mobile phase after the extraction step.

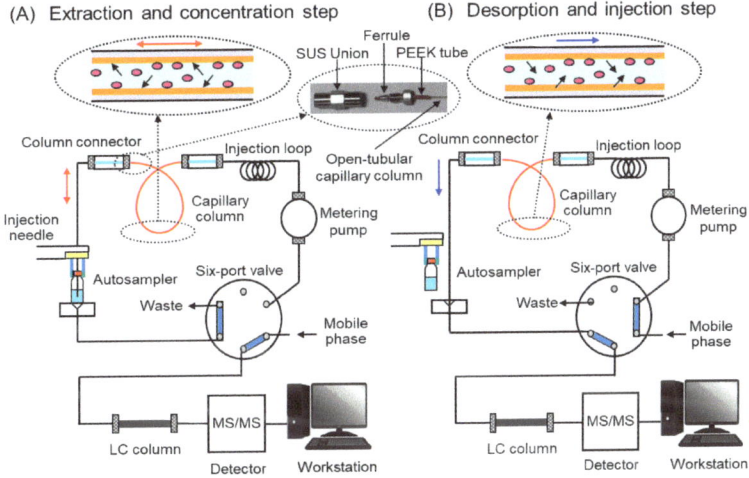

Figure 1. On-line IT-SPME LC–MS/MS system.

To develop a simultaneous analytical method for biomarkers using on-line automated IT-SPME LC–MS/MS conditions including the length and coating type of the capillary, number, flow-rate of drawings/ejections, and pH of the sample solution were optimized using a standard mixture containing 5.0 ng mL^{-1} Nic, 1.0 ng mL^{-1} Cot, 0.2 ng mL^{-1} CRT, 2.0 ng mL^{-1} 5-HT, 1.0 ng mL^{-1} MEL, 20 ng mL^{-1} DA, and 50 ng mL^{-1} OXT. The enrichment factors of each biomarker were calculated as the peak height ratio obtained by IT-SPME compared to direct injection (10 µL) of the standard mixture. Five commercially available GC capillary columns were evaluated for biomarker enrichment. OXT showed high enrichment on all capillaries, while Nic, Cot, CRT, and DA showed lower enrichment on CP-Sil 5CB, CP-Sil 19, CB, and CP-Wax 52CB, which are coated with thin film liquid phase type adsorbents. Among the capillaries tested, Supel-Q PLOT and Carboxen 1006 (Supelco Bellefonte, PA, USA), coated with porous adsorbents, showed higher enrichment due to their large adsorption surface areas and thick films (Figure 2). Since Carboxen 1006 tends to detach its coating from the capillary inner wall, Supel-Q PLOT was chosen as the best extraction device for IT-SPME in this study. Longer capillaries increase the sample load, but they also broaden the peaks and require more time for extraction. Therefore, considering the amount of analyte extracted and peak broadening, it was optimal to use a capillary with a length of 60 cm and an inner diameter of 0.32 mm (approximately 48 µL) and a load sample of 40 µL, which does not exceed this volume. Furthermore, the number of draw/eject cycles, flow rate, and sample pH influence the amounts of compounds extracted and the extraction time. As shown in Figure 3, all seven biomarkers were efficiently extracted into the Supel-Q PLOT capillary by the repeated drawing/ejecting of 40 µL samples more than 20 times. A flow rate of 0.2 mL min^{-1} was chosen as optimum; too slow a flow rate extends extraction time, while too fast a flow rate reduces extraction efficiency. Regarding the sample solution's pH, Nic and CRT can be efficiently extracted at a weakly acidic pH. Among the pH 3–8 buffers tested, acetate buffer (pH 4) was the most effective.

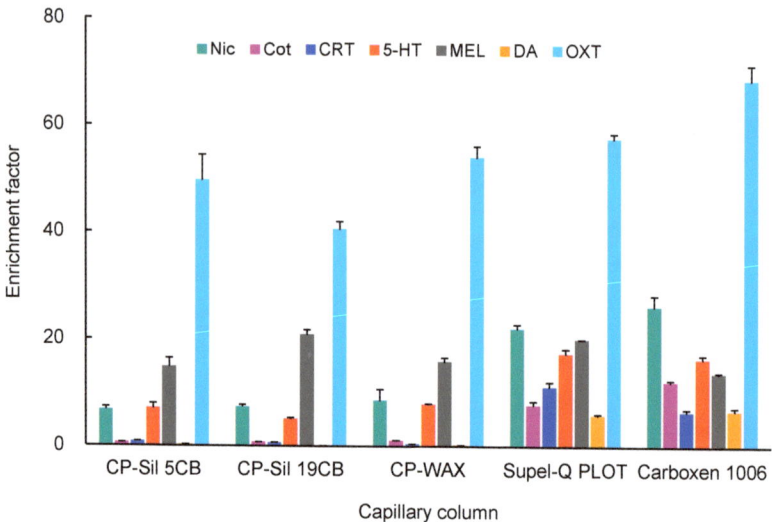

Figure 2. Effects of capillary coatings on IT-SPME of seven biomarkers. Extraction was performed by 25 draw/eject cycles of 40 µL of standard solution.

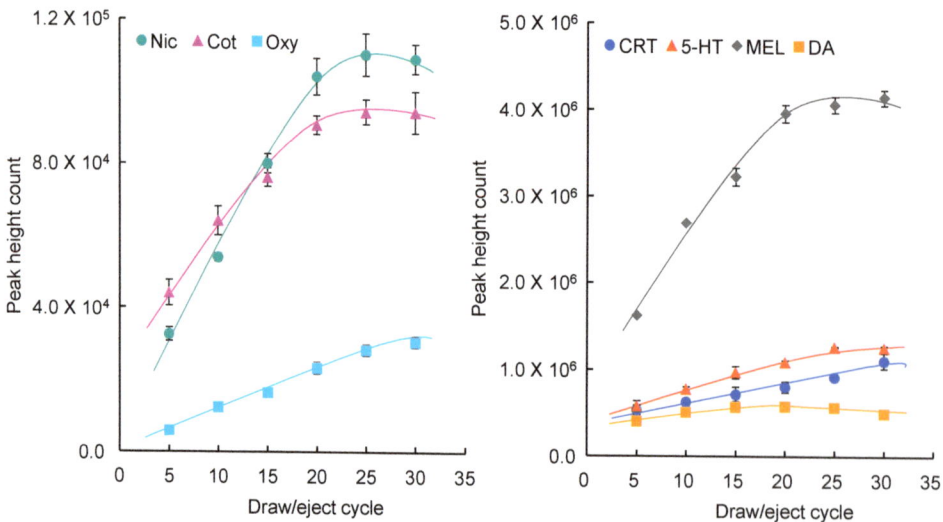

Figure 3. Effects of draw/eject cycles on IT-SPME of seven biomarkers. Extraction was performed with Supel-Q PLOT capillary by draw/eject cycles of 40 µL of standard solution.

The biomarkers extracted into the stationary phase in the capillary were almost completely desorbed and introduced directly into the LC column by switching the valve and flowing the mobile phase into the capillary by column switching (Figure 1B). The absolute amount of biomarker extracted into the capillary tube was calculated by comparing the peak area obtained under optimized IT-SPME conditions using a Supel-Q PLOT capillary with 25 drawings/ejections of 40 µL standard solution at 0.2 mL min^{-1} flow rate with that obtained by direct injection (10 µL) of the standard mixture. Absolute extraction yields were 42, 49, 45, 79, 52, 36, and 40% for Nic, Cot, CRT, 5-HT, MET, DA, and OXT, respectively, but the reproducibility and quantitation of the IT-SPME method were good due to the autosampler and internal standard.

2.2. LC–MS/MS Analysis of Biomarkers

MS/MS operating parameters for seven biomarkers and their stable isotope-labeled compounds were optimized using API 4000 tuning software 1.6.2. These compounds showed high sensitivity in ESI positive ionization mode. The protonated ion $[M + H]^+$ of each compound and the most intense fragment ion produced by cleavage of $[M + H]^+$ were selected as the precursor ion (Q1 mass) and product ion (Q3 mass), respectively. The optimized parameters and MRM transitions for each compound are listed in Table 1. The results were consistent with previously reported data [19,20,50–52,60–64,66–71].

Chromatographic conditions were optimized using LC columns that allowed separation with short retention times, considering the matrix's influence and peak shape. Among the LC columns tested, the Kinetex Biphenyl column (100 mm × 3.0 mm) showed good separation and peak shape for each biomarker. As shown in Figure 4A, the seven biomarkers and their stable isotope-labeled compounds were selectively detected as good peaks within 6 min at 0.05% formic acid solution/acetonitrile (60/40, v/v) and a flow rate of 0.2 mL min^{-1}. With the developed on-line IT-SPME LC–MS/MS system, the analysis time per sample was approximately 24 min, and overnight operation enabled automated analysis of about 60 samples per day.

2.3. Validation of the Developed IT-SPME LC–MS/MS Method

To evaluate the analytical performance of the proposed method, linearity, sensitivity, and intra- and inter-day precisions were validated. As shown in Table 2, the calibration curves constructed from the peak height ratios of biomarkers to each IS showed a linear relationship in the range of 0.01 to 25 ng mL^{-1} (triplicate analyses of each compound at six concentrations) with correlation coefficients above 0.9974 (n = 18). The LOD at signal-to-noise (S/N) ratios of 3 for each biomarker ranged 0.22–3.7 pg mL^{-1}, 21–53 times more sensitive than the direct injection method (10 µL injection). The intra- and inter-day precision (RSD, %) rates were 1.3–7.9% and 2.2–14.6%, respectively, with acceptable precision for quantitative analysis (Table 3).

Table 2. Linearity and sensitivity of the IT-SPME LC–MS/MS method for biomarkers.

Biomarker	Linearity		LOD [2] (pg mL^{-1})		LOQ [3] (pg mL^{-1})
	Range (ng mL^{-1})	CC [1]	Direct Injection	IT-SPME	IT-SPME
Nic	0.05–2.5	0.9977	7.0	0.22	7.7
Cot	0.01–0.5	0.9999	2.6	0.12	4.0
CRT	0.2–10	0.9997	23	0.78	27
5-HT	0.02–1.0	1.0000	18	0.39	13
MEL	0.01–0.5	0.9994	24	0.45	16
DA	0.2–10	0.9993	31	1.4	48
OXT	0.5–25	0.9998	138	3.7	124

[1] Correlation coefficient (n = 18). [2] Limits of detection: pg mL^{-1} sample solution (S/N = 3). [3] Limits of quantification: pg mL^{-1} saliva sample (S/N = 10).

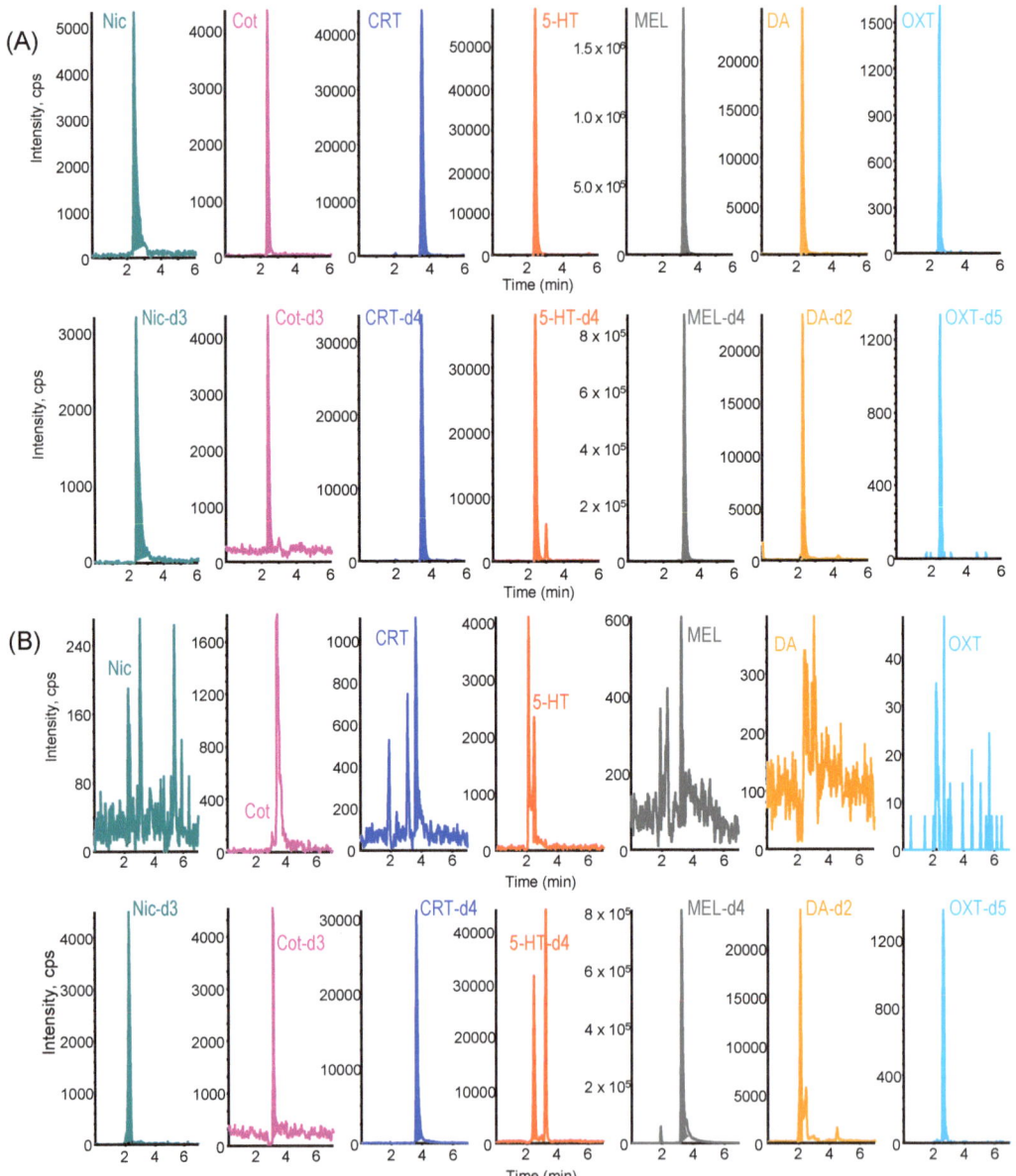

Figure 4. MRM chromatograms obtained from (**A**) a standard solution containing 5.0 ng mL^{-1} Nic, 1.0 ng mL^{-1} Cot, 0.20 ng mL^{-1} CRT, 2.0 ng mL^{-1} 5-HT, 1.0 ng mL^{-1} MEL, 20 ng mL^{-1} DA, 50 ng mL^{-1} OXT, and their stable isotope-labeled internal standard (IS) compounds and (**B**) 0.05 mL of saliva sample. IT-SPME LC-MS/MS conditions are described in the Materials and Methods section.

Table 3. Precision of the IT-SPME LC–MS/MS method for biomarkers.

Compound	Concentration (ng mL^{-1})	Precision (RSD [1] %), (n = 5)	
		Intra–Day	Inter–Day
Nic	0.25	5.7	9.5
	1	5.9	9.9
	2.5	5.8	9.4
Cot	0.05	3.4	7.3
	0.2	6.8	13.5
	0.5	3.2	5.0
CRT	1	6.1	6.7
	4	2.6	6.8
	10	1.6	3.8
5-HT	0.5	6.7	14.6
	2	3.0	5.7
	5	1.9	2.8
MEL	0.05	4.8	5.5
	0.2	3.0	2.3
	0.5	1.3	2.2
DA	1	5.0	7.8
	4	3.9	4.7
	10	3.4	2.9
OXT	2.5	7.9	8.0
	10	3.7	4.0
	25	3.6	3.7

[1] RSD, relative standard deviation.

2.4. Analysis of Saliva Samples

Saliva is a suitable biological material for noninvasive sampling, but the salivary concentrations of target biomarkers are affected by circadian rhythms [27]. In this study, saliva samples were collected with Salisoft® (Assist, Tokyo, Japan) between 14:00 and 16:00, when changes in the concentrations of these biomarkers are relatively small. Saliva samples were ultrafiltrated using Amicon Ultra® (Millipore, Tullagreen, Ireland) immediately after collection to remove macromolecular components. The ultrafiltrate can be stored stably in a −20 °C freezer if not analyzed immediately. The matrix effect of samples on the IT-SPME LC–MS/MS analysis could be corrected by adding the respective stable isotope-labeled IS, which behaves in ways similar to the target biomarker, to the saliva sample. As shown in Figure 5B, the salivary biomarkers were selectively detected, although background noise was noticeable in samples with low content. The LOQ (S/N = 10) of the seven biomarkers, except for OXT, were less than several tens of pg mL^{-1} saliva (Table 2), comparable to or superior to the sensitivities of previously reported LC–MS/MS methods (Table 1) [51,60–64]. Furthermore, the overall recoveries of biomarkers spiked to pooled saliva samples ranged from 82.1–106.6% with RSDs of 0.5–5.7% (Table 4). These results demonstrate that the proposed IT-SPME LC–MS/MS method is capable of accurately and quantitatively analyzing saliva samples.

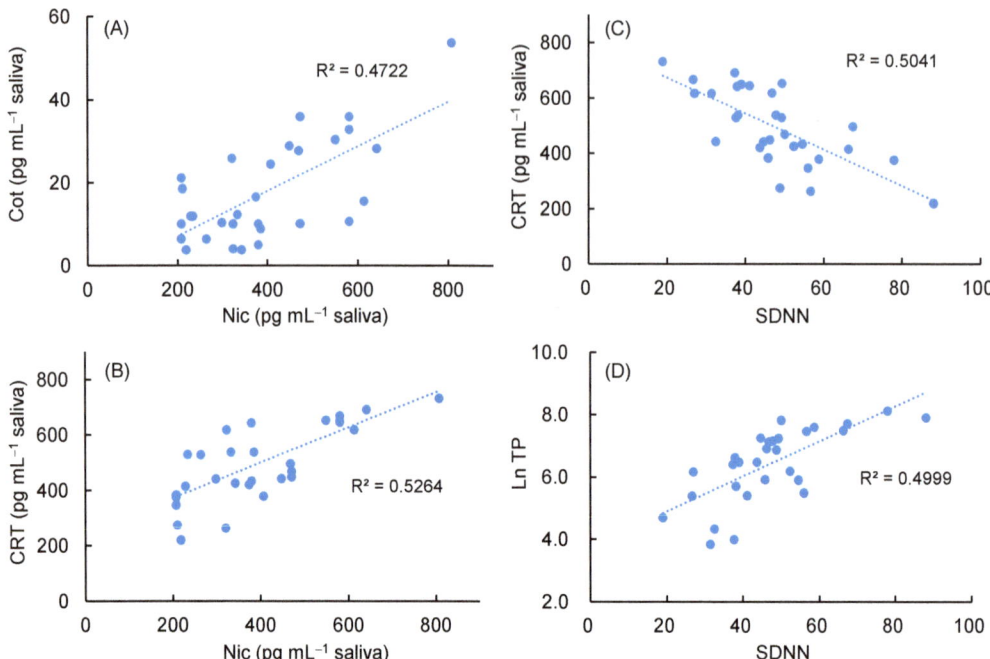

Figure 5. Correlation between salivary biomarker concentrations and HRV indicators in 30 non-smokers. (**A**) Nic concentration vs. Cot concentration, (**B**) Nic concentration vs. CRT concentration, (**C**) SDNN vs. CRT concentration, (**D**) SDNN vs. Ln TP.

Table 4. Recoveries of biomarkers spiked into saliva samples.

Compound	Spiked (ng mL^{-1} Saliva) (ng mL^{-1} Saliva)	Recovery ± SD (%), (n = 3) (n = 3)
Nic	2.5 10 25	105.4 ± 2.5 96.8 ± 0.6 98.2 ± 0.9
Cot	0.5 2 5	105.4 ± 4.0 99.1 ± 0.7 100.4 ± 0.4
CRT	10 40 100	99.1 ± 1.4 97.9 ± 4.4 101.0 ± 1.2
5-HT	5 20 50	98.2 ± 2.2 98.9 ± 1.5 99.2 ± 2.6
MEL	0.5 2 5	96.9 ± 0.5 96.2 ± 0.7 102.9 ± 5.9
DA	10 40 100	82.3 ± 2.6 96.7 ± 3.9 82.1 ± 1.2
OXT	25 100 250	93.1 ± 3.7 106.6 ± 1.1 93.8 ± 2.5

2.5. Analysis of the Relationship between Stress and Tobacco Smoke Exposure by Measurements of Salivary Biomarkers and HRV

Salivary biomarkers were measured in thirty nonsmokers using the developed method, and HRV analysis was performed simultaneously by fingertip pulse wave measurement using TAS9VIEW. As shown in Table 5, despite non-smoking participants, Nic was detected in saliva at an average of 395 pg mL^{-1}, and Cot was also detected at less than 1/20 of that level. These results indicate that nonsmokers are unconsciously exposed to environmental tobacco smoke [19,20,66]. In contrast, CRT, 5-HT, DA, and OXT were detected at levels ranging from tens to hundreds of pg mL^{-1}, while MEL, a sleep hormone, was below the LOQ because it was a daytime saliva sample. Additionally, 5-HT was detected at a maximum level of several ng mL^{-1} with a wide range of concentrations due to dietary influences. Since salivary CRT is a known stress biomarker [9,11,27] and SDNN, LF/HF ratio, and TP in heart rate variability analysis have also been reported to be indicators of stress assessment [41–43], the correlation among these parameters was analyzed. As shown in Figure 5A,B, salivary Nic concentrations showed a modest positive correlation with Cot and CRT concentrations, but no significant correlation with other biomarker concentrations. In contrast, HRV analysis showed that SDNN values in the proper range (36–106) with an average of 47, and the Ln LF/Ln HF ratio averaged 1.0, indicating a good sympathetic–parasympathetic balance. As shown in Figure 5C,D, SDNN values showed a modest negative correlation with salivary CRT content and a modest positive correlation with Ln TP. These results suggest that individuals with higher SDNN levels tend to have lower CRT levels, higher levels of autonomic activity, and lower stress states. However, since salivary biomarker contents and HRV indices represent levels at the time of measurement, and there are individual differences, it is important to measure them periodically to assess fluctuations in their levels.

Table 5. Salivary biomarker contents and HRV indicators in thirty nonsmokers.

		Content [1] (pg mL^{-1} Saliva)						HRV Indicator		
	Nic	Cot	CRT	5-HT	MEL	DA	OXT	SDNN [3]	Ln LF/Ln HF [4]	Ln TP [5]
Max.	807	54	732	3706	<LOQ [2]	224	691	88	1.7	9.1
Med.	377	12	482	164	<LOQ	44	114	47	0.9	6.6
Min.	207	4	220	23	<LOQ	<LOQ	<LOQ	19	0.7	3.8
Ave.	395	18	497	457	<LOQ	64	178	47	1.0	6.4
SD	153	12	134	792	–	46	162	15	0.2	1.2

[1] The salivary biomarker content was obtained from the average of three independent measurements for each subject. [2] LOQ, limit of quantification; [3] SDNN, standard deviation of the normal-to-normal intervals; [4] LF, low frequency band, HF, high frequency band; [5] TP, total power.

To evaluate the effect of passive smoking on stress levels, salivary biomarker measurements and HRV analysis were performed before and after tobacco smoke exposure in one nonsmoker as a pilot study. Figure 6A,B shows the increase and decrease in salivary biomarker concentrations before and after exposure as a ratio of concentration, with the pre-exposure concentration as 1. Nic concentration increased 160-fold immediately after 30 min of tobacco smoke exposure and then halved within 30 min, while Cot concentration gradually increased. In contrast, CRT concentrations increased over time, 5-HT concentrations decreased over time, but DA concentrations increased transiently immediately after exposure. On the other hand, among the HRV indices (Figure 6C), SDNN and TP decreased transiently immediately after exposure but increased thereafter. In contrast, the LF/HF ratio increased immediately after exposure and then decreased but remained sympathetically dominant. These results suggest that exposure to tobacco smoke causes a slight increase in DA concentration with an increase in Nic concentration, but this increase cannot be sustained, resulting in an increase in CRT and a decrease in 5-HT concentration, indicating a state of stress. In contrast, HRV analysis showed a transient stress state followed by recovery. This may be because the heart rate-based stress response was observed simultaneously with tobacco smoke exposure, whereas the biomarkers were secreted later after the stress stimulus [27]. The variation of these parameters due to passive smoking confirms the modest correlations between each parameter in Figure 5.

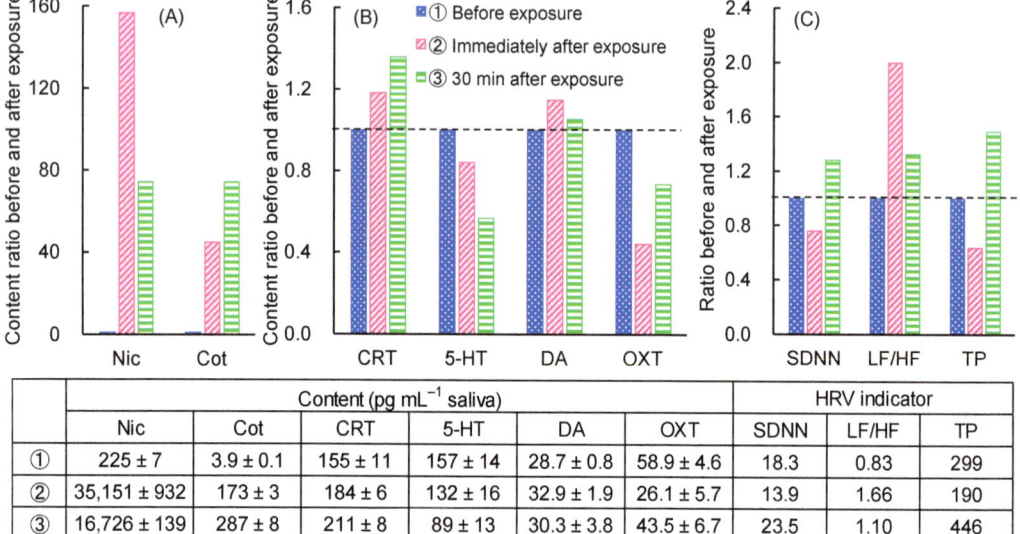

	Content (pg mL^{-1} saliva)						HRV indicator		
	Nic	Cot	CRT	5-HT	DA	OXT	SDNN	LF/HF	TP
①	225 ± 7	3.9 ± 0.1	155 ± 11	157 ± 14	28.7 ± 0.8	58.9 ± 4.6	18.3	0.83	299
②	35,151 ± 932	173 ± 3	184 ± 6	132 ± 16	32.9 ± 1.9	26.1 ± 5.7	13.9	1.66	190
③	16,726 ± 139	287 ± 8	211 ± 8	89 ± 13	30.3 ± 3.8	43.5 ± 6.7	23.5	1.10	446

Figure 6. Variation in salivary biomarker concentrations and HRV indicators before and after tobacco smoke exposure in nonsmoker. (**A**) Tobacco smoke exposure biomarkers, (**B**) Stress and relaxation biomarkers and (**C**) HRV indicators.

3. Materials and Methods

3.1. Reagents and Materials

Nic and its metabolite Cot were purchased from Sigma-Aldrich Japan (Tokyo, Japan); CRT, 5-HT, MEL, and DA from Sigma-Aldrich Japan (Tokyo, Japan); and OXT from Peptide Institute Inc. (Osaka, Japan). Internal standards (IS) used were Nic-d$_3$ (isotopic purity 98.4%), Cot-d$_3$ (isotopic purity 99.9%), CRT-d$_4$ (isotopic purity 99%), MEL-d$_4$ (isotopic purity 98.7%), DA-d$_2$ hydrochloride (isotopic purity 99.5%), and OXT-d$_5$ trifluoroacetate (isotopic purity 98.9%) from Toronto Research Chemicals Inc. (Toronto, ON, Canada), and 5-HT-d$_4$ (isotopic purity >99%) from Cayman Chemical (Ann Arbor, MI, USA). Structures of these biomarkers and their stable isotope-labeled IS compounds are shown in Figure 7.

Each standard and IS except OXT and OXT-d$_5$ were dissolved at 0.5 mg mL^{-1} in LC–MS grade methanol. OXT and OXT-d$_5$ were dissolved at 0.1 mg mL^{-1} in LC–MS grade distilled water. Stock solutions were tightly capped and stored at −20 °C. Working standard mixtures were prepared by diluting stock solutions with distilled water to the required concentrations before use and stored at 4 °C until use. LC–MS grade methanol, acetonitrile, and distilled water were purchased from Kanto Chemical (Tokyo, Japan).

GC capillary columns (60 cm × 0.32 mm i.d.) used as extraction devices for IT-SPME included CP-Sil 5CB (100% polydimethylsiloxane, 5 μm film thick), CP-Sil 19CB (14% cyanopropyl phenyl methyl silicone, 1.2 μm film thick), and CP-Wax 52CB (polyethylene glycol, 1.2 μm film thick) from Varian Inc. (Lake Forest, CA); and Supel-Q PLOT (divinylbenzene polymer, 17 μm film thick) and Carboxen 1006 PLOT (Carboxen molecularsives, 15 μm film thick) from Supelco (Bellefonte, PA, USA).

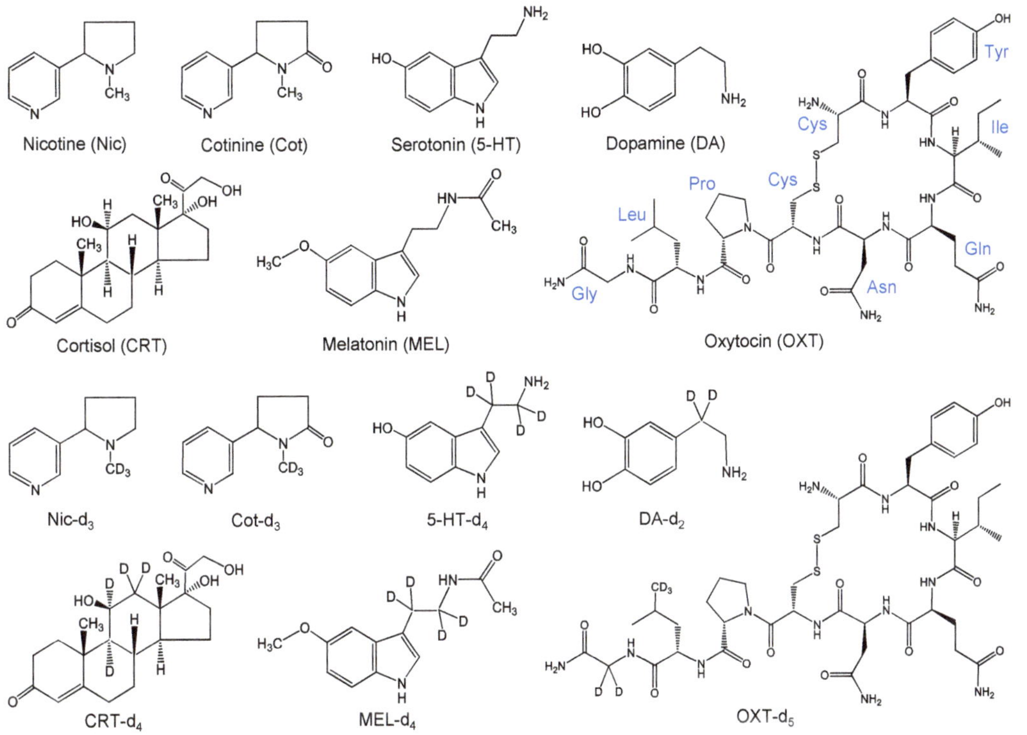

Figure 7. Structures of target biomarkers and their stable isotope-labeled internal standards.

3.2. Sampling, Preparation and Analysis of Saliva Samples

This study was approved by the ethics committee of Shujitsu University and informed consent was obtained from all participants. Saliva samples were collected from 7 healthy males and 23 healthy females aged 20–26 years. Each participant rinsed their mouth with water 30 min before saliva collection, refrained from eating or drinking, and provided approximately 1 mL of saliva in a collection tube using a Salisoft® kit (Assist, Tokyo, Japan) containing a polypropylene–polyethylene swab. Saliva was donated between 14:00 and 16:00, when circadian rhythms of target biomarkers are less variable [27]. After removal of insoluble material by centrifugation at 2500× g for 1 min, the supernatant saliva sample was transferred into polyethylene tubes with caps, stored at −20 °C if not immediately used for analysis, and thawed spontaneously just prior to analysis. To each 0.05 mL of collected saliva sample, an aliquot of IS mixture solution and distilled water were added to make a total volume of 0.5 mL, and ultrafiltered at 15,000 rpm for 20 min using an Amicon Ultra® 0.5-mL 3K regenerated cellulose 3000 molecular weight cut-off centrifugal filter device (Millipore, Tullagreen, Ireland). The filtrate (ca. 0.4 mL) was taken into a 2.0-mL autosampler vial with a septum. Then, 0.05 mL of 0.2 M acetate buffer (pH 4) was added, and the solution was made up to a total volume of 0.5 mL with distilled water. The vial was then placed in the autosampler for IT-SPME LC–MS/MS analysis. The salivary concentrations of the target biomarkers were calculated from the ratio of the peak heights of each biomarker to the IS compound using a calibration curve.

3.3. LC–MS/MS Conditions

An Agilent 1100 series LC (Agilent Technologies, Boeblingen Germany) coupled to an API 4000 triple quadrupole mass spectrometer (AB SCIEX, Foster City, CA, USA) was used for LC–MS/MS analysis. For LC separation, a Kinetex Biphenyl column (100 mm × 3.0 mm, particle size 2.6 µm; Phenomenex Inc., Torrance, CA, USA) and 0.05% formic acid/acetonitrile (60/40, v/v) as the mobile phase were used at a column temperature of 40 °C and a flow rate of 0.2 mL min^{-1}. For electrospray ionization (ESI)-MS/MS, turbo ion spray voltage and temperature were 4200 V and 600 °C, respectively. Additionally, the flow rates of the ion-source gasses GS1 and GS2 were 50 L min^{-1} and 80 L min^{-1}, respectively, and the flow rates of the curtain and collision gasses were 30 L min^{-1} and 4 L min^{-1}, respectively. Multiple reaction monitoring (MRM) transitions of target biomarkers and their stable isotope-labeled compounds in positive ion mode and other parameters are shown in Table 6. Each compound was quantified by MRM of the protonated precursor molecular ions [M + H]$^+$ (Q1) and their related product ions (Q3). LC–MS/MS data were processed using Analyst Software 1.6.2 (AB SCIEX).

Table 6. MRM transitions and setting parameters for target biomarkers and their stable isotope-labeled compounds.

Compound	Mass Transition (m/z)	DP [1] (V)	EP [2] (V)	CE [3] (V)	CXP [4] (V)
Nicotine (Nic)	163.1 → 132.1	70	10	20	10
Cotinine (Cot)	177.1 → 80.2	75	10	30	15
Cortisol (CRT)	363.0 → 120.9	70	10	30	10
Serotonin (5-HT)	177.2 → 160.2	25	5	15	3
Melatonin (MEL)	233.1 → 174.1	20	9	20	12
Dopamine (DA)	154.2 → 91.1	50	4	30	8
Oxytocin (OXT)	1008.3 → 724.5	60	9	40	12
Nic-d$_3$	166.1 → 132.1	70	10	20	10
Cot-d$_3$	180.1 → 80.2	75	10	30	15
CRT-d$_4$	367.1 → 121.4	70	10	30	10
5-HT-d$_4$	181.2 → 164.3	25	5	15	3
MEL-d$_4$	237.1 → 178.1	20	9	20	12
DA-d$_2$	156.2 → 93.1	50	4	30	8
OXT-d$_5$	1013.3 → 724.5	60	9	40	12

[1] Declustering potential (V). [2] Entrance potential (V). [3] Collision energy (V). [4] Collision cell exit potential (V).

3.4. IT-SPME Procedure and On-Line Coupling with LC–MS/MS

As shown in Figure 1, IT-SPME, using a capillary column as the extraction device, is incorporated into the on-line coupling system with LC–MS/MS. Both ends of a GC capillary (60 cm × 0.32 mm i.d. d.; 48 µL internal volume) were threaded through a 2.5 cm sleeve of a 1/16-inch polyetheretherketone tube with a 330 µm internal diameter, and a standard 1/16-inch stainless-steel nut, ferrule, and stainless-steel union connector were used to connect the injection needle and the injection loop of the autosampler. The injection loop was kept in the system to avoid contamination of the metering pump by the sample.

For extraction and concentration, a 2 mL screw cap autosampler vial with a silicone/polytetrafluoroethylene septum containing 0.5 mL of sample solution was placed in the sample tray in the autosampler. Additionally, three vials (1.5 mL methanol, 1.5 mL water, and another blank) were placed in the autosampler. Prior to sample extraction, the capillary column was washed and conditioned with two repeated drawings/ejections of 40 µL methanol, by drawing 50 µL air from the blank vial, and with two repeated drawings/ejections of 40 µL distilled water at each flow rate of 0.2 mL min^{-1} with the 6-port valve in the LOAD position (Figure 1A). The target biomarkers and the IS compounds in the sample were then extracted onto the capillary coating with 25 repeated drawings/ejections of 40 µL samples at a flow rate of 0.2 mL min^{-1} with the 6-port valve in the LOAD position (Figure 1A). After extraction, the tip of the injection needle was washed with one drawing/ejection of 2 µL methanol from another autosampler vial. For desorption and injection, the extracted compounds in the capillary column were desorbed with the mobile phase by valve switching to the INJECT position (Figure 1B), transported to the LC column, and detected by the MS/MS system. These IT-SPME steps including conditioning, extraction, desorption, and injection were fully automated by the autosampler software (Analyst 1.6.2, AB SCIEX).

3.5. Method Validation Study

The linearity, limit of detection (LOD), limit of quantification (LOQ), and precision of the developed method were evaluated. Linearities were validated by triplicate analyses of six concentrations of Nic (0.05, 0.1, 0.25, 0.5, 1.0, and 2.5 ng mL^{-1}), six concentrations each of Cot and MEL (0.01, 0.02, 0.05, 0.1, 0.2, and 0.5 ng mL^{-1}), six concentrations each of CRT and DA (0.2, 0.4, 1.0, 2.0, 4.0, and 10 ng mL^{-1}), six concentrations of 5-HT (0.02, 0.04, 0.1, 0.2, 0.4, and 1.0 ng mL^{-1}), and six concentrations of OXT (0.5, 1.0, 2.5, 5.0, 10, and 25 ng mL^{-1}) in the presence of the IS mixture containing 1.0 ng mL^{-1} Nic-d$_3$, 0.2 ng mL^{-1} Cot-d$_3$, 4.0 ng mL^{-1} CRT-d$_4$, 0.4 ng mL^{-1} 5-HT-d$_4$, 0.2 ng mL^{-1} MEL-d$_4$, 4.0 ng mL^{-1} DA-d$_2$, and 10 ng mL^{-1} OXT-d$_5$. Calibration curves were constructed from the ratio of the peak heights of each compound to the IS compound at each concentration. The LOD and LOQ were calculated from the signal-to-noise ratio (S/N) of 3 and 10, respectively. For each analyte, the intra-day and inter-day precision was verified from five analyses using low, medium, and high concentration solutions, respectively, and expressed as relative standard deviations (RSD, %).

3.6. HRV Analysis by Fingertip Pulse Wave Measurement

HRV was measured using the acceleration pulse wave measuring device Pulse Analyzer Plus View (TAS9VIEW, YKC Corp., Tokyo, Japan). This device is designed to noninvasively evaluate the state of stress by placing a sensor between the index fingers of the left hand. It extracts the pulse with high accuracy from the interval peaks of pulse wave height for 150 sec at rest and analyzes changes in the pulse rate [39,44].

In time domain analysis, the RR interval of the HRV is analyzed, and the standard deviation of the normal-to-normal intervals (SDNN) is calculated. In frequency domain analysis, the very low frequency band (VLH, 0.003–0.04 Hz), the low frequency band (LH, 0.04–0.15 Hz) based on both sympathetic and parasympathetic functions, and the high frequency band (HF, 0.15–0.4 Hz) based on parasympathetic function are analyzed. The ratio of LF to HF or the ratio of the natural logarithm of each was used as an index of stress [37,40–43].

SDNN represents the degree of activity of the autonomic nervous system, with higher values indicating better physical condition. Additionally, an LF/HF or Ln LF/Ln HF ratio close to 1 indicates a state of autonomic balance and no stress, while a ratio below 1 indicates parasympathetic dominance and a relaxed state. Furthermore, the sum of the three spectral bands VLF, LF, and HF represents the total power (TP) of the RR interval variation, and a decrease in this value indicates decreased autonomic activity. These parameter indices were measured for thirty nonsmokers who provided saliva samples.

3.7. Tobacco Smoke Exposure Assessment by Biomarker and HRV Analyses

Salivary biomarkers and HRV were analyzed in one nonsmoker before and after passive smoking to evaluate the stress response to tobacco smoke exposure. According to a previously reported method [20], the nonsmoker was exposed for 30 min by nasal breathing to secondhand smoke generated by burning a single cigarette in a plastic smoke exposure chamber (W40 × D40 × H40 cm). Saliva samples were collected before, immediately after, and 30 min after the 30-min exposure trial. The content of tobacco smoke exposure biomarkers and stress- and relaxation-related biomarkers were measured according to the method in Section 3.5. Additionally, the fingertip pulse wave was measured using TAS9VIEW simultaneously with saliva collection, as described in Section 3.6.

4. Conclusions

In this study, a novel analytical method was developed to noninvasively and simultaneously measure biomarkers for tobacco smoke exposure and stress- and relaxation-related biomarkers, clarifying the relationship between passive smoking and stress. The proposed IT-SPME LC–MS/MS method can selectively analyze seven biomarkers with high sensitivity and accuracy by simply ultrafiltrating a small amount of saliva. This method is environmentally friendly, as it does not use organic solvents and is fully automated from extraction and concentration of sample solutions to separation, detection, and data analysis, enabling unattended nighttime operations and reducing labor costs.

A unique feature of this method is its ability to objectively analyze tobacco smoke exposure levels and stress levels in a single analysis using the same sample. Additionally, more accurate analysis of stress and relaxation states is possible by measuring the balance of autonomic nervous system activity based on heart rate variability in conjunction with biomarker level measurements. These analyses indicate that tobacco smoke exposure is a stressor in nonsmokers.

The biomarker analysis method developed in this study is expected to be a useful tool not only for analyzing the effects of passive smoking on stress, but also for the early diagnosis and prevention of related health issues.

Author Contributions: Conceptualization, H.K.; methodology, H.K., S.M. and K.E.; software, H.K.; validation, H.K., S.M. and K.E.; formal analysis, H.K., S.M. and K.E.; investigation, H.K., S.M., and K.E.; resources, H.K.; data curation, H.K.; writing—original draft preparation, H.K.; writing—review and editing, H.K. and K.E.; visualization, H.K. and S.M.; supervision, H.K.; project administration, H.K.; funding acquisition, H.K. All authors have read and agreed to the published version of the manuscript.

Funding: This research was funded by a JSPS KAKENHI (Grant Number JP23K06091) and Smoking Research Foundation, 2024.

Institutional Review Board Statement: The study was conducted according to the guidelines of the Declaration of Helsinki and approved by the ethics committee of Shujitsu University (approval code 283; 2 October 2023).

Informed Consent Statement: Informed consent was obtained from all subjects involved in the study.

Data Availability Statement: The data presented in this study are available on request from the corresponding author.

Conflicts of Interest: The authors declare no conflict of interest.

References

1. Hu, Q.; Hou, H. (Eds.) *Tobacco Smoke Exposure Biomarkers*; CRC Press, Taylor & Francis Group: Boca Raton, FL, USA, 2015.
2. Talhout, R.; Schulz, T.; Florek, E.; van Benthem, J.; Wester, P.; Opperhuizen, A. Hazardous compounds in tobacco smoke. *Int. J. Environ. Res. Public Health* **2011**, *8*, 613–628. [CrossRef] [PubMed]
3. Sikorska-Jaroszyńska, M.H.; Mielnik-Błaszczak, M.; Krawczyk, D.; Nasiłowska-Barud, A.; Błaszczak, J. Passive smoking as an environmental health risk factor. *Ann. Agric. Environ. Med.* **2012**, *19*, 547–550. [PubMed]
4. Mattes, W.; Yang, X.; Orr, M.S.; Richter, P.; Mendrick, D.L. Biomarkers of tobacco smoke exposure. *Adv. Clin. Chem.* **2014**, *67*, 1–45. [CrossRef]
5. Ni, X.; Xu, N.; Wang, Q. Meta-analysis and systematic review in environmental tobacco smoke risk of female lung cancer by research type. *Int. J. Environ. Res. Public Health* **2018**, *15*, 1348. [CrossRef]
6. Torres, S.; Merino, C.; Paton, B.; Correig, X.; Ramírez, N. Biomarkers of exposure to secondhand and thirdhand tobacco smoke: Recent advances and future perspectives. *Int. J. Environ. Res. Public Health* **2018**, *15*, 2693. [CrossRef]
7. Hori, M.; Tanaka, H.; Wakai, K.; Sasazuki, S.; Katanoda, K. Secondhand smoke exposure and risk of lung cancer in Japan: A systematic review and meta-analysis of epidemiologic studies. *Jpn. J. Clin. Oncol.* **2016**, *46*, 942–951. [CrossRef] [PubMed]
8. Kurahashi, N.; Inoue, M.; Liu, Y.; Iwasaki, M.; Sasazuki, S.; Sobue, T.; Tsugane, S. Passive smoking and lung cancer in Japanese non-smoking women: A prospective study. *Int. J. Cancer* **2008**, *122*, 653–657. [CrossRef]
9. Dhama, K.; Latheef, S.K.; Dadar, M.; Samad, H.A.; Munjal, A.; Khandia, R.; Karthik, K.; Tiwari, R.; Yatoo, M.I.; Bhatt, P.; et al. Biomarkers in Stress Related Diseases/Disorders: Diagnostic, Prognostic, and Therapeutic Values. *Front. Mol. Biosci.* **2019**, *6*, 91. [CrossRef]
10. Vinkers, C.H.; Kuzminskaite, E.; Lamers, F.; Giltay, E.J.; Penninx, B.W. An integrated approach to understand biological stress system dysregulation across depressive and anxiety disorders. *J. Affect. Disord.* **2021**, *283*, 139–146. [CrossRef]
11. Chojnowska, S.; Ptaszyńska-Sarosiek, I.; Kępka, A.; Knaś, M.; Waszkiewicz, N. Salivary Biomarkers of Stress, Anxiety and Depression. *J. Clin. Med.* **2021**, *10*, 517. [CrossRef]

12. Wekenborg, M.K.; von Dawans, B.; Hill, L.K.; Thayer, J.F.; Penz, M.; Kirschbaum, C. Examining reactivity patterns in burnout and other indicators of chronic stress. *Psychoneuroendocrinology* **2019**, *106*, 195–205. [CrossRef] [PubMed]
13. Strain, J.J. The psychobiology of stress, depression, adjustment disorders and resilience. *World J. Biol. Psychiatry* **2018**, *19* (Suppl. S1), S14–S20. [CrossRef]
14. Kivimaki, M.; Steptoe, A. Effects of stress on the development and progression of cardiovascular disease. *Nat. Rev. Cardiol.* **2018**, *15*, 215–229. [CrossRef]
15. Chami, R.; Monteleone, A.M.; Treasure, J.; Monteleone, P. Stress hormones and eating disorders. *Mol. Cell. Endocrinol.* **2019**, *497*, 110349. [CrossRef]
16. Tomiyama, A.J. Stress and Obesity. *Annu. Rev. Psychol.* **2019**, *70*, 703–718. [CrossRef] [PubMed]
17. Koob, G.; Kreek, M.J. Stress, dysregulation of drug reward pathways, and the transition to drug dependence. *Am. J. Psychiatry* **2007**, *164*, 1149–1159. [CrossRef]
18. Gould, G.S.; Havard, A.; Lim, L.L.; Kumar, R. Exposure to tobacco, environmental tobacco smoke and nicotine in pregnancy: A pragmatic overview of reviews of maternal and child outcomes, effectiveness of interventions and barriers and facilitators to quitting. *Int. J. Environ. Res. Public Health* **2020**, *17*, 2034. [CrossRef]
19. Kataoka, H.; Kaji, S.; Moai, M. Risk Assessment of Passive Smoking Based on Analysis of Hair Nicotine and Cotinine as Exposure Biomarkers by In-Tube Solid-Phase Microextraction Coupled On-Line to LC-MS/MS. *Molecules* **2021**, *26*, 7356. [CrossRef] [PubMed]
20. Kataoka, H.; Ohshima, H.; Ohkawa, T. Simultaneous analysis of multiple steroidal biomarkers in saliva for objective stress assessment by on-line coupling of automated in-tube solid-phase microextraction and polarity-switching LC-MS/MS. *Talanta Open* **2023**, *7*, 100177. [CrossRef]
21. Wadsworth, M.E.; Broderick, A.V.; Loughlin-Presnal, J.E.; Bendezu, J.J.; Joos, C.M.; Ahlkvist, J.A.; Perzow, S.E.D.; McDonald, A. Co-activation of SAM and HPA responses to acute stress: A review of the literature and test of differential associations with preadolescents' internalizing and externalizing. *Dev. Psychobiol.* **2019**, *61*, 1079–1093. [CrossRef]
22. Bleker, L.S.; van Dammen, L.; Leeflang, M.M.G.; Limpens, J.; Roseboom, T.J.; de Rooij, S.R. Hypothalamic-pituitary-adrenal axis and autonomic nervous system reactivity in children prenatally exposed to maternal depression: A systematic review of prospective studies. *Neurosci. Biobehav. Rev.* **2020**, *117*, 243–252. [CrossRef]
23. Mueller, B.; Figueroa, A.; Robinson-Papp, J. Structural and functional connections between the autonomic nervous system, hypothalamic-pituitary-adrenal axis, and the immune system: A context and time dependent stress response network. *Neurol. Sci.* **2022**, *43*, 951–960. [CrossRef] [PubMed]
24. Dorsey, A.; Scherer, E.; Eckhoff, R.; Furberg, R. *Measurement of Human Stress: A Multidimensional Approach*; RTI Press: Research Triangle Park, NC, USA, 2022. [CrossRef]
25. Stefano, G.B.; Fricchione, G.L.; Esch, T. Relaxation: Molecular and physiological significance. *Med. Sci. Monit.* **2006**, *12*, HY21–HY31. [PubMed]
26. Moberg, K.U.; Handlin, L.; Petersson, M. Neuroendocrine mechanisms involved in the physiological effects caused by skin-to-skin contact—With a particular focus on the oxytocinergic system. *Infant Behav. Dev.* **2020**, *61*, 101482. [CrossRef] [PubMed]
27. Giacomello, G.; Scholten, A.; Parr, M.K. Current methods for stress marker detection in saliva. *J. Pharm. Biomed. Anal.* **2020**, *191*, 113604. [CrossRef]
28. Łoś, K.; Waszkiewicz, N. Biological Markers in Anxiety Disorders. *J. Clin. Med.* **2021**, *10*, 1744. [CrossRef]
29. Noushad, S.; Ahmed, S.; Ansari, B.; Mustafa, U.H.; Saleem, Y.; Hazrat, H. Physiological biomarkers of chronic stress: A systematic review. *Int. J. Health Sci.* **2021**, *15*, 46–59.
30. Kataoka, H. Application of In-Tube SPME to Analysis of Stress-Related Biomarkers. In *Evolution of SPME Technology*; Pawliszyn, J., Ed.; Royal Society of Chemistry: Cambridge, UK, 2023; Chapter 14; pp. 419–440. ISBN 978-1-83916-680-8.
31. Ali, N.; Nater, U.M. Salivary Alpha-Amylase as a Biomarker of Stress in Behavioral Medicine. *Int. J. Behav. Med.* **2020**, *27*, 337–342. [CrossRef]
32. Obayashi, K. Salivary mental stress proteins. *Clin. Chim. Acta* **2013**, *425*, 196–201. [CrossRef]
33. Carter, C.S.; Pournajafi-Nazarloo, H.; Kramer, K.M.; Ziegler, T.E.; White-Traut, R.; Bello, D.; Schwertz, D. Oxytocin: Behavioral associations and potential as a salivary biomarker. *Ann. N. Y. Acad. Sci.* **2007**, *1098*, 312–322. [CrossRef]
34. Steckl, A.; Ray, P. Stress Biomarkers in Biological Fluids and Their Point-of-Use Detection. *ACS Sens.* **2018**, *3*, 2025–2044. [CrossRef]
35. Zamkah, A.; Hui, T.; Andrews, S.; Dey, N.; Shi, F.; Sherratt, R.S. Identification of Suitable Biomarkers for Stress and Emotion Detection for Future Personal Affective Wearable Sensors. *Biosensors* **2020**, *10*, 40. [CrossRef] [PubMed]
36. Singh, N.K.; Chung, S.; Chang, A.-Y.; Wang, J.; Hall, D.A. A non-invasive wearable stress patch for real-time cortisol monitoring using a pseudoknot-assisted aptamer. *Biosens. Bioelectron.* **2023**, *227*, 115097. [CrossRef] [PubMed]
37. Shaffer, F.; Ginsberg, J.P. An Overview of Heart Rate Variability Metrics and Norms. *Front. Public Health* **2017**, *5*, 258. [CrossRef]
38. Sieciński, S.; Kostka, P.S.; Tkacz, E.J. Heart Rate Variability Analysis on Electrocardiograms, Seismocardiograms and Gyrocardiograms on Healthy Volunteers. *Sensors* **2020**, *20*, 4522. [CrossRef]
39. Natarajan, A.; Pantelopoulos, A.; Emir-Farinas, H.; Natarajan, P. Heart rate variability with photoplethysmography in 8 million individuals: A cross-sectional study. *Lancet Digit. Health* **2020**, *2*, e650–e657. [CrossRef]
40. Nayak, S.K.; Pradhan, B.; Mohanty, B.; Sivaraman, J.; Ray, S.S.; Wawrzyniak, J.; Jarzębski, M.; Pal, K. A Review of Methods and Applications for a Heart Rate Variability Analysis. *Algorithms* **2023**, *16*, 433. [CrossRef]

41. Kim, H.-G.; Cheon, E.-J.; Bai, D.-S.; Lee, Y.H.; Koo, B.-H. Stress and Heart Rate Variability: A Meta-Analysis and Review of the Literature. *Psychiatry Investig.* **2018**, *15*, 235–245. [CrossRef]
42. Cao, R.; Rahmani, A.M.; Lindsay, K.L. Prenatal stress assessment using heart rate variability and salivary cortisol: A machine learning-based approach. *PLoS ONE* **2022**, *17*, e0274298. [CrossRef]
43. Immanuel, S.; Teferra, M.N.; Baumert, M.; Bidargaddi, N. Heart Rate Variability for Evaluating Psychological Stress Changes in Healthy Adults: A Scoping Review. *Neuropsychobiology* **2023**, *82*, 187–202. [CrossRef]
44. Elgendi, M. On the Analysis of Fingertip Photoplethysmogram Signals. *Curr. Cardiol. Rev.* **2012**, *8*, 14–25. [CrossRef] [PubMed]
45. Chen, H.-K.; Hu, Y.-F.; Lin, S.-F. Methodological considerations in calculating heart rate variability based on wearable device heart rate samples. *Comput. Biol. Med.* **2018**, *102*, 396–401. [CrossRef] [PubMed]
46. Liu, I.; Ni, S.; Peng, K. Happiness at Your Fingertips: Assessing Mental Health with Smartphone Photoplethysmogram-Based Heart Rate Variability Analysis. *Telemed. J. e-Health* **2020**, *26*, 1483–1491. [CrossRef] [PubMed]
47. Matsumoto, A.; Ino, T.; Ohta, M.; Otani, T.; Hanada, S.; Sakuraoka, A.; Matsumoto, A.; Ichiba, M.; Hara, M. Enzyme-linked immunosorbent assay of nicotine metabolites. *Environ. Health Prev. Med.* **2010**, *15*, 211–216. [CrossRef] [PubMed]
48. Da Fonseca, B.M.; Moreno, I.E.; Magalhães, A.R.; Barroso, M.; Queiroz, J.A.; Ravara, S.; Calheiros, J.; Gallardo, E. Determination of biomarkers of tobacco smoke exposure in oral fluid using solid-phase extraction and gas chromatography-tandem mass spectrometry. *J. Chromatogr. B Analyt. Technol. Biomed. Life Sci.* **2012**, *889–890*, 116–122. [CrossRef]
49. Shaik, F.B.; Nagajothi, G.; Swarnalatha, K.; Kumar, C.S.; Maddu, N. Quantification of Nicotine and Cotinine in Plasma, Saliva, and Urine by HPLC Method in Chewing Tobacco Users. *Asian Pac. J. Cancer Prev.* **2019**, *20*, 3617–3623. [CrossRef]
50. Kataoka, H.; Inoue, R.; Yagi, K.; Saito, K. Determination of nicotine, cotinine, and related alkaloids in human urine and saliva by automated in-tube solid-phase microextraction coupled with liquid chromatography-mass spectrometry. *J. Pharm. Biomed. Anal.* **2009**, *49*, 108–114. [CrossRef]
51. Miller, E.I.; Norris, H.R.; Rollins, D.E.; Tiffany, S.T.; Moore, C.M.; Vincent, M.J.; Agrawal, A.; Wilkins, D.G. Identification and quantification of nicotine biomarkers in human oral fluid from individuals receiving low-dose transdermal nicotine: A preliminary study. *J. Anal. Toxicol.* **2010**, *34*, 357–366. [CrossRef]
52. Chang, Y.J.; Muthukumaran, R.B.; Chen, J.L.; Chang, H.Y.; Hung, Y.C.; Hu, C.W.; Chao, M.R. Simultaneous determination of areca nut- and tobacco-specific alkaloids in saliva by LC-MS/MS: Distribution and transformation of alkaloids in oral cavity. *J. Hazard. Mater.* **2022**, *426*, 128116. [CrossRef]
53. Tahara, Y.; Huang, Z.; Kiritoshi, T.; Onodera, T.; Toko, K. Development of Indirect Competitive Immuno-Assay Method Using SPR Detection for Rapid and Highly Sensitive Measurement of Salivary Cortisol Levels. *Front. Bioeng. Biotechnol.* **2014**, *2*, 15. [CrossRef]
54. Pritchard, B.T.; Stanton, W.; Lord, R.; Petocz, P.; Pepping, G.J. Factors Affecting Measurement of Salivary Cortisol and Secretory Immunoglobulin A in Field Studies of Athletes. *Front. Endocrinol.* **2017**, *8*, 168. [CrossRef] [PubMed]
55. Ben Halima, H.; Bellagambi, F.G.; Brunon, F.; Alcacer, A.; Pfeiffer, N.; Heuberger, A.; Hangouët, M.; Zine, N.; Bausells, J.; Errachid, A. Immuno field-effect transistor (ImmunoFET) for detection of salivary cortisol using potentiometric and impedance spectroscopy for monitoring heart failure. *Talanta* **2023**, *257*, 123802. [CrossRef]
56. Casals, G.; Ballesteros, M.A.; Zamora, J.; Martínez, I.; Fernández-Varo, G.; Mora, M.; Hanzu, F.A.; Morales-Ruiz, M. LC-HRMS and GC-MS Profiling of Urine Free Cortisol, Cortisone, 6beta-, and 18-Hydroxycortisol for the Evaluation of Glucocorticoid and Mineralocorticoid Disorders. *Biomolecules* **2024**, *14*, 558. [CrossRef] [PubMed]
57. De Palo, E.F.; Antonelli, G.; Benetazzo, A.; Prearo, M.; Gatti, R. Human saliva cortisone and cortisol simultaneous analysis using reverse phase HPLC technique. *Clin. Chim. Acta* **2009**, *405*, 60–65. [CrossRef] [PubMed]
58. Abujaber, F.; Corps Ricardo, A.I.; Ríos, Á.; Guzmán Bernardo, F.J.; Rodríguez Martín-Doimeadios, R.C. Ionic liquid dispersive liquid-liquid microextraction combined with LC-UV-Vis for the fast and simultaneous determination of cortisone and cortisol in human saliva samples. *J. Pharm. Biomed. Anal.* **2019**, *165*, 141–146. [CrossRef]
59. Saracino, M.A.; Iacono, C.; Somaini, L.; Gerra, G.; Ghedini, N.; Raggi, M.A. Multimatrix assay of cortisol, cortisone and corticosterone using a combined MEPS-HPLC procedure. *J. Pharm. Biomed. Anal.* **2014**, *88*, 643–648. [CrossRef]
60. Fustinoni, S.; Polledri, E.; Mercadante, R. High-throughput determination of cortisol, cortisone, and melatonin in oral fluid by on-line turbulent flow liquid chromatography interfaced with liquid chromatography/tandem mass spectrometry. *Rapid Commun. Mass Spectrom.* **2013**, *27*, 1450–1460. [CrossRef]
61. Cao, Z.T.; Wemm, S.E.; Han, L.; Spink, D.C.; Wulfert, E. Noninvasive determination of human cortisol and dehydroepiandrosterone sulfate using liquid chromatography-tandem mass spectrometry. *Anal. Bioanal. Chem.* **2019**, *411*, 1203–1210. [CrossRef]
62. Bakusicm, J.; De Nys, S.; Creta, M.; Godderis, L.; Duca, R.C. Study of temporal variability of salivary cortisol and cortisone by LC-MS/MS using a new atmospheric pressure ionization source. *Sci. Rep.* **2019**, *9*, 19313. [CrossRef]
63. Gregory, S.; Denham, S.G.; Lee, P.; Simpson, J.P.; Homer, N.Z.M. Using LC-MS/MS to determine salivary steroid reference intervals in a European older adult population. *Metabolites* **2023**, *13*, 265. [CrossRef]
64. Lanfermeijer, M.; van Winden, L.J.; Starreveld, D.E.J.; Razab-Sekh, S.; van Faassen, M.; Bleiker, E.M.A.; van Rossum, H.H. An LC-MS/MS-based method for the simultaneous quantification of melatonin, cortisol and cortisone in saliva. *Anal. Biochem.* **2024**, *689*, 115496. [CrossRef]
65. Kataoka, H. In-tube solid-phase microextraction: Current trends and future perspectives. *J. Chromatogr. A* **2021**, *1636*, 461787. [CrossRef]

66. Inukai, T.; Kaji, S.; Kataoka, H. Analysis of nicotine and cotinine in hair by on-line in-tube solid-phase microextraction coupled with liquid chromatography-tandem mass spectrometry as biomarkers of exposure to tobacco smoke. *J. Pharm. Biomed. Anal.* **2018**, *156*, 272–277. [CrossRef] [PubMed]
67. Kataoka, H.; Ehara, K.; Yasuhara, R.; Saito, K. Simultaneous determination of testosterone, cortisol and dehydroepiandrosterone in saliva by stable isotope dilution on-line in-tube solid-phase microextraction coupled with liquid chromatography–tandem mass spectrometry. *Anal. Bioanal. Chem.* **2013**, *405*, 331–340. [CrossRef] [PubMed]
68. Moriyama, E.; Kataoka, H. Automated analysis of oxytocin by on-line in-tube solid-phase microextraction coupled with liquid chromatography-tandem mass spectrometry. *Chromatography* **2015**, *2*, 382–391. [CrossRef]
69. Ishizaki, A.; Uemura, A.; Kataoka, H. A sensitive method to determine melatonin in saliva by automated online in-tube solid-phase microextraction coupled with stable isotope-dilution liquid chromatography-tandem mass spectrometry. *Anal. Methods* **2017**, *9*, 3134–3140. [CrossRef]
70. Kataoka, H.; Nakayama, D. Online in-tube solid-phase microextraction coupled with liquid chromatography-tandem mass spectrometry for automated analysis of four sulfated steroid metabolites in saliva samples. *Molecules* **2022**, *27*, 3225. [CrossRef]
71. Hitomi, T.; Kataoka, H. Development of Noninvasive Method for the Automated Analysis of Nine Steroid Hormones in Saliva by Online Coupling of In-Tube Solid-Phase Microextraction with Liquid Chromatography–Tandem Mass Spectrometry. *Analytica* **2024**, *5*, 233–249. [CrossRef]

Disclaimer/Publisher's Note: The statements, opinions and data contained in all publications are solely those of the individual author(s) and contributor(s) and not of MDPI and/or the editor(s). MDPI and/or the editor(s) disclaim responsibility for any injury to people or property resulting from any ideas, methods, instructions or products referred to in the content.

MDPI AG
Grosspeteranlage 5
4052 Basel
Switzerland
Tel.: +41 61 683 77 34

Molecules Editorial Office
E-mail: molecules@mdpi.com
www.mdpi.com/journal/molecules

Disclaimer/Publisher's Note: The title and front matter of this reprint are at the discretion of the Guest Editor. The publisher is not responsible for their content or any associated concerns. The statements, opinions and data contained in all individual articles are solely those of the individual Editor and contributors and not of MDPI. MDPI disclaims responsibility for any injury to people or property resulting from any ideas, methods, instructions or products referred to in the content.

www.ingramcontent.com/pod-product-compliance
Lightning Source LLC
LaVergne TN
LVHW072327090526
838202LV00019B/2367